METHODS IN MOLECULAR BIOLOGY

For further volumes:
http://www.springer.com/series/7651

For over 35 years, biological scientists have come to rely on the research protocols and methodologies in the critically acclaimed *Methods in Molecular Biology* series. The series was the first to introduce the step-by-step protocols approach that has become the standard in all biomedical protocol publishing. Each protocol is provided in readily-reproducible step-by-step fashion, opening with an introductory overview, a list of the materials and reagents needed to complete the experiment, and followed by a detailed procedure that is supported with a helpful notes section offering tips and tricks of the trade as well as troubleshooting advice. These hallmark features were introduced by series editor Dr. John Walker and constitute the key ingredient in each and every volume of the *Methods in Molecular Biology* series. Tested and trusted, comprehensive and reliable, all protocols from the series are indexed in PubMed.

The Oral Microbiome

Methods and Protocols

Edited by

Guy R. Adami

College of Dentistry, University of Illinois at Chicago, Chicago, IL, USA

Humana Press

Editor
Guy R. Adami
College of Dentistry
University of Illinois at Chicago
Chicago, IL, USA

ISSN 1064-3745 ISSN 1940-6029 (electronic)
Methods in Molecular Biology
ISBN 978-1-0716-1517-1 ISBN 978-1-0716-1518-8 (eBook)
https://doi.org/10.1007/978-1-0716-1518-8

This Humana imprint is published by the registered company Springer Science+Business Media, LLC part of Springer Nature.
The registered company address is: 1 New York Plaza, New York, NY 10004, U.S.A.

Preface

Attention has turned to the oral microbiome and oral and systemic health. The mouth has a variety of microbes second only to the gut in complexity of human sites. It offers easy access to samples, but the variety of surfaces in the oral cavity and oral pharynx, and the different types of saliva, for example, stimulated versus not stimulated, make sampling decisions difficult. The book begins with analytical techniques for identifying and measuring oral bacteria, which can be distinct from those used for gut bacteria. Guidance is provided in matching sample collection methods and data analysis with the goals of the specific oral microbiome project. Strategies for controlling common sources of variability in oral microbiome are also outlined and methods for viral and fungal analysis are described. Methods and study strategies, from oral DNA and RNA samples of oral microbiome, to identifying molecular pathways relevant to disease are included. Mouse models are included, with methods to study mouse teeth microbiome and a well-validated approach to antibiotic purging of systemic bacteria. Finally, methods of metaproteomic analysis of various oral samples are covered, along with approaches to simultaneous noninvasive measurement of mucosal host miRNA and mucosal bacteria from a single site.

16s rRNA amplicon sequencing is the most common method of taxa identification and quantitation due to the simplicity of sample processing, the low sequencing depth required, and the sophisticated and reproducible tools of taxa identification. Methods including both comprehensive bacterial libraries and oral-specific libraries are included. Chapter 14 offers an alternative approach to 16s rRNA gene sequencing, instead using the 16s–23s intergenic region, ITR. ITR amplicon sequencing allows relatively rapid, accurate strain level identification of taxa not possible with 16s rRNA gene sequencing even while using an oral microbiome-specific library.

The oral microbiome is of course not just bacteria. Techniques that take advantage of high-throughput DNA sequencing can be used for analysis of oral virus. While shotgun sequencing of DNA or RNA for analysis of virus in theory provides a simple approach to measuring virus present, in operation the method is anything but simple, high levels of non-virus sequence make it necessary to sequence at great depth, increasing the cost. Then, the large amount of sequence makes computational requirements for the sequence alignment to viral genomes extreme. Isolation of virus particles prior to sequencing using flow cytometry approaches is not yet optimized [1]. An alternative strategy that is much more practical for those interested in quantitation of known virus families, using NGS, is the Bait Capture approach. With this method the initial work is to select and make probes using a bait library that can be purchased from one of several suppliers. After the probe library is constructed, the steps to screen additional samples for the same virus are facilitated. This approach is also applicable to the study of bacteriophage genomes and cellular RNA and genes.

Functional genomics, a mainstay of molecular biology approaches to gene characterization, has been used to explore the oral microbiome. Chapter 4 has a method to isolate bacteriophage based on the ability to target any bacteria of interest that can be maintained in bacterial mono-culture. Chapter 3 describes a proven method to screen metagenes from any oral source, in this case saliva bacteria, for genes that offer antibiotic resistance, though any selectable trait can be studied. While the selection assay is in vitro, the method is applicable to all taxa present in the oral microbiome.

Metagenomics is the future of oral microbiome studies. It offers identification of metagenes and thus a window to the molecular pathways that reveal cell functions. The mature nature of the field of gut microbiome research simplifies further analysis at that site, with it being routine to infer function based on taxa identity [2, 3]. The oral cavity offers a host of surfaces, variable pH, and variable oxygen accessibility suggesting taxa and metagenes will be distinct functionally in many ways, versus the gut microbiome. Thus, there is great need to do metagenomic DNA sequence analysis.

Examples of microbiome analysis using shotgun metagenomic sequencing are included. Chapter 7 provides an example of shotgun sequencing in the case of precious low yield, samples where major effort must be undertaken to avoid contamination during sample discovery, collection, and DNA isolation. This is applicable to not just ancient teeth but any low yield samples. Sequencing of modern-day high yield oral samples has different challenges, such as excess levels of nonbacterial DNA. Methods for experimental reduction of nonbacterial DNA are helpful. Chapter 6 discusses a range of solutions, with the authors providing a protocol for the elimination of nonbacterial cells from fresh, and more importantly, frozen samples. Their protocol allows acquisition of samples at a clinical site with, following addition of a preservative, freezing and storage of the sample. After thawing, host cell depletion is done in the laboratory by exposure of a DNA degrading chemical to the contents of mammalian cells, but not that of the bacteria. It has been validated on saliva samples and has been shown to enrich for bacterial DNA by 4- to 80-fold. This approach, which is unique in that has been shown to work with frozen samples, results in greatly reduced sequencing depth requirements per sample, and so substantial cost-saving, and it reduced computational needs when performing DNA sequence alignment and metagene identification. An example of transcriptome analysis is also included. Oral sampling where microbiome can be easily harvested directly from their niches, unlike samples from the gut, allows rapid preservation of RNA constituent as found in vivo. As a result, transcriptomic studies offer a great opportunity to understand the system. For the same reason, peptide analysis is also a key tool to better understand how the oral microbiome may affect oral physiology/health and that of the entire body.

Two chapters focus on taking the step from metagenes to understanding what bacteria are doing. One approach described in Chapter 9 starts with a hypothesis that taxa associated with nitrogen-reducing activity, either identified earlier experimentally or by previous metagenomic analysis, can be measured to determine correlation of their levels with a clinical measure, in this case systolic blood pressure. Chapter 10 presents a method for identification of biosynthetic gene clusters that uses computational tools AntiSmash and BigScape, which use what is known about biosynthetic clusters in the database, to allow the identification of the same, related, or new biosynthetic gene clusters in new sample types.

Chicago, IL, USA *Guy R. Adami*

References

1. Martinez Martinez J, Martinez-Hernandez F, Martinez-Garcia M (2020) Single-virus genomics and beyond. Nat Rev Microbiol. https://doi.org/10.1038/s41579-020-00444-0
2. Iwai S, Weinmaier T, Schmidt BL, Albertson DG, Poloso NJ, Dabbagh K et al (2016) Piphillin: improved prediction of metagenomic content by direct inference from human microbiomes. PLoS One 11(11):e0166104. https://doi.org/10.1371/journal.pone.0166104
3. Douglas GM, Maffei VJ, Zaneveld JR, Yurgel SN, Brown JR, Taylor CM et al (2020) PICRUSt2 for prediction of metagenome functions. Nat Biotechnol 38(6):685–688. https://doi.org/10.1038/s41587-020-0548-6

Contents

Contributors

GUY R. ADAMI • *Department of Oral Medicine and Diagnostic Sciences, College of Dentistry, University of Illinois at Chicago, Chicago, IL, USA*

DANIELLE E. ANDERSON • *Programme in Emerging Infectious Diseases, Duke-NUS Medical School, Singapore, Singapore*

MICHAEL J. ANG • *Department of Oral Medicine and Diagnostic Sciences, College of Dentistry, University of Illinois at Chicago, Chicago, IL, USA*

RAMY K. AZIZ • *Department of Microbiology and Immunology, Faculty of Pharmacy, Cairo University, Cairo, Egypt; Center for Genome and Microbiome Research, Cairo University, Cairo, Egypt*

JONATHON L. BAKER • *Genomic Medicine Group, J. Craig Venter Institute, La Jolla, CA, USA*

BRUNO BOHN • *Division of Epidemiology and Community Health, School of Public Health, University of Minnesota, Minneapolis, MN, USA*

TEAGAN BROWN • *Quadram Institute Bioscience, Norfolk, UK*

CRAIG D. BYRON • *Department of Biology, Mercer University, Macon, GA, USA*

STUART DASHPER • *Melbourne Dental School, Faculty of Medicine, Dentistry & Health Science, The University of Melbourne, Melbourne, VIC, Australia*

RYAN T. DEMMER • *Division of Epidemiology and Community Health, School of Public Health, University of Minnesota, Minneapolis, MN, USA; Department of Epidemiology, Mailman School of Public Health, Columbia University, New York, NY, USA*

ANNA EDLUND • *Genomic Medicine Group, J. Craig Venter Institute, La Jolla, CA, USA; Department of Pediatrics, University of California at San Diego, La Jolla, CA, USA*

CHARLENE E. GOH • *Faculty of Dentistry, National University of Singapore, Singapore, Singapore*

DIVYA GOPINATH • *Oral Diagnostic & Surgical Sciences Department, School of Dentistry, International Medical University, Kuala Lumpur, Malaysia*

STEFAN J. GREEN • *Research Resources Center, University of Illinois at Chicago, Chicago, IL, USA*

HEBA HUSSEIN • *Oral Medicine, Diagnosis, and Periodontology Department, Faculty of Dentistry, Cairo University, Cairo, Egypt; Department of Oral Medicine and Diagnostic Sciences, College of Dentistry, University of Illinois at Chicago, Chicago, IL, USA*

MWILA KABWE • *Department of Pharmacy and Biomedical Sciences, La Trobe Institute for Molecular Science, La Trobe University, Bendigo, VIC, Australia*

THOMAS KAISER • *Division of Basic and Translational Research, Department of Surgery, University of Minnesota, Minneapolis, MN, USA; BioTechnology Institute, University of Minnesota, St. Paul, MN, USA*

ELISSA M. KIM • *Department of Oral Medicine and Diagnostic Sciences, College of Dentistry, University of Illinois at Chicago, Chicago, IL, USA*

ROB KNIGHT • *Department of Pediatrics, University of California San Diego, La Jolla, CA, USA; Center for Microbiome Innovation, University of California San Diego, La Jolla, CA, USA; Department of Computer Science and Engineering, University of California San Diego, La Jolla, CA, USA*

HENG KU • *Department of Pharmacy and Biomedical Sciences, La Trobe Institute for Molecular Science, La Trobe University, Bendigo, VIC, Australia*

EUGENE J. LEYS • *Division of Biosciences, College of Dentistry, The Ohio State University, Columbus, OH, USA*

CLARISSE MAROTZ • *Department of Pediatrics, University of California San Diego, La Jolla, CA, USA*

ROHIT KUNNATH MENON • *Clinical Dentistry (Prosthodontics), School of Dentistry, International Medical University, Kuala Lumpur, Malaysia*

NICOLE E. MOORE • *Department of Anthropology, The Pennsylvania State University, University Park, PA, USA*

CHIRANJIT MUKHERJEE • *Jet Propulsion Laboratory, California Institute of Technology, Pasadena, CA, USA*

JALEES A. NASIR • *M.G. DeGroote Institute for Infectious Disease Research, Department of Biochemistry and Biomedical Sciences, DeGroote School of Medicine, McMaster University, Hamilton, ON, Canada*

ALEXANDER RABE • *Department of Functional Genomics, Interfaculty Institute for Genetics and Functional Genomics, University Medicine Greifswald, Greifswald, Germany*

LIAM J. REYNOLDS • *UCD School of Biomolecular and Biomedical Science, UCD Earth Institute and UCD Conway Institute, University College Dublin, Dublin, Ireland*

ADAM P. ROBERTS • *Department of Tropical Disease Biology, Liverpool School of Tropical Medicine, Liverpool, UK*

MANUELA GESELL SALAZAR • *Department of Functional Genomics, Interfaculty Institute for Genetics and Functional Genomics, University Medicine Greifswald, Greifswald, Germany*

LEA M. SEDGHI • *Department of Orofacial Sciences, School of Dentistry, University of California San Francisco, San Francisco, CA, USA*

DAVID J. SPEICHER • *Department of Laboratory Medicine, St. Joseph's Healthcare Hamilton, Hamilton, ON, Canada; M.G. DeGroote Institute for Infectious Disease Research, Department of Biochemistry and Biomedical Sciences, DeGroote School of Medicine, McMaster University, Hamilton, ON, Canada; Menzies Health Institute Queensland, Griffith University, Gold Coast, QLD, Australia*

CHRISTOPHER STALEY • *Division of Basic and Translational Research, Department of Surgery, University of Minnesota, Minneapolis, MN, USA; BioTechnology Institute, University of Minnesota, St. Paul, MN, USA*

SUPATHEP TANSIRICHAIYA • *Department of Clinical Dentistry, Faculty of Health Sciences, UiT the Arctic University of Norway, Tromsø, Norway*

JOSEPH TUCCI • *Department of Pharmacy and Biomedical Sciences, La Trobe Institute for Molecular Science, La Trobe University, Bendigo, VIC, Australia*

UWE VÖLKER • *Department of Functional Genomics, Interfaculty Institute for Genetics and Functional Genomics, University Medicine Greifswald, Greifswald, Germany*

TAKAYASU WATANABE • *Department of Chemistry, Nihon University School of Dentistry, Tokyo, Japan*

LAURA S. WEYRICH • *Department of Anthropology, The Pennsylvania State University, University Park, PA, USA; School of Biological Sciences, University of Adelaide, Adelaide, SA, Australia*

LIVIA ZARAMELA • *Department of Pediatrics, University of California San Diego, La Jolla, CA, USA*

KARSTEN ZENGLER • *Department of Pediatrics, University of California San Diego, La Jolla, CA, USA; Center for Microbiome Innovation, University of California San Diego, La Jolla, CA, USA*
PENG ZHOU • *CAS Key Laboratory of Special Pathogens, Wuhan Institute of Virology, Center for Biosafety Mega-Science, Chinese Academy of Sciences, Wuhan, China*
ZHIGANG ZHU • *Division of Basic and Translational Research, Department of Surgery, University of Minnesota, Minneapolis, MN, USA; BioTechnology Institute, University of Minnesota, St. Paul, MN, USA*
CRISTAL ZUNIGA • *Department of Pediatrics, University of California San Diego, La Jolla, CA, USA*

Chapter 1

Increasing Reproducibility in Oral Microbiome Research

Divya Gopinath and Rohit Kunnath Menon

Abstract

Evidence on the role of the oral microbiome in health and disease is changing the way we understand, diagnose, and treat ailments. Numerous studies on diseases affecting the oral cavity have revealed a large amount of data that is invaluable for the advancements in diagnosing and treating these diseases. However, the clinical translation of most of these exploratory data is stalled by variable methodology between studies and non-uniform reporting of the data.

Understanding the key areas that are gateways to bias in microbiome studies is imperative to overcome this challenge faced by oral microbiome research. Bias can be multifactorial and may be introduced in a microbiome research study during the formulation of the study design, sample collection and storage, or the sample processing protocols before sequencing. This chapter summarizes the recommendations from literature to eliminate bias in the microbiome research studies and to ensure the reproducibility of the microbiome research data.

Key words Microbiome, Oral microbiome, 16S rRNA, DNA sequencing, Human microbiome

1 Introduction

The human microbiome is progressively attaining significance in the field of translational medicine and its potential as a diagnostic adjunct and therapeutic intervention is being established [1]. The resident microbiome is considered a key player in the homeostasis of human metabolism and microbial dysbiosis and has been associated with various human diseases, including oral diseases. The oral microbiome is considered the second largest collection of microbiota, next to the gut, harboring diverse microbes consisting of over 700 species of bacteria and other microorganisms like fungi, protozoa, and archaea. Until recently, most data on the specific effects of oral microbes were derived from in vitro studies [2, 3]. However, the influence of specific or single bacterial species on human cells studied in the secluded environment of duplicated culture plates cannot fully reflect the complexity of the human body. Next-generation sequencing (NGS) technologies offered a new perspective to our concept of the role of oral microbes in

Guy R. Adami (ed.), *The Oral Microbiome: Methods and Protocols*, Methods in Molecular Biology, vol. 2327,
https://doi.org/10.1007/978-1-0716-1518-8_1,

health and disease and to the understanding of the continuous cross-talk between epithelial cells and millions of bacterial strains that occur in health and disease. Studies of microbial pathogenesis have shifted from consideration of individual pathogens to a paradigm, which considers the whole compositional and functional attributes of a microbial community, i.e., of the microbiome. Earlier studies investigating the microbial communities utilized the original DNA sequencing techniques, such as Sanger sequencing, and could provide the genetic code of only a few DNA fragments per sample. On the other hand, recent NGS techniques extend this provision across millions of segments in a very short time. The 16S ribosomal RNA (16S rRNA) gene harbors DNA sequences, which are highly conserved among prokaryotes, and they have short tracks of hypervariable sequences (V1–V9 regions) that can provide specific signatures useful for taxonomic and phylogenetic characterization to identify different bacteria. This is a relatively cost-effective technique and can identify rare bacteria at the genus level. With numerous databases available, this has been the most common approach for bacterial community identification. Recently, whole-genome, so-called shotgun metagenomic, sequencing allows the study of the phylogenetic characteristics, as well as of the genes encoding for the functions of these communities [4], whereas metatranscriptomics describes active gene expression.

Increased understanding of the structure and functions of complex and intricate bacterial communities and their influence on human health have revolutionized this field of research. However, the diverse ways in which oral microbiome is integrated into several facets of the oral disease spectrum pose equally distinct challenges in designing oral microbiome studies. Given the peculiar composition of oral cavity niches and heterogeneous clinical phenotypes of oral diseases, unbiased characterization of oral microbial diversity requires stringent monitoring of the measures in planning the study design, sample collection methodology, and downstream bioinformatics analysis. This chapter addresses the challenges in designing clinical microbiome studies and focuses on the recommendations from the literature on the microbiome studies, pertaining to study design, sample collection, storage, and sample processing protocols before sequencing.

2 Study Design

Any microbiome study should begin with defining a clear research question with a well-defined objective that determines the choice of the sequencing strategy. Careful consideration of the population to be targeted is necessary to select a representative sample set and to ensure that the results are generalizable [5–7]. Inclusion and exclusion criteria should be well defined to lower the interindividual heterogeneity. A crucial aspect of microbiome studies includes the calculation of effect sizes with adequate statistical power to detect relevant differences between the groups being studied. The effect size calculation is still a matter of debate for microbiome studies in general [8, 9]. Anticipated effect size from a previous study with similar outcomes can be used to calculate the sample size, and depending upon the study objective, the study parameter may be chosen [8, 10]. However, calculations based on a pilot study can also be considered, if previous studies are not available. Numerous web-based applications are also available for sample size calculation [9, 11].

The microbiome can be studied cross-sectionally as well as longitudinally. Cross-sectional studies, especially case-control studies, are usually simpler to plan; however, they can only validate associations as outcomes and cannot define cause-and-effect relationships. Such studies are limited by samples collected at one point. However, well-defined case-control studies can help to identify microbial differences between certain pathological conditions as compared with healthy controls. A proper definition of cases and controls, with inclusion and exclusion criterion, is necessary before data collection. Use of systemic antibiotics, antifungals or antivirals, within 6 months of sampling is a recommended exclusion criterion by the NIH Human Microbiome Project. The influence of antibiotics on the oral microbiome can vary substantially between individuals [12–14]. Large multicentric case-control studies are encouraged to elucidate the microbial profile of chronic diseases owing to the huge interindividual variations of the microbiome [15]. They may help to offer insights to design more comprehensive longitudinal studies. For investigating causality of the microbiome in oral diseases, a longitudinal cohort study is considered superior to cross-sectional studies. Repeated sampling can help to study the temporal dynamics of the microbial community, which in turn can help to elucidate the progressive severity of a disease, treatment response, and relapse. As these kind of studies are intensive and demanding, very few cohort studies have been reported till now and results are inconclusive owing to several factors, including inadequate power, dropouts, improper time-frame, and the resilience of the oral microbiome [12, 15]. Regardless of the study design, following guidelines outlined in "The Strengthening of the Reporting of Observational Studies in Epidemiology Statement" (STROBE) can ensure uniform reporting and, hence, help in the interpretation of data linked from different microbiome studies [6].

Careful selection of controls is often a challenging matter and has to be defined by the objective of the study [16, 17]. Controls for cross-sectional studies to find discriminating microbiota should be carefully selected with a distinct phenotype from the cases and studied with matching sociodemographic and other relevant factors to avoid confounding bias. Multiple groups as controls could offer additional insights into the confounding factors that can have an impact on the microbiome [6]. For longitudinal studies, either the same individual can be used as control if the research question is based on a treatment or different individuals can act as controls if separate outcomes are expected after the longitudinal study. When the oral microbiome is being studied for monitoring the progression of caries or periodontitis or the efficacy of a drug or procedure, a subset of patients with less severe disease or response could be identified as controls.

For oral microbiome studies, collection of information on additional clinical variables that can impact the microbiome including the periodontal parameters, caries, tooth loss, and systematic diseases should be considered as these can provide invaluable insights. This data collectively known as metadata (covariates) for the sample can help to ensure consistency in downstream statistical analysis [10]. Furthermore, statistical analysis can be more simplified when the metadata is more consistent across groups and unnecessary variables are curtailed.

3 Sample Collection

Despite significant endeavors in streamlining the sequencing and data analysis, the variability in sample collection, preservation, and storage has been shown to influence microbial profiles [18, 19]. The oral cavity is a distinct microenvironment as it encompasses several sub-ecological niches including teeth, gingiva, sulcus, tongue, buccal mucosa, the floor of the mouth, lip, retromolar trigone, hard palate, and soft palate, all of which are colonized by distinct microbial communities [20]. Moreover, close proximity of the oral cavity with the nasopharynx and oropharynx can result in overlap of the microbiome. Based on the research objective, it is important to decide whether to collect samples from individual niches or the whole microbiome of the oral cavity. The latter can be saliva or mouth gargles, which can reflect the local bacterial alterations of the microbes from the oral sub niches. Sampling site can have an impact on the experimental design, specifically with regard to the number of subjects and the number of samples to be taken. Moreover, sample collection method will also vary according to the type of oral sample to be collected.

As a biological fluid, saliva is comparatively infrequently used in laboratory analysis in contrast to blood and urine despite being easy to collect [21] and considered a feasible alternative to study oral and general health. Saliva has been shown to demonstrate temporal stability in microbial diversity and structure over 1 year of sampling [22]. There are conflicting results in the literature regarding the variation introduced by different sample collection methodologies and their impact on biomolecules in saliva [23]. Moreover, saliva can be collected as unstimulated saliva, stimulated saliva, or as rinses or gargles and these may differ in their compositional profile. Microbial diversity of stimulated saliva was found to be higher than unstimulated saliva [24]. However, there was no difference in bacterial salivary profiles according to another study on a different population that compared unstimulated and stimulated saliva [24]. However, sample collection methodology and downstream bioinformatic analysis were different in both studies, highlighting the need for a standardized methodology to deliver comparable results. It has been shown that mouthwash samples perform similarly to saliva samples for analysis of the oral microbiome [25, 26]. For saliva collection, the subjects should be asked to refrain from drinking/eating for an hour preceding the sample collection. Saliva can be either collected directly using falcon tubes or commercial saliva collection and stabilization kits, such as OMNIgene (DNA Genotek) and GeneFiX (Isohelix). The use of commercial kits eases the collection and storage, thus eliminating the need for cold chains that often complicates the collection process, especially in remote locations. Moreover, the incorporation of a funnel design helps in noninvasive collection, even in patients with neuromuscular disorders. A compromise between these two extremes is the usage of a preservative, which is added to samples upon collection. The difference in salivary collection methods does not have a significant influence on the oral microbial profile in general, though changes in measured levels of specific taxa can occur [27].

In an attempt to replace invasive sampling of tissues, oral swabs can be considered to collect the mucosa-associated microbiome. However, significant differences have been found in microbial populations when comparing oral swabs and biopsy samples from pathological specimens [28]. Oral swabs can be utilized for collecting microbiome samples from hard tissue, as well as soft tissue surfaces. Examples of commercial oral swabs include Becton-Dickinson, BBL CultureSwab EZ II, and Isohelix Swab SK-2S. The relative performance of the swabs regarding microbial DNA yield has not been studied yet. The preferred method would be to stroke the target surface multiple times and immediately replace the seal in a collection tube or wrapper subsequently for safe containment. Minimizing contamination during oral swab collection can be critical for oral microbiome studies. It is recommended to pool the DNA from one or two swabs to increase your DNA content if shotgun sequencing is planned. If there are no room temperature stabilizing buffers, samples should be immediately stored in the −80 °C freezer until DNA extraction. Several commercial companies (Norgen Biotek, Iso helix, DNA Genotek) are marketing swab collection devices with stabilization buffers, which help with storage at room temperature.

Oral mucosal tissue samples should be collected aseptically from tissue at the time of surgery and should either be flash-frozen or should be stored in appropriate sample collection media. Special care should be taken to record the spatial dimension of the tissue collected, as deeper areas might harbor different organisms due to the variable microenvironment [18].

Paper points and curettes were routinely used for the collection of subgingival plaque and to analyze the microbiota long before the advent of microbiome studies [29, 30]. Certain factors prior to sampling can impact the microbiome collected, including plaque control as well as scaling and polishing. However, clinical researchers are in agreement that supragingival plaque should be removed before subgingival sampling [31]. Nevertheless, the extent of this cleansing procedure is still a matter of debate. Retrieving adequate microbial DNA for sequencing from a single subgingival niche can be challenging. Pooling of samples is the currently practiced method [31–33]. Ideally, analysis of single site ecology should be performed as site-specific microbiome might play a role in the etiology of diseases of the periodontium [34, 35].

4 Sample Storage

The microbial composition of a sample will begin to alter immediately after its collection and thus retaining the integrity of collected samples is a key challenge. Stabilizing the sample from the point of collection is mandatory to avoid bias introduced from microbial growth. It is not feasible to extract DNA from samples as soon as they are collected; therefore, samples are to be stored prior to DNA extraction. Sample storage may not be consistent throughout the world owing to resource limitations. Current evidence on the impact of different storage conditions alone on the oral microbiome is inconsistent [36–38]. However, storage of the fresh specimen at −80 °C and processing in one batch is considered the best approach in microbiome studies [39–41]. This practice may not be achievable at remote sites where immediate access to low-temperature storage is inaccessible. In such situations, room temperature stabilization buffers/kits can come in handy.

Various preservatives have been utilized by different studies, including RNA later, Tris-EDTA, and commercial nucleic acid-preserving reagents, to maintain the integrity of clinical samples when immediate freezing is not available [42]. Studies have indicated that preservatives alone have minimal effect on microbial compositions when compared to the downstream computational microbiota analyses [43, 44]. The duration of storage has been shown to have minimal overall impact on the microbial community structure [43]. Regardless of the method used for preservation, it is more crucial to confirm that all the study samples are stored in a similar manner and storage duration should be recorded [45]. Studies focusing on RNA and the metabolome are more stringent in storage requirements; DNA is considered to be relatively stable; however, different storage conditions and repeated thawing and re-freezing have been shown to have an effect on the abundances of certain microbial taxa [46, 47].

5 Negative and Positive Controls

Following the principles of good scientific practice, controls must be included at all stages of sample collection and processing. As the microbiome field is evolving rapidly with newer findings published quite frequently, publishing 16S rRNA gene sequences identified from positive and negative controls may become mandatory, thereby assisting the readers to interpret results from each study.

Bacterial DNA contamination can occur in various stages of sample collection or sample processing. Contamination may be due to various factors, including the laboratory environment, the DNA extraction process, and the PCR reaction. It has been reported that contaminants in the DNA extractions kit, which has been referred to as the "kitome," varied among manufacturers [48]. Even if the researcher had worked in a completely sterile workspace with appropriate infection control policies, the results can still be affected by the so-called "kitome." Contamination with microbes or microbial DNA from the kit can lead to the detection of microbes not present in the samples and can hinder the outcome, especially in clinical samples [49]. So, it is important to sequence the reagent samples as controls alongside the study samples. If the reagent controls are not included in sequencing along with the samples, there could be bias that can lead to the misinterpretation of critical results. Moreover, it is a good practice to process all samples together in the lab using the same batch of reagents [44] and record the batches used if several kits were used. It is often difficult to distinguish the microbiome composition of samples with low biomass from those with single negative controls. Hence, it is now recommended that multiple negative controls can help in these scenarios to distinguish between contaminants and real microbes in the sample.

Several practices have been recommended for removing the contaminating DNA from kit and lab reagents, including UV and gamma radiation [50, 51], enzymatic treatments [51, 52], caesium chloride density gradient centrifugation [53], and silica-membrane filtration [54]. These methods vary by decontamination capacities and the effect of these on the quality of the reagent is still inconclusive. The use of primer extension PCR has been recommended to avoid the amplification of contaminant DNA in the PCR [40].

If contaminants are identified in the negative controls, most of the time these can be removed from the samples, as well as in the downstream analysis. However, if the contaminating microbes are biologically plausible in the samples and could not be separated in the downstream analysis, alternative approaches have been identified [47]. Statistical approaches have been suggested that predict the likelihood of reads that could have originated from contaminants or that combine quantitative PCR with relative abundance [40, 55]. These can be applied by using software packages that have been developed to aid in identifying and removing contaminant DNA sequences from the sequences output for each sample. These methods work on the principle that taxa at relative higher levels in negative controls and low concentration samples are likely to be contaminants introduced after sample collection, possibly during library preparation or sequencing [56, 57]. Moreover, for shotgun metagenomics of bacteria, strain-specific genes or phylogenetic information on marker genes may help in discrimination.

The majority of the oral microbiome research reported till now has not utilized positive controls [37]. Positive bacterial controls, in the form of communities, are commercially available. The manufacturers have carefully selected relevant gram-positive and -negative bacterial species commonly prevalent in the environment and the human body. Currently, many scientific companies provide these mock controls, including BEI Resources and ATCC, which market bacteria only, while ZymoResearch has options for bacteria and fungi.

Ideally, for an appropriate positive control, the knowledge of expected microbes in the working sample is required. As this is often not the case in real clinical scenarios, the selection of a diverse mock community is recommended as a positive control [36]. If the mock communities targeting the species involved are not available commercially, creating lab-based mock communities can also be considered. However, standardized protocols to construct such controls are currently lacking. PCR for identification of the species-specific genes must be performed to ensure that the microbe is present in the mock community [58, 59]. Unexpected contamination can also happen when creating a lab-based mock community and special care needs to be taken. Owing to the exact microenvironment under investigation, it is important to consider whether the positive control selected is a valid representation for the specific research question being investigated.

Downstream use of sequencing reads from the reagents controls is still under debate. However, there is consensus among microbiome researchers that including reagent controls is useful for the quality control of data. Thus, the inclusion of reagent controls is advised when planning a 16S microbiome research project.

6 DNA/RNA Extraction

Following sample collection, the next step is the extraction of DNA and/or RNA. The majority of microbiome studies have focused on the genome; therefore we will discuss DNA extraction in detail. DNA extraction is a critical step that can lead to significant bias that can distort the microbial profiles. Hence, uniformity is urged to facilitate comparison across oral microbiology studies. Standardization of the DNA extraction method that is quality controlled and suitable for all organisms remains a challenge. Commercially available DNA extraction kits are usually preferred over manual extraction because they are superior in reproducibility, quality control, and automation [60]. The kits rely on a variety of principles for DNA purification, including solid-phase protocols of DNA-adsorbing materials, such as silica-membranes or ion exchange columns, that rely on binding and release with an appropriate buffer or may be entirely chemical-based and depend on differential precipitation [61]. Studies that have compared the different commercial kits report that most of the kits produced DNA of sufficient quantity and quality for sequencing [62, 63]. The choice of kit can be based on the type of clinical samples and the manufacturer's report on efficacy with different microbial groups. No bacterial DNA extraction kit is equally effective for all bacteria, which results in a distortion of the composition of the microbiota. As this field is progressing, many commercial companies have come up with DNA kits specifically designed for the extraction of the microbiome (Merck, Qiagen, Norgen, and others). However, the relative efficacy of all these kits has not been established.

Regardless of the DNA extraction kit used, the most crucial step is the adequate lysis of the bacterial cells to expose DNA [64]. Gram-positive bacteria cell wall consists of several layers of peptidoglycans, which cannot be easily destroyed [24, 25]. DNA extraction will be most effective if these bonds are broken either mechanically or chemically. Mechanical lysis with bead beating is considered one of the best methods for destroying the cell wall [55]. Alternatively, chemical lysis with detergents like sodium dodecyl sulfate or enzymatic lysis with lysozyme or specific proteinases have also been suggested [65]. Chemical lysis is more effective when the bacteria are known; for example, treatment with lysostaphin is a standard method for the disruption of *Staphylococcus aureus*. However, various genera might be present in most clinical specimens, and thus methods that can be universally applied are desired. Care should be also taken to add only the specified quantity as recommended, as residual enzyme can affect the downstream PCR.

Both bead beating and sonication effectively disrupt bacteria in microbiome studies [66], with small beads the most effective for cracking bacteria. Zirconia or silica beads of 100μm or less are considered cost-effective for disrupting bacteria. Bead beating for clinical samples can be performed in microtubes half-filled with sample and beads. The clinical samples can be homogenized or resuspended in a suitable homogenization/lysis buffer. It is recommended to load the tubes before adding samples [67]. Beating the sample for short durations is optimal for homogenization in clinical samples. HT Homogenizer (OPS Diagnostics) and Geno/Grinder (SPEX Sample Prep) are the most popular homogenizers, which can be used for processing large number of clinical samples. However, several oral microbiome studies have highlighted that the bead-beating step resulted in a higher representation of various gram-positive genera and species present in the oral cavity [40, 44, 45, 68, 69]. The disadvantage of bead beating is that even when done optimally some fragmentation of DNA occurs, and when done to excess degradation can occur. The usual low amounts of fragmentation that occur have no effects on 16S rRNA gene measurement or that of standard shotgun metagenomic analysis.

Generation of DNA fragments from samples by sonication is performed by exposing the bacteria mixture in a microcentrifuge tube for brief periods. Sonication is conducted for a varying number of 10-s bursts using the maximum output and continuous power [29]. Exact conditions for sonication should be empirically determined for a given DNA sample before a preparative sonication is performed. The sonication treatment disrupts cell surface structures to release DNA, yet does not disintegrate bacterial cells into a clear lysate. Therefore, most cell debris can be eliminated by centrifugation, leaving a relatively clean supernatant with primarily nucleic acids and soluble components, such as proteins. Different sonic power settings impact the DNA yield. Longer treatment on high power slightly decreases the size of the DNA and might lead to complete lysing of some gram-negative bacterial stains. Shorter treatment results in lower DNA yield but does not produce any significantly larger DNA fragments of easily lysed gram-negative species [70, 71]. Contrarily, under-lysis can underestimate gram-positive bacteria that may not lyse sufficiently [28]. After DNA extraction, the yield and quality of DNA should be assessed using various methods, including agarose gel electrophoresis, absorbance, and the use of fluorescent dyes.

The optimization of the enzymatic lysis method and DNA extraction is recommended to ensure optimal DNA yield of superior quality for16S rRNA gene sequencing. For metatranscriptomic studies, extraction of mRNA, which is more technique-sensitive than DNA, is required.

7 Conclusion

The introduction of NGS has offered us a complex picture of the microbiome and currently, we have more questions than answers. Profiling of the oral microbiome can help in the detection of the onset of an oral disease or estimate individual risk for oral disease. The most recent technologies, like nanopore sequencing, allow rapid sequencing on benchtop platforms and can help clinical translation from bench to chairside with rapid diagnostic preventive and therapeutic strategies.

However, the reliability of the obtained data remains a concern in microbiome studies. It is now well known that bias can be introduced into microbiota studies at all methodological stages from sampling to bioinformatic analysis. Hence, it is imperative that consistent methodology be employed throughout a microbiome study and investigators stay well informed regarding recent practices to reduce potential bias and improve reproducibility. There is no consensus on a unique recipe for all microbiome studies. Consistency in every step and comprehensive recording of data related to subjects, experiments, and bioinformatic analysis is pivotal to ensure reproducibility. Publication of all metadata as well as submission of the sequence files to the public databases is mandatory. The use of two cohorts as a discovery cohort and validation cohort can help to validate taxa identified in the former and can help in illustrating reproducibility in biological conclusions [72, 73].

Oral microbiome research until now has been comprised mostly of exploratory studies with no attempts at replications to confirm the reproducibility of the results. It is an accepted fact that results must be reproducible, but they need to be replicable and generalizable as well [74]. Similar to microarray technology, international collaborative efforts engaging scientific communities in different countries are encouraged to establish better standards and guidelines for oral microbiome research [75]. To date, there is no specific method to apply for all oral microbiome studies due to the complexity of the microflora, as different microbes preferentially favor different intraoral habitats. Thus, it is crucial to align the study design to the objectives and to assess and validate methods suitable for each specific study.

References

1. Falony G, Vandeputte D, Caenepeel C, Vieira-Silva S, Daryoush T, Vermeire S et al (2019) The human microbiome in health and disease: hype or hope. Acta Clin Belg 74(2):53–64
2. Salli KM, Ouwehand AC (2015) The use of *in vitro* model systems to study dental biofilms associated with caries: a short review. J Oral Microbiol 7(1):26149
3. Darrene L-N, Cecile B (2016) Experimental models of oral biofilms developed on inert substrates: a review of the literature. Biomed Res Int 2016:7461047

4. Bikel S, Valdez-Lara A, Cornejo-Granados F, Rico K, Canizales-Quinteros S, Soberón X et al (2015) Combining metagenomics, metatranscriptomics and viromics to explore novel microbial interactions: towards a systems-level understanding of human microbiome. Comput Struct Biotechnol 13:390–401
5. Mallick H, Ma S, Franzosa EA, Vatanen T, Morgan XC, Huttenhower C (2017) Experimental design and quantitative analysis of microbial community multiomics. Genome Biol 18(1):228
6. Varoni EM, Bavarian R, Robledo-Sierra J, Porat Ben-Amy D, Wade WG, Paster B et al (2019) World workshop on oral medicine VII: targeting the microbiome for oral medicine specialists—Part 1. A methodological guide. Oral Dis 25(S1):12–27
7. Keiding N, Louis TA (2016) Perils and potentials of self-selected entry to epidemiological studies and surveys. J R Stat Soc A 179 (2):319–376
8. Kelly BJ, Gross R, Bittinger K, Sherrill-Mix S, Lewis JD, Collman RG et al (2015) Power and sample-size estimation for microbiome studies using pairwise distances and PERMANOVA. Bioinformatics 31(15):2461–2468
9. La Rosa PS, Brooks JP, Deych E, Boone EL, Edwards DJ, Wang Q et al (2012) Hypothesis testing and power calculations for taxonomic-based human microbiome data. PLoS One 7 (12):e52078
10. Debelius J, Song SJ, Vazquez-Baeza Y, Xu ZZ, Gonzalez A, Knight R (2016) Tiny microbes, enormous impacts: what matters in gut microbiome studies? Genome Biol 17(1):217
11. Mattiello F, Verbist B, Faust K, Raes J, Shannon WD, Bijnens L et al (2016) A web application for sample size and power calculation in case-control microbiome studies. Bioinformatics 32(13):2038–2040
12. Menon RK, Gomez A, Brandt BW, Leung YY, Gopinath D, Watt RM et al (2019) Long-term impact of oral surgery with or without amoxicillin on the oral microbiome-a prospective cohort study. Sci Rep 9(1):18761
13. Abeles SR, Jones MB, Santiago-Rodriguez TM, Ly M, Klitgord N, Yooseph S et al (2016) Microbial diversity in individuals and their household contacts following typical antibiotic courses. Microbiome 4(1):39
14. Jakobsson HE, Jernberg C, Andersson AF, Sjölund-Karlsson M, Jansson JK, Engstrand L (2010) Short-term antibiotic treatment has differing long-term impacts on the human throat and gut microbiome. PLoS One 5(3):e9836
15. Acosta N, Whelan FJ, Somayaji R, Poonja A, Surette MG, Rabin HR et al (2017) The evolving cystic fibrosis microbiome: a comparative cohort study spanning 16 years. Ann Am Thorac Soc 14(8):1288–1297
16. Gopinath D, Menon RK (2018) Comments on "Compositional and functional variations of oral microbiota associated with the mutational changes in oral cancer" by Yang et al. Oral Oncol 78:216–217
17. Nayfach S, Pollard KS (2016) Toward accurate and quantitative comparative metagenomics. Cell 166(5):1103–1116
18. Gopinath D, Menon RK, Banerjee M, Su Yuxiong R, Botelho MG, Johnson NW (2019) Culture-independent studies on bacterial dysbiosis in oral and oropharyngeal squamous cell carcinoma: a systematic review. Crit Rev Oncol Hematol 139:31–40
19. Fricker AM, Podlesny D, Fricke WF (2019) What is new and relevant for sequencing-based microbiome research? A mini-review. J Adv Res 19:105–112
20. Segata N, Haake S, Mannon P, Lemon KP, Waldron L, Gevers D et al (2012) Composition of the adult digestive tract bacterial microbiome based on seven mouth surfaces, tonsils, throat and stool samples. Genome Biol 13(6): R42
21. Belstrøm D (2020) The salivary microbiota in health and disease. J Oral Microbiol 12 (1):1723975
22. Cameron SJS, Huws SA, Hegarty MJ, Smith DPM, Mur LAJ (2015) The human salivary microbiome exhibits temporal stability in bacterial diversity. FEMS Microbiol Ecol 91(9): fiv091
23. Pfaffe T, Cooper-White J, Beyerlein P, Kostner K, Punyadeera C (2011) Diagnostic potential of saliva: current state and future applications. Clin Chem 57(5):675–687
24. Gomar-Vercher S, Simón-Soro A, Montiel-Company JM, Almerich-Silla JM, Mira A (2018) Stimulated and unstimulated saliva samples have significantly different bacterial profiles. PLoS One 13(6):e0198021
25. Fan X, Peters BA, Min D, Ahn J, Hayes RB (2018) Comparison of the oral microbiome in mouthwash and whole saliva samples. PLoS One 13(4):e0194729
26. Cabral DJ, Wurster JI, Flokas ME, Alevizakos M, Zabat M, Korry BJ et al (2017) The salivary microbiome is consistent between subjects and resistant to impacts of short-term hospitalization. Sci Rep 7(1):11040

27. Lim Y, Fukuma N, Totsika M, Kenny L, Morrison M, Punyadeera C (2018) The performance of an oral microbiome biomarker panel in predicting oral cavity and oropharyngeal cancers. Front Cell Infect Microbiol 8:267
28. Liu A-Q, Vogtmann E, Shao D-T, Abnet CC, Dou H-Y, Qin Y et al (2019) A comparison of biopsy and mucosal swab specimens for examining the microbiota of upper gastrointestinal carcinoma. Cancer Epidemiol Biomark Prev 28 (12):2030–2037
29. Hartroth B, Seyfahrt I, Conrads G (1999) Sampling of periodontal pathogens by paper points: evaluation of basic parameters. Oral Microbiol Immunol 14(5):326–330
30. Sahl EF, Henkin JM, Angelov N (2014) Recovery of putative periodontal pathogens from curette sampling at different depths of periodontal lesions: an in vivo cross-sectional clinical study. J Int Acad Periodontol 16 (3):78–85
31. Santigli E, Trajanoski S, Eberhard K, Klug B (2016) Sampling modification effects in the subgingival microbiome profile of healthy children. Front Microbiol 7:2142
32. Nowicki EM, Shroff R, Singleton JA, Renaud DE, Wallace D, Drury J et al (2018) Microbiota and metatranscriptome changes accompanying the onset of gingivitis. mBio 9(2): e00575–e00518
33. Belstrøm D, Sembler-Møller ML, Grande MA, Kirkby N, Cotton SL, Paster BJ et al (2017) Microbial profile comparisons of saliva, pooled and site-specific subgingival samples in periodontitis patients. PLoS One 12(8):e0182992
34. Teles R, Teles F, Frias-Lopez J, Paster B, Haffajee A (2013) Lessons learned and unlearned in periodontal microbiology. Periodontol 2000 62(1):95–162
35. Socransky SS, Haffajee AD (2005) Periodontal microbial ecology. Periodontol 2000 38 (1):135–187
36. Lim Y, Totsika M, Morrison M, Punyadeera C (2017) The saliva microbiome profiles are minimally affected by collection method or DNA extraction protocols. Sci Rep 7(1):8523
37. Menon RK, Gopinath D (2018) Eliminating bias and accelerating the clinical translation of oral microbiome research in oral oncology. Oral Oncol 79:84–85
38. Kim Y, Koh I, Rho M (2015) Deciphering the human microbiome using next-generation sequencing data and bioinformatics approaches. Methods 79–80:52–59
39. Bahl MI, Bergström A, Licht TR (2012) Freezing fecal samples prior to DNA extraction affects the firmicutes to bacteroidetes ratio determined by downstream quantitative PCR analysis. FEMS Microbiol Lett 329 (2):193–197
40. Pollock J, Glendinning L, Wisedchanwet T, Watson M (2018) The madness of microbiome: attempting to find consensus "best practice" for 16S microbiome studies. Appl Environ Microbiol 84(7):e02627-17
41. Tsilimigras MCB, Fodor AA (2016) Compositional data analysis of the microbiome: fundamentals, tools, and challenges. Ann Epidemiol 26(5):330–335
42. Zhou X, Nanayakkara S, Gao J-L, Nguyen K-A, Adler CJ (2019) Storage media and not extraction method has the biggest impact on recovery of bacteria from the oral microbiome. Sci Rep 9(1):14968
43. Goodrich JK, Di Rienzi SC, Poole AC, Koren O, Walters WA, Caporaso JG et al (2014) Conducting a microbiome study. Cell 158(2):250–262
44. Kim D, Hofstaedter CE, Zhao C, Mattei L, Tanes C, Clarke E et al (2017) Optimizing methods and dodging pitfalls in microbiome research. Microbiome 5(1):52
45. Luo T, Srinivasan U, Ramadugu K, Shedden KA, Neiswanger K, Trumble E et al (2016) Effects of specimen collection methodologies and storage conditions on the short-term stability of oral microbiome taxonomy. Appl Environ Microbiol 82(18):5519–5529
46. Bhattarai KR, Kim H-R, Chae H-J (2018) Compliance with saliva collection protocol in healthy volunteers: strategies for managing risk and errors. Int J Med Sci 15(8):823–831
47. Kuczynski J, Lauber CL, Walters WA, Parfrey LW, Clemente JC, Gevers D et al (2012) Experimental and analytical tools for studying the human microbiome. Nat Rev Genet 13 (1):47–58
48. Salter SJ, Cox MJ, Turek EM, Calus ST, Cookson WO, Moffatt MF et al (2014) Reagent and laboratory contamination can critically impact sequence-based microbiome analyses. BMC Biol 12(1):87
49. Robinson CK, Brotman RM, Ravel J (2016) Intricacies of assessing the human microbiome in epidemiologic studies. Ann Epidemiol 26 (5):311–321
50. Humphrey B, McLeod N, Turner C, Sutton JM, Dark PM, Warhurst G (2015) Removal of contaminant DNA by combined UV-EMA

treatment allows low copy number detection of clinically relevant bacteria using pan-bacterial real-time PCR. PLoS One 10(7):e0132954

51. Corless CE, Guiver M, Borrow R, Edwards-Jones V, Kaczmarski EB, Fox AJ (2000) Contamination and sensitivity issues with a real-time universal 16S rRNA PCR. J Clin Microbiol 38(5):1747–1752
52. Klaschik S, Lehmann LE, Raadts A, Hoeft A, Stuber F (2002) Comparison of different decontamination methods for reagents to detect low concentrations of bacterial 16S DNA by real-time-PCR. Mol Biotechnol 22 (3):231–242
53. Rand KH, Houck H (1990) Taq polymerase contains bacterial DNA of unknown origin. Mol Cell Probes 4(6):445–450
54. Mohammadi T, Reesink HW, Vandenbroucke-Grauls CMJE, Savelkoul PHM (2005) Removal of contaminating DNA from commercial nucleic acid extraction kit reagents. J Microbiol Methods 61(2):285–288
55. Lazarevic V, Gaïa N, Girard M, Schrenzel J (2016) Decontamination of 16S rRNA gene amplicon sequence datasets based on bacterial load assessment by qPCR. BMC Microbiol 16 (1):73
56. Davis NM, Proctor DM, Holmes SP, Relman DA, Callahan BJ (2018) Simple statistical identification and removal of contaminant sequences in marker-gene and metagenomics data. Microbiome 6(1):226
57. Hornung BVH, Zwittink RD, Kuijper EJ (2019) Issues and current standards of controls in microbiome research. FEMS Microbiol Ecol 95(5):fiz045
58. Fouhy F, Clooney AG, Stanton C, Claesson MJ, Cotter PD (2016) 16S rRNA gene sequencing of mock microbial populations-impact of DNA extraction method, primer choice and sequencing platform. BMC Microbiol 16(1):123
59. Parada AE, Needham DM, Fuhrman JA (2016) Every base matters: assessing small subunit rRNA primers for marine microbiomes with mock communities, time series and global field samples. Environ Microbiol 18 (5):1403–1414
60. Wu J (2014) Comparison of DNA extraction methods for human oral microbiome research. Br J Med Med Res 4(10):1980–1991
61. Tan SC, Yiap BC (2009) DNA, RNA, and protein extraction: the past and the present. J Biomed Biotechnol 2009:574398
62. Lim MY, Song E-J, Kim SH, Lee J, Nam Y-D (2018) Comparison of DNA extraction methods for human gut microbial community profiling. Syst Appl Microbiol 41(2):151–157
63. Rosenbaum J, Usyk M, Chen Z, Zolnik CP, Jones HE, Waldron L et al (2019) Evaluation of oral cavity DNA extraction methods on bacterial and fungal microbiota. Sci Rep 9(1):1531
64. Sohrabi M, Nair RG, Samaranayake LP, Zhang L, Zulfiker AHM, Ahmetagic A et al (2016) The yield and quality of cellular and bacterial DNA extracts from human oral rinse samples are variably affected by the cell lysis methodology. J Microbiol Methods 122:64–72
65. Shehadul Islam M, Aryasomayajula A, Selvaganapathy P (2017) A review on macroscale and microscale cell lysis methods. Micromachines 8 (3):83
66. Starke R, Jehmlich N, Alfaro T, Dohnalkova A, Capek P, Bell SL et al (2019) Incomplete cell disruption of resistant microbes. Sci Rep 9 (1):5618
67. Goldberg S (2008) Mechanical/physical methods of cell disruption and tissue homogenization. In: Posch A (ed) 2D PAGE: sample preparation and fractionation, Methods in molecular biology, vol 424. Humana, Totowa, NJ
68. de Boer R, Peters R, Gierveld S, Schuurman T, Kooistra-Smid M, Savelkoul P (2010) Improved detection of microbial DNA after bead-beating before DNA isolation. J Microbiol Methods 80(2):209–211
69. Abusleme L, Hong B-Y, Dupuy AK, Strausbaugh LD, Diaz PI (2014) Influence of DNA extraction on oral microbial profiles obtained via 16S rRNA gene sequencing. J Oral Microbiol 6(1):23990
70. Salonen A, Nikkilä J, Jalanka-Tuovinen J, Immonen O, Rajilić-Stojanović M, Kekkonen RA et al (2010) Comparative analysis of fecal DNA extraction methods with phylogenetic microarray: effective recovery of bacterial and archaeal DNA using mechanical cell lysis. J Microbiol Methods 81(2):127–134
71. Walker AW, Martin JC, Scott P, Parkhill J, Flint HJ, Scott KP (2015) 16S rRNA gene-based profiling of the human infant gut microbiota is strongly influenced by sample processing and PCR primer choice. Microbiome 3(1):26
72. Sabino J, Vieira-Silva S, Machiels K, Joossens M, Falony G, Ballet V et al (2016) Primary sclerosing cholangitis is characterised by intestinal dysbiosis independent from IBD. Gut 65(10):1681–1689
73. Schmidt BL, Kuczynski J, Bhattacharya A, Huey B, Corby PM, Queiroz ELS et al (2014) Changes in abundance of oral

microbiota associated with oral cancer. PLoS One 9(6):e98741
74. Schloss PD (2018) Identifying and overcoming threats to reproducibility, replicability, robustness, and generalizability in microbiome research. mBio 9(3):e00525-18
75. Scott AJ, Alexander JL, Merrifield CA, Cunningham D, Jobin C, Brown R et al (2019) International cancer microbiome consortium consensus statement on the role of the human microbiome in carcinogenesis. Gut 68 (9):1624–1632

Chapter 2

Oral Sampling Techniques

Heba Hussein

Abstract

The human oral cavity is a major point of entry for microorganisms, many of which live and multiply in the mouth. In addition, it provides an accessible site for sampling compared to other parts of the body; however, caution should be taken during oral sampling as many factors contribute to the microbial diversity in a site-dependent manner. The accessibility of the oral cavity and its microbial diversity emphasize the crucial need to avoid cross-contamination during the sampling procedure. In this chapter, we describe various detailed oral sampling procedures. These methods include supragingival dental plaque sampling, subgingival dental plaque sampling, oral mucosal sampling, and endodontic sampling methods for extracted teeth or in the patient's mouth. The proposed protocols provide tips to avoid contamination between different oral sources of bacteria and possible alternatives to the tools used.

Key words Supragingival plaque, Subgingival plaque, Microbiome, Oral cavity, Sampling, Endodontic sample, Oral mucosal sample, Oral sampling, Teeth

1 Introduction

The oral cavity harbors one of the most diverse microbial populations in the human body [1]. This microbial population includes about 1000 indigenous bacterial species, 300–2000 viral genotypes, and 20 fungal types [2–4]. Previous studies have shown that oral bacteria are not only connected to oral diseases but also systemic diseases [5].

Proctor and Relman [6] underscored the concept of the spatial ecology of the microbial populations in the human body based on the theory of landscape ecology, which simply explains spatial heterogeneity rather than assuming a homogenous space. For example, in the oral cavity, biomass accumulation varies by tooth surface site; toothbrushing removes biofilms fairly well from the cheek (buccal) and tongue (lingual) surfaces (Fig. 1) of the teeth and less well from the biting (occlusal) surfaces of the posterior teeth, thus permitting higher biomass accumulation at the biting surfaces. Another factor contributing to the microbial differences between the buccal and lingual surfaces of the tooth is the proximity of the

Guy R. Adami (ed.), *The Oral Microbiome: Methods and Protocols*, Methods in Molecular Biology, vol. 2327, https://doi.org/10.1007/978-1-0716-1518-8_2, © Springer Science+Business Media, LLC, part of Springer Nature 2021

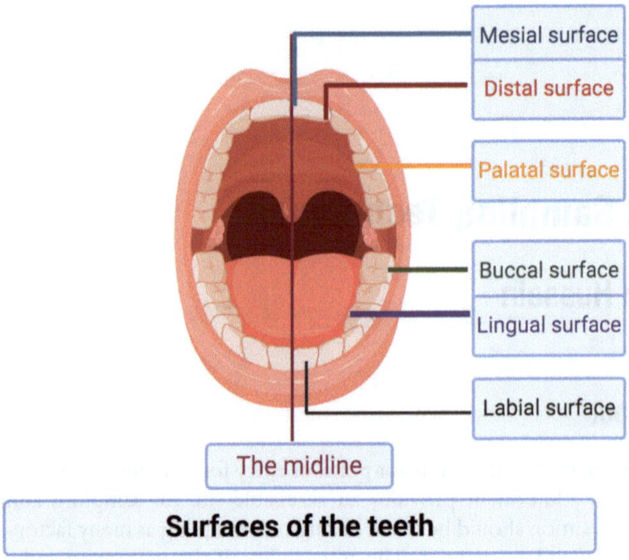

Fig. 1 Diagram representing different tooth surfaces. (Created with BioRender.com)

surfaces to the nearest salivary gland. The minor salivary glands in the labial, buccal, and palatal mucosa release viscous secretions with poor buffering capacity [7, 8]. An additional example of spatial heterogeneity in the oral cavity is the difference between bacteria found in plaque above the gum line (supragingival) and that in plaque below the gum line (subgingival) (Fig. 2). This difference is because the anaerobic bacteria in the mouth prefer areas where oxygen is limited, as in the subgingival crevice [6].

The salivary glands not only contribute to the spatial heterogeneity of the hard tissues but also that of the soft tissues, as the three major salivary glands (the parotid, the submandibular, and the sublingual) differ in the secretory rates and the salivary composition [9], suggesting that microbial communities of the soft tissues may vary along moisture or pH gradient [10]. Another example of this space-related colonization heterogenicity is the difference of the community structure between the dorsal surface of the tongue and the lateral or ventral (lower) tongue surfaces [11]. These differences are likely due to the different keratinization and different tongue papillae arrangement of these sites [12].

In this chapter, we present simplified protocols for oral sampling methods. Also, we present the oral and dental terms and tools related to those sampling techniques. The techniques presented are supragingival dental plaque sampling, subgingival dental plaque sampling, oral mucosal sampling, and two types of endodontic sampling methods, one on extracted teeth and the second inside the patient's mouth. While these methods are presented for the collection of bacterial DNA, they also can be used to collect fungus

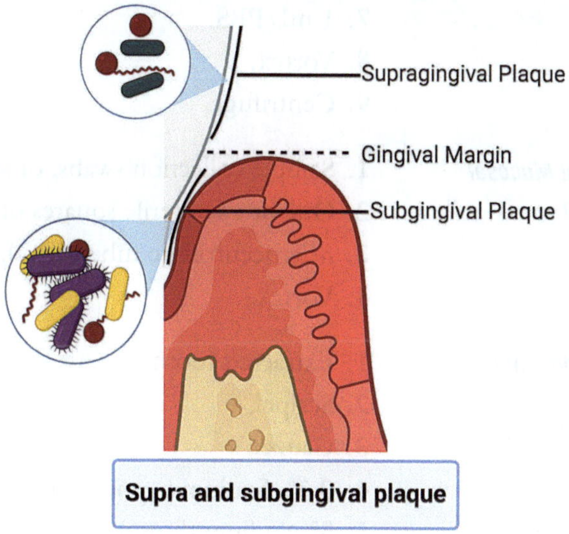

Fig. 2 Diagram representing supragingival and subgingival plaque. (Created with BioRender.com)

and virus samples. If RNA is to be collected, a key difference is that soon after collection, the sample must be immersed in an RNA preservative to maintain RNA profiles identical to that found in vivo.

2 Materials

2.1 Supragingival Dental Plaque Sampling

1. Surgical mask and eye protection (goggles or a face shield).
2. Sterile dental cotton rolls.
3. Sampling tool. A sterile Teflon spatula, sterile dental excavator, or a sterile Gracey curette.
4. 1.5 mL Sterile centrifuge tube.
5. Suspending solution: RNAprotect reagent or 1 mL TE buffer: 10 mM Tris–HCl, 1 mM EDTA, pH 8.0, or 1 mL PBS.
6. Centrifuge.
7. Air-water syringe.
8. Vortex.

2.2 Subgingival Dental Plaque Sampling

1. Surgical mask and eye protection (goggles or a face shield).
2. Sterile dental cotton rolls.
3. Sterile dental scaler.
4. Sterile dental tweezers.
5. Sterile, size 30 paper points.
6. Tube with 750μL Tris EDTA buffer, pH 8.0.

7. 1 mL PBS.
8. Vortex.
9. Centrifuge.

2.3 Oral Mucosal Sampling

1. Sample collection swabs, or small cotton-tipped applicators.
2. One or two sterile squares of 2″ × 2″ gauze.
3. Microcentrifuge tube with 0.8 mL TE.
4. Vortex.

2.4 Endodontic Samples from Extracted Teeth

1. Extracted tooth.
2. Scalpel.
3. Gauze.
4. 0.5% Sodium hypochlorite.
5. 5% Sodium thiosulfate.
6. Two sterilized diamond disks.
7. RNA stabilization solution.
8. Lysis buffer.
9. 2 mL Screw-capped tubes.
10. Freezer mill.
11. Centrifuge.

2.5 Endodontic Sample from a Patient

1. Rubber dam.
2. 30% Hydrogen peroxide solution (H_2O_2).
3. 2.5% Sodium hypochlorite solution (NaOCl).
4. High-speed sterile carbide bur.
5. 5% Sodium thiosulfate.
6. Gates-Glidden burs.
7. K- and/or H-type files.
8. ProTaper retreatment files.
9. Tube containing 0.75 mL reduced transport fluid (RTF).
10. Cryotube containing 0.75 mL RTF.
11. K files through size 30.
12. Sterile saline solution.
13. Sterile #10 K-type hand file.
14. Endodontic irrigation syringes with 27 G needles.
15. Sterile paper points.
16. Surgical scissors.
17. Apex locator.

3 Methods

3.1 Supragingival Dental Plaque Sampling

This sampling technique is ideally performed by a dentist or a dental hygienist (*see* **Note 1**). Sampling should be performed by one examiner to maintain uniformity.

1. Wear a surgical mask and eye protection (goggles or a face shield) to avoid the splatter of dental plaque, calculus, or saliva.
2. Request the patients refrain from brushing teeth for 24 h before the appointment, but not more than 48 h (*see* **Note 2**).
3. Isolate the sampling sites with sterile dental cotton rolls (Fig. 3) to prevent contamination with saliva, and dry with a gentle stream of air from an air-water syringe (*see* **Note 3**).
4. Collect supragingival plaque by scraping the tooth or teeth (*see* **Note 4**) using one of the following tools: A sterile Teflon spatula of a suitable size [13], a curved spatula is preferred; a sterile dental excavator [14]; or a Sterile Gracey curette [15] (Fig. 4), as it has a curved end that acts as a scooper.
5. Whatever tool is used to remove the plaque, it needs to be immediately immersed and agitated in a solution to transfer sample to a suspension. This can be done using 500–1000μL of nucleic acid preservative, which also preserves bacteria structure, such as RNAprotect cell reagent. Alternatively, place the plaque-covered end of the tool in a labeled 1.5 mL sterile centrifuge tube containing 1 mL TE buffer comprised of 10 mM Tris–HCl and 1 mM EDTA, pH 8.0, or 1 mL PBS, agitate for 4 or 5 s, and scrape as much plaque as possible into the tube. Depending on the experimental design, you may pool like samples in one tube.
6. Store on ice until transport to the laboratory. Process or freeze the clinical samples within 2 h, then store at −80 °C.
7. Before or after freezing, vortex the sample (3 × 30 s), and centrifuge at 14,000 × *g* for 10 min at 4 °C to obtain the pellet that contains the bacteria with the DNA. Freeze the cell pellets at − 80 °C until DNA extraction is performed (*see* **Note 5**).

3.2 Subgingival Dental Plaque Sampling

This sampling technique needs to be performed by a dentist or a dental hygienist (*see* **Note 1**). Sampling should be performed by one examiner to maintain sample uniformity. Subgingival samples are most easily collected from patients with periodontal disease, as the sulcus is enlarged (*see* **Note 6**).

1. Wear a surgical mask and eye protection (goggles or a face shield) to avoid contact with the splatter of dental plaque, calculus, or saliva.

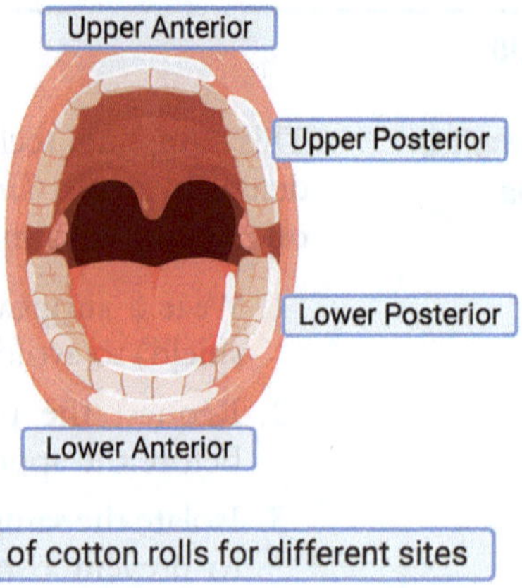

Fig. 3 Diagram representing cotton roll placement for different areas in the mouth. (Created with BioRender.com)

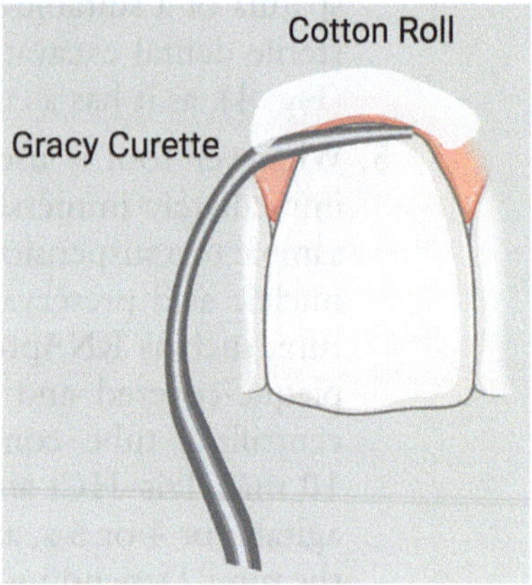

Fig. 4 Diagram representing plaque collection for an upper anterior tooth with a Gracey curette. (Created with BioRender.com)

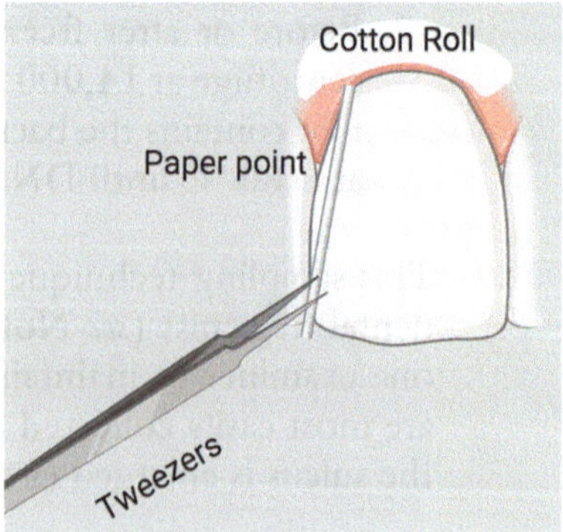

Fig. 5 Diagram representing plaque collection for an upper tooth with a paper point. (Created with BioRender.com)

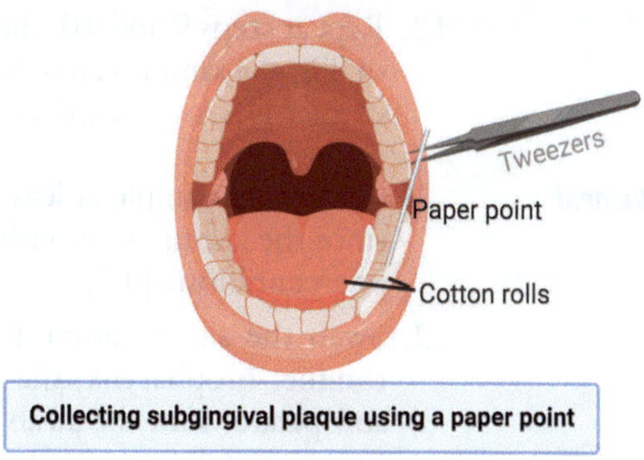

Fig. 6 Diagram representing plaque collection for a lower posterior tooth with a paper point. (Created with BioRender.com)

2. Determine the sampling sites. Some [16] have sampled the subgingival plaque from the buccal sulcus of the first molar in each quadrant, though any tooth can be sampled and especially those with deep pockets (*see* **Note 6**).
3. Isolate the sites from the buccal mucosa (*see* **Note 3**) by placing sterile dental cotton rolls in the buccal vestibule related to the tooth. Note that the buccal vestibule is the junction between the teeth and the interior of the cheeks or lips.
4. Before collecting the samples, remove the supragingival plaque or tooth surface biofilm above the gingiva to avoid cross-contamination with the subgingival plaque sample. This step must be done by a dentist or a dental hygienist.
5. For each site to be sampled, insert a sterile, size 30 paper point (Figs. 5 and 6), held with a dental tweezer in the buccal (facial/outer/external) surface of the tooth as deep as possible into the gingival sulcus, the thin space between the tooth and the surrounding gingiva. Start at the nearest part of the sulcus to the jaw midline (mesiobuccal) and slowly pass to the farthest part of the sulcus away from the midline (distobuccal), as shown in Fig. 1 (*see* **Notes 7** and **8**).
6. Place the paper points from all the sampled teeth into one tube with 750μL Tris-EDTA buffer, pH 8.0, and store at −80 °C.
7. When ready for DNA extraction, thaw the sample, then vortex it at full speed to dislodge the dental plaque from the paper point.
8. Centrifuge at 15,000 × *g* for 5 min at 4 °C.
9. Remove the supernatant.
10. Replace with 1 mL PBS, resuspend as best possible with vortexing. Note that not all clumps will disappear.
11. Repeat centrifugation.

12. Repeat **steps 9** and **10**, then resuspend the pellet in 200μL PBS or TE; the sample can be stored frozen or you may immediately extract DNA via standard methods (*see* **Note 5**).

3.3 Oral Mucosal Sample

1. Collect the sample at least 1 h after the last food intake, 1 day since the last use of mouthwash, and 1 month since the last use of an antibiotic [17].
2. Swab the site of interest, which can include buccal mucosa, tongue, oropharynx, the floor of the mouth, hard palate, or soft palate. Run the swab over the area of interest for 5–10 s, being careful to cover all surfaces of the swab. To ease access to sites of interest, it is important to use one or two sterile squares of 2″ × 2″ gauze to grasp the tongue and move it out of the way. Use light to aid in collecting samples from the back of the mouth (soft palate) and adjust the patient's chair (*see* **Notes 9–11**).
3. Place the swab in the microcentrifuge tube with 0.8 mL TE. B-reak the shaft so that it fits in the tube and then close. Store on ice for up to 1 h before transfer to the freezer for long-term storage.
4. Upon thawing, vortex vigorously 10 s with a swab in tube and remove what you need. 100μL should produce a yield of about 1 to several μg of total DNA. One can also aliquot the remainder after pushing the sides of the swab against the tube to drain the liquid.

3.4 Endodontic Sample from Extracted Teeth

All steps for an endodontic sample from extracted teeth [18] must be conducted under strict aseptic conditions.

1. After extraction, remove all tissue with a scalpel and then scrub the tooth surface vigorously with gauzes soaked in 0.5% sodium hypochlorite.
2. Wipe off the tooth with a gauze soaked in 5% sodium thiosulfate.
3. Decoronate the tooth (by removing the crown) using a sterilized diamond disk under saline cooling to split the tooth perpendicular to the long axis at the cementum enamel border.
4. Split the remaining root into an apical and a coronal segment with another sterile diamond disk under saline cooling, 2–6 mm from the apex.
5. Store the segments at −80 °C in a 15 mL conical tube until further processing.
6. Cryo-pulverize the root segments under aseptic conditions using a freezer mill operated at −80 °C and store at −80 °C in a 5 mL RNA stabilization solution (*see* **Note 12**).

7. Thaw the powdered root segments and centrifuge for 5 min at 4500 × *g*, remove the supernatant, resuspend the powdered root segment in 200μL lysis buffer suitable for DNA isolation with the chosen method.

3.5 Endodontic Sample from a Patient

Endodontic sample from a patient [19] should be done by a single experienced endodontist to maintain uniformity in samples.

Some studies have focused on teeth with evidence of periapical disease, such as the widening of the apical periodontal ligament space or radiolucent lesion shown in radiological examinations (*see* **Note 13**).

Exclusion criteria of the reference study are antibiotic treatment in the previous 3 months (*see* **Note 14**) or have teeth that are difficult to isolate with a rubber dam. Care must be taken to evaluate the tooth for coronal leakage, due to an inadequate crown, and periodontal disease before sampling.

1. Clean the tooth with pumice and isolate it with a rubber dam.
2. Decontaminate the tooth and disinfect it with a 30% hydrogen peroxide solution (H_2O_2) and then with 2.5% sodium hypochlorite solution (NaOCl).
3. Prepare the access cavity (of the previously treated root canal) with a high-speed sterile carbide bur, and before the pulp chamber exposure, clean the tooth and rubber dam as described in **step 1** of this section.
4. Quench the NaOCl solution with sodium thiosulfate to avoid interference with the bacterial sample (*see* **Note 15**).
5. Check the sterility of the operating field by taking samples of the disinfected tooth crown using two pellets of cotton wool and transfer it to a tube containing RTF. If growth detected after 72 h of anaerobic incubation, exclude the sample.
6. Remove the gutta-percha (the root canal obturation material) in the coronal canal using Gates-Glidden burs and the gutta-percha in the apical portion of the root with K- and/or H-type files and ProTaper retreatment files. Don't use a solvent (*see* **Note 16**).
7. Remove the root filling material from the apical portion of the canals and transfer it into a tube containing 0.75 mL RTF.
8. Take radiographs to ensure that all the root filling had been removed (It is a routine step in root canal re-treatment).
9. Now, proceed with a regular root canal filling re-treatment. Obtain apical patency and establish the working length with an electronic apex locator and subsequent radiographic control. File the canal walls again, but gently, at full working length with K-files until size 30. Use a sterile saline solution as the irrigant (*see* **Note 17**).

10. After use, stir the active portion of each instrument in a cryotube with 0.75 mL RTF to obtain the dentine debris.
11. After using size 30 hand files, place 0.2 mL of sterile saline solution into the root canal using an endodontic irrigation syringe with a 27 G needle. Absorb the contents of the root canal into four consecutive sterile paper points. Hold each paper point in place within the canal at working length for 1 min and transfer it into the same tube with 0.75 mL RTF in which the active portions of the files had been rinsed. Cut the terminal 4 mm of the paper point using a sterile pair of surgical scissors.
12. Process both samples obtained in **steps 10** and **11** within 2 h for DNA extraction (*see* **Note 18**) and complete the root canal treatment of each tooth.

4 Notes

1. The collection of dental plaque samples (supragingival, subgingival) should be collected by someone with a license to practice health care. It could be a dentist, a hygienist, a nurse, or a physician, all of whom should be licensed in their clinical field, but all this is based on state law, and it may be possible the research protocol allows for other personnel within a university research setting when approved by the university Institutional Review Board. The collection of subgingival plaque must be done by dental personnel to avoid injury to the gingiva.
2. Not brushing the teeth for at least 24 h and not more than 48 h will leave a thin film of soft plaque on teeth surfaces and will not allow it to harden into calculus (tartar). Ask the participants to note the moment of the last brushing.
3. Place the sterile cotton rolls in the buccal vestibule for upper teeth, and in both buccal/labial and lingual vestibule for lower teeth as in Fig. 3. The lingual surface of lower anterior teeth and the buccal surface of upper posterior teeth are opposite to salivary gland ducts. Some patients have gagging with placement of the cotton roll, so the cotton rolls should not be bulky.
4. Usually, one or two scrapes are enough to collect a good amount of dental plaque, as the dental plaque is soft, though multiple strokes are possible to ensure maximal yield. Sample carefully enough to not harm the gingival papilla, but adequately until you can see material on the scraper. Harvest the dental plaque from one side of the tooth surface to the other site (mesial to distal or distal to mesial). It is crucial this step must be done in a consistent manner for all subjects with a similar number of scrapes and amount of pressure.

5. Sample can be stored frozen in TE or PBS before DNA isolation. Though if the supernatant is removed after thaw and centrifugation there is the theoretical risk of loss of some bacterial DNA due to lysis after the freeze/thaw. Thus, for the most stringent protocols, samples are centrifuged and the supernatant removed before the pellet is stored frozen at −80 °C. Alternatively, a preservative, such as RNAprotect Cell/Bacteria Reagent, can be added to the sample on collection, though this may cause other differences [20–22]. Whatever method is chosen, it should be used consistently.
6. Although any tooth can be sampled for subgingival plaque, start by trying the posterior teeth as they are more likely to have deeper pockets than the anteriors.
7. Avoid touching any tooth or gingival surfaces above the gum line with the paper point. A key component of sampling subgingival microbiome is to ensure that contaminating bacteria from saliva or supragingival tooth surfaces are not mistakenly sampled. Any contaminated paper point should be discarded.
8. Alternatively, an individual sterile Gracey curette can be used to sample subgingival plaque after **step 4**, of the Subheading 3.2 instead of a paper point. A Gracey curette is used by dental personnel. If the curette is used, insert the curette into the deepest point of the pocket, and move from one surface of the tooth to the other with a single stroke on the tooth surface with constant pressure to aid in reproducibility. As with all sample taking, it is best to use a single operator. Plaque is scraped into a 1.5 mL tube and resuspended in TE (this topic is covered by Gopinath and Menon in Chapter 1).
9. It is important to choose sites that can be reproducibly swabbed in different people. For example, the tongue has distinct surfaces depending on the position, dorsum versus ventral versus lateral border of the tongue and base of the tongue, that may have distinct bacterial profiles. For that reason, effort must be made to sample the same site, making self-collected swabs of less value.
10. To avoid contamination with the teeth microbes, avoid touching the teeth.
11. Care must be taken in collecting oropharyngeal samples in order not to induce a gag reflex.
12. Alternatively, five post cryo pulverization samples can be frozen as a powder at −80 °C for storage. Then resuspended in lysis solution depending on the method of DNA isolation chosen.

13. In that there needs to be a reason to perform re-treatment on the tooth to allow the in situ sampling, typically there will be pathology in the tooth.
14. Exclude patients who have taken antibiotics for the last 3 months because the composition of bacteria in the tooth pulp may have been affected. This length of time is chosen out of utmost caution, as it is not clear what the effects of antibiotic use cessation are on bacteria turnover in the tooth pulp.
15. 5% sodium thiosulfate is used to neutralize the effect of NaOCl solution, so that NaOCl would not affect the character/kill the bacteria that is present apically. This solution is originally used in swimming pools to neutralize the whitening effect of chlorine.
16. Again, to avoid killing the bacteria, don't use a solvent.
17. Use sterile saline as an irrigant so as to not kill the bacteria.
18. Sampling from a tooth in situ is likely to be less representative of what is present in the canal as compared to samples from extracted teeth.

References

1. The Human Microbiome Project Consortium (2012) Structure, function and diversity of the healthy human microbiome. Nature 486:207–214
2. Dewhirst F, Chen T, Izard J, Paster BJ, Tanner ACR, Yu W-H (2010) The human oral microbiome. J Bacteriol 192(19):5002–5017
3. Pride D, Salzman J, Haynes M, Rowher F, Davis-Long C, White RA et al (2012) Evidence of a robust resident bacteriophage population revealed through analysis of the human salivary viromes. ISME J 6(5):915–926
4. Ghannoum MA, Jurevic RJ, Mukherjee PK, Cui F, Sikaroodi M, Naqvi A et al (2010) Characterization of the oral fungal microbiome (mycobiome) in healthy individuals. PLoS Pathog 6(1):e1000713
5. Wade WG (2013) The oral microbiome in health and disease. Pharmacol Res 69 (1):137–143
6. Proctor DM, Relman DA (2017) The landscape ecology and microbiota of the human nose, mouth, and throat. Cell Host Microbe 21(4):421–432
7. Dawes C, Wood CM (1973) The composition of human lip mucous gland secretions. Arch Oral Biol 18(3):343–350
8. Sato Y, Yamagishi J, Yamashita R, Shinozaki N, Ye B, Yamada T et al (2015) Inter-individual differences in the oral bacteriome are greater than intra-day fluctuations in individuals. PLoS One 10(6):e0131607
9. Schneyer LH, Levin LK (1955) Rate of secretion by individual salivary gland pairs of man under conditions of reduced exogenous stimulation. J Appl Physiol 7(5):508–512
10. Wolff M, Kleinberg I (1998) Oral mucosal wetness in hypo- and normosalivators. Arch Oral Biol 43(6):455–462
11. Aas JA, Paster BJ, Stokes LN, Olsen I, Dewhirst FE (2005) Defining the normal bacterial flora of the oral cavity. J Clin Microbiol 43(11):5721–5732
12. Aufdemorte TB, Cameron IL (1981) The relation of keratinization to bacterial colonization on the baboon tongue as demonstrated by scanning electron microscopy. J Dent Res 60 (6):1008–1014
13. Keijser BJF, van den Broek TJ, Slot DE, van Twillert L, Kool J, Thabuis C et al (2018) The impact of maltitol-sweetened chewing gum on the dental plaque biofilm microbiota composition. Front Microbiol 9:381
14. Shi W, Tian J, Xu H, Zhou Q, Qin M (2018) Distinctions and associations between the microbiota of saliva and supragingival plaque of permanent and deciduous teeth. PLoS One 13(7):e0200337

15. Caselli E, Fabbri C, D'Accolti M, Soffritti I, Bassi C, Mazzacane S et al (2020) Defining the oral microbiome by whole-genome sequencing and resistome analysis: the complexity of the healthy picture. BMC Microbiol 20(1):120
16. Al-Kamel A, Baraniya D, Abdulnaser Al-Hajj W, Halboub E, Abdulrab S, Chen T et al (2019) Subgingival microbiome of experimental gingivitis: shifts associated with the use of chlorhexidine and *N*-acetyl cysteine mouthwashes. J Oral Microbiol 11 (1):1608141
17. Zaura E, Brandt BW, Teixeira de Mattos MJ, Buijs MJ, Casper MPM, Rashid MU et al (2015) Same exposure but two radically different responses to antibiotics: resilience of the salivary microbiome versus long-term microbial shifts in feces. mBio 6(6):e01693–e01615
18. Persoon IF, Buijs MJ, Özok AR, Crielaard W, Krom BP, Zaura E et al (2017) The mycobiome of root canal infections is correlated to the bacteriome. Clin Oral Investig 21 (5):1871–1881
19. Sánchez-Sanhueza G, Bello-Toledo H, González-Rocha G, Gonçalves AT, Valenzuela V, Gallardo-Escárate C (2018) Metagenomic study of bacterial microbiota in persistent endodontic infections using next-generation sequencing. Int Endod J 51(12):1336–1348
20. Zhou X, Nanayakkara S, Gao J, Nguyen K, Adler CJ (2019) Storage media and not extraction method has the biggest impact on recovery of bacteria from the oral microbiome. Sci Rep 9(1):14968
21. Adler CJ, White A, Bockmann M, Browne GV, Townsend G, Hughes T (2018) VMG II transport medium stabilises oral microbiome samples for next-generation sequencing. J Microbiol Methods 144:91–98
22. Luo T, Sirnivasan U, Ramadugu K, Shedden KA, Neiswanger K, Trumble E et al (2016) Effects of specimen collection methodologies and storage conditions on the short-term stability of oral microbiome taxonomy. Appl Environ Microbiol 82(18):5519–5529

Chapter 3

Functional Metagenomic Screening for Antimicrobial Resistance in the Oral Microbiome

Supathep Tansirichaiya, Liam J. Reynolds, and Adam P. Roberts

Abstract

A large proportion of bacteria, from a multitude of environments, are not yet able to be grown in the laboratory, and therefore microbiological and molecular biological investigations of these bacteria are challenging. A way to circumvent this challenge is to analyze the metagenome, the entire collection of DNA molecules that can be isolated from a particular environment or sample. This collection of DNA molecules can be sequenced and assembled to determine what is present and infer functional potential, or used as a PCR template to detect known target DNA and potentially unknown regions of DNA nearby those targets; however assigning functions to new or conserved hypothetical, functionally cryptic, genes is difficult. Functional metagenomics allows researchers to determine which genes are responsible for selectable phenotypes, such as resistance to antimicrobials and metabolic capabilities, without the prerequisite needs to grow the bacteria containing those genes or to already know which genes are of interest. It is estimated that a third of the resident species of the human oral cavity is not yet cultivable and, together with the ease of sample acquisition, makes this metagenome particularly suited to functional metagenomic studies. Here we describe the methodology related to the collection of saliva samples, extraction of metagenomic DNA, construction of metagenomic libraries, as well as the description of functional assays that have previously led to the identification of new genes conferring antimicrobial resistance.

Key words Functional metagenomics, Functional screening, Oral metagenome, High-throughput screening, Antimicrobial resistance genes, Oral microbiome, Antibiotic resistance, Antiseptic resistance, AMR

1 Introduction

Many bacterial species in microbial communities from different environments have been identified as uncultivable, or yet-to-be cultivated, in the laboratory due to specific physical, chemical, and biological conditions required by each bacteria for growth such as nutrient availability, temperature, and secondary metabolites from other members in the community [1, 2]. The human oral cavity is the second-most complex microbial community in the human body and it is composed of various distinct microbial habitats such as the surface of teeth, cheek, supra, and sub-gingival, which have

Guy R. Adami (ed.), *The Oral Microbiome: Methods and Protocols*, Methods in Molecular Biology, vol. 2327,
https://doi.org/10.1007/978-1-0716-1518-8_3,

different properties such as pH, oxygen level, and nutrients [3]. Approximately 200–300 bacterial taxa can be found per mouth, and more than 700 bacterial species have been listed in the human oral microbiome database, of which more than one-third are considered unculturable bacteria [4–8]. Relying on a culture-dependent method is, therefore, not enough to study entire functional potential of the microbiome, such as metabolism, antimicrobial resistance (the resistome), and mobile genetic elements (the mobilome), as functions of new genes or gene families would not be able to be assigned based solely on the conserved motifs or available sequences in the database [9].

Functional metagenomics is a term used to describe experimental approaches which match genes, regardless of their source, to phenotypes. These experiments are powerful in that they can be used to identify genes of interest from uncultured bacteria, which confer abilities on a surrogate bacterial host, and these methodologies have been applied to various environmental metagenomic samples such as soil, sediment, wastewater, and also human microflora [10–13]. Total community DNA or metagenomic DNA can be extracted directly from environmental samples and used to construct a metagenomic library by ligating the DNA into cloning vectors and introducing these constructs into surrogate hosts. Clones with genes conferring the function of interest can be selected by screening the metagenomic library in a suitable assay, for example, on selective media containing antibiotics. Function-based metagenomics has a number of advantages including (1) high-throughput screening that can investigate bacterial genes from multiple species at one time, (2) the analysis of genes from both culturable and unculturable bacteria, and (3) the potential to identify novel genes as it relies on the function of the genes and no prior sequence knowledge is required.

For the human oral cavity, metagenomic DNA can be extracted from saliva samples, which is easily collected and often used to represent the oral microbiome and even to identify diagnostic markers for several diseases [14–16]. With the high level of diversity in the oral microbiome, functional metagenomics has the potential to discover various genes depending on research purposes that can be designed through a selection of surrogate host(s), cloning vectors, and screening methods. In this chapter, we described functional metagenomic protocols used to identify several novel antimicrobial resistance genes conferring resistance to antibiotics, such as tetracycline, β-lactams, and sulfonamide, and antiseptics, such as cetyltrimethylammonium bromide (CTAB) and triclosan [17–22]. Throughout the chapter, we have indicated where methodological alterations can be considered to widen the functional assays to investigate additional aspects of biology.

2 Materials

2.1 Saliva Sample Collection and DNA Extraction

1. Sterilized saliva collection tubes (*see* **Note 1**).
2. Gentra Puregene Yeast/Bact. Kit (Qiagen) (*see* **Note 2**).
3. Isopropanol.
4. 70% Ethanol in distilled sterile water.

2.2 Construction of a Metagenomic Library

1. Restriction enzyme *Hin*dIII (20 U/μL and 1 U/μL diluted in dH_2O).
2. Alkaline phosphatase, Calf intestinal: 1 U/μL.
3. pCC1BAC vector [23] (*see* **Notes 3–5**).
4. PCR purification kit.
5. Fast-Link DNA Ligation Kit (Lucigen).
6. 0.1-cm-gap electroporation cuvettes.
7. TransforMax EPI300 Electrocompetent *E. coli* (Lucigen) (*see* **Notes 3–5**).
8. SOC medium.
9. 70% Ethanol in distilled sterile water and 100% ethanol.
10. 3 M Sodium acetate (NaOAc): Add about 70 mL of sterile distilled water to a 100-mL glass bottle. Weigh 12.3 g NaOAc and transfer to the glass bottle. Mix by using a magnetic stirrer and adjust the pH to 5.2 by adding glacial acetic acid. Add water up to 100 mL and filter-sterilize.
11. Sterile toothpicks.

2.3 Screening and Characterization of Antimicrobial Resistance Genes

1. Antimicrobial compounds for screening: Dissolve the compounds as described in the manufacturer's MSDS (*see* **Note 6**).
2. QIAprep Spin Miniprep Kit (Qiagen).
3. Restriction enzyme *Hin*dIII (10 U/μL).
4. Template Generation System II Kit (Thermo Fisher Scientific).
5. 1000× CopyControl Induction Solution (Lucigen) (*see* **Note 4**).
6. pCC1-F (Forward primer): 5′- GGATGTGCTGCAAGGC GATTAAGTTGG-3′.
7. pCC1-R (Reverse primer): 5′- CTCGTATGTTGTGTG GAATTGTGAGC-3′.
8. PCR master mix.

2.4 DNA Quality Control

1. 50× Tris-acetate-EDTA (TAE): Dissolve 242 g Tris Base and 18.6 g disodium EDTA in 700 mL dH_2O. Add 57 mL glacial acetic acid and top up with dH_2O to 1 L.

2. Agarose gel running buffer (1× TAE): Dilute 20 mL of 50× stock solution with 980 mL of distilled sterile water to a final volume of 1 L (1:50 dilution).
3. 10 mg/mL Ethidium bromide.
4. 1% Agarose gel: Prepare in TAE buffer, add ethidium bromide at the final concentration of 0.5μg/mL while still molten (*see* **Note 7**).
5. 6× Loading dye: Add 25 mg bromophenol blue, 3 mL glycerol, and 7 mL dH_2O.
6. Standard DNA ladder with sizes ranging from 0.5 to 50 kb.
7. Qubit fluorometer and Qubit dsDNA HS assay kit (*see* **Note 8**).
8. Nanodrop spectrophotometer (*see* **Note 8**).
9. Gel electrophoresis apparatus.
10. Gel imager system.

2.5 Bacterial Cultures

1. Lysogeny Broth (LB): Add 10 g NaCl, 10 g tryptone, and 5 g yeast extract in 950 mL dH_2O. Mix well to dissolve and adjust pH to 7.2. Top up with dH_2O to 1 L and sterilize by autoclave.
2. LB agar: Add 10 g NaCl, 10 g tryptone, 5 g yeast extract, and 15 g agar in 950 mL dH_2O. Mix well to dissolve and adjust pH to 7.2. Top up with dH_2O to 1 L and sterilize by autoclave. Sterilize by autoclaving and pour approximately 20–25 mL per plate when the molten agar is at 50 °C.
3. 100 mM Isopropyl-β-D-thiogalactopyranoside (IPTG) stock solution: Dissolve in dH_2O and filter sterilize.
4. 20 mg/mL 5-Bromo-4-chloro-3-indolyl-β-D-galactopyranoside (X-Gal) stock solution: Dissolve in DMSO.
5. 12.5 mg/mL Chloramphenicol stock solution: dissolve chloramphenicol in 70% ethanol, stored at −20 °C.
6. 20 mg/mL. Kanamycin stock solution: dissolve kanamycin in sterile water and filter sterilize.
7. LB–chloramphenicol broth: add 1μL chloramphenicol stock solution per 1 mL sterile LB broth.
8. LB–chloramphenicol agar: add 1μL of chloramphenicol stock solution per 1 mL molten LB agar.
9. LB–chloramphenicol IPTG/X-Gal agar: add 1μL of chloramphenicol stock solution and 1μL of IPTG and 2μL X-Gal solution per 1 mL molten LB agar (*see* **Note 9**).
10. LB–chloramphenicol–kanamycin broth: add 1μL chloramphenicol stock solution and 1μL kanamycin stock solution per 1 mL sterile LB broth.

11. LB–chloramphenicol–kanamycin agar: add 1μL chloramphenicol stock solution and 1μL kanamycin stock solution per 1 mL molten LB agar.
12. 40% Glycerol solution: Mix 40 mL glycerol and 60 mL dH_2O and sterilize by autoclaving.

2.6 Equipment, Consumables, and Instruments

1. Basic laboratory material including 1.5 mL microcentrifuge tubes, 0.2-mL PCR tubes, 50-mL centrifuge tubes, Petri dishes, L-shaped spreaders, 96-well sterile microdilution plates, mixed cellulose ester sterile filters (0.22μm pore size).
2. Centrifuges for 1.5 and 50 mL tubes.
3. Class II biosafety cabinet.
4. Incubator with orbital shaker (37 °C).
5. Heat block.
6. Electroporator.
7. 96-Well microplate replicator.
8. Spectrophotometer.
9. PCR thermal cycler.

3 Methods

This protocol outlines the identification of genes conferring antimicrobial resistance by using functional metagenomics (Fig. 1). To identify genes conferring other selectable traits, different approaches to screen the metagenomic library can be used. For example:

- Screening for clones with specific metabolic activity: Grow the metagenomic library on medium containing an enzyme substrate that change appearance (e.g., color) following enzymatic activity.
  - Novel glycosyl hydrolases were discovered from a cast of earthworms by selecting for colonies with intense yellow color on medium containing *p*-nitrophenyl-β-D-glucopyranoside and *p*-nitrophenyl-α-L-arabinopyranoside [24].
  - Novel carboxyl-ester hydrolase was found from bovine rumen microbiome by screening for clones that can degrade Tween20 and Impranil and resulting in halo [25]
- Screening for clones that grow under selective conditions: Grow the metagenomic library on medium with nutrient deficiency, supplemented with antibiotics, or construct the library with mutant host strains.
  - Novel prebiotic degradation pathways were found in human gut metagenomic library by screening on a minimum

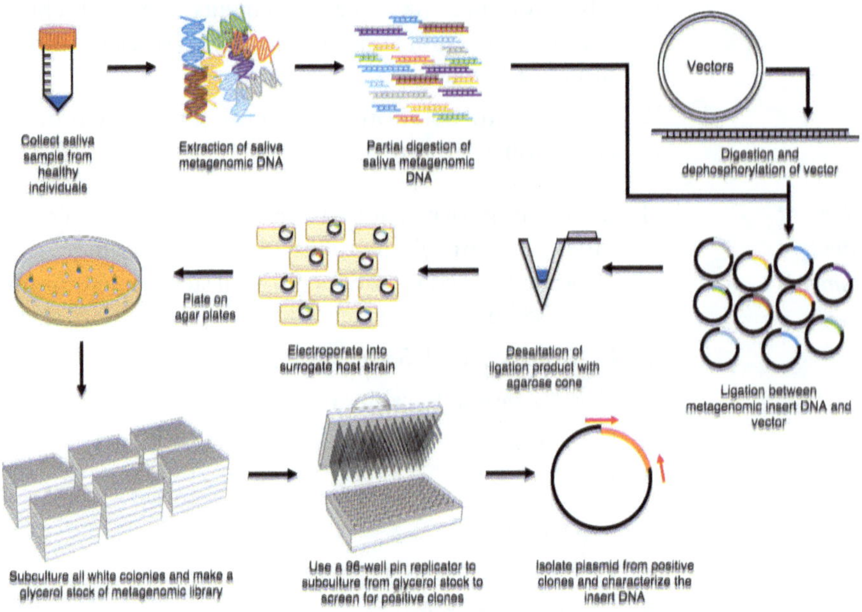

Fig. 1 Schematic representation of functional metagenomics to screen for microbial genes in the human saliva metagenome. Metagenomic DNA, extracted from the collected saliva sample, is partially digested and ligated to digested and dephosphorylated vector. The ligation products between metagenomic fragments and vector are desalted with an agarose cone and electroporated into a surrogate host strain. The white transformants are subcultured and prepared as glycerol stocks of the metagenomic library in 96-well microdilution plates. Clones with phenotypes of interest can be screened for by subculturing the glycerol stocks with a 96-well pin replicator into new plates and performing screening assay on the subculture plates. Plasmids from the positive clones are extracted, and gene(s) conferring functions of interest can be characterized by sequencing, transposon mutagenesis, and subcloning

medium that contains prebiotic oligosaccharides as the only carbon source [26].

  - Novel DNA polymerase I genes were discovered from glacial ice samples by constructing the library in a cold-sensitive *Escherichia coli polA* mutant and screening for target genes by growing at temperature below 20 °C [27].

- Screening clones producing antimicrobial substances: Grow the metagenomic library overnight on agar plates and screen against another bacterial species by pouring a thin layer of agar mixed with the bacterial suspension on top of the metagenomic library plates. Then, select for a colony with a zone of growth inhibition surrounding it.
  - Antibacterial enzymes were found by screening a soil metagenomic library, constructed in *Ralstonia metallidurans*, against a lawn of *Bacillus subtilis* for colony with a zone of growth inhibition surrounding it [28].
- Reporter-based screening: Screen for clones with a change in the expression level of a reporter gene in response to a substrate of interest or a gene's product encoding by metagenomics DNA insert.

- Substrate-induced gene expression (SIGEX) has been used to identify salicylate oxygenase genes in an aromatic hydrocarbon-contaminated soil by cloning metagenomic DNA upstream of a green fluorescent protein reporter gene and using fluorescence associated cell sorting (FACS) to screen for clones with an increased expression of the reporter gene in response to aromatic compounds [29].

Carry out all procedures at room temperature unless otherwise specified. Standard aseptic techniques should be used to perform all procedures involving saliva sample preparation, bacterial culture, screening, and media preparation. All standard laboratory materials should be clean and sterile before usage, and all centrifugations are carried out at the 15,700 × *g* and 4500 × *g* for 1.5-mL and 50-mL tubes, respectively. The protocol should be modified accordingly to the manufacturer's instruction when different enzymes, reagents, and kits are used.

3.1 Saliva Sample Collection and DNA Extraction

1. Collect 2 mL of saliva (not including bubbles) from healthy volunteers, who have not had antibiotics for at least 3 months and have not eaten, drank, rinsed, or cleaned their mouth for at least 1 h, by spitting into the provided collection tubes (*see* **Notes 10** and **11**).
2. Transfer the samples to the laboratory on ice before proceeding with DNA extraction as soon as possible (*see* **Note 12**).
3. Transfer 750μL saliva into 1.5-mL microcentrifuge tube within a Class II biosafety cabinet and pellet the cells by centrifuge for 5 min.
4. Carefully remove the supernatant using a pipette without disturbing the pellet.
5. Extract DNA from the pellet using the Gentra Puregene Yeast/Bact. Kit following the DNA purification protocol for Gram-positive bacteria.
6. Check the concentration, purity, and integrity of the extracted DNA using Qubit, Nanodrop, and gel electrophoresis on 1% agarose gel, respectively (*see* **Notes 8** and **13**). Store DNA at −20 °C.

3.2 Construction of Saliva Metagenomic DNA Library

3.2.1 Partial Digestion of Saliva Metagenome

1. Set up two partial digestion reactions of the saliva metagenomic DNA by mixing the following components in a 20μL reaction volume in each in 1.5-mL microcentrifuge tube (*see* **Note 14**): 11μL 50 ng/μL Saliva metagenomic DNA, 2μL 10× Restriction enzyme buffer, 1μL 1 U/μL *Hin*dIII (*see* **Note 15**), and 6μL Molecular Grade Water. Incubate one reaction for 1 min and the other for 2 min at 37 °C.

2. Add 100% ethanol to stop the reaction and mix by pipetting, then add 2μL 3 M NaOAc to allow precipitation of partially digested DNA and mix again by pipetting (*see* **Note 16**).
3. Incubate the tubes on ice for 30 min and centrifuge for 15 min.
4. Discard the supernatant and rinse the DNA pellet by adding 1 mL 70% ethanol.
5. Centrifuge for 15 min, remove supernatant, and air-dry the pellet for 5 min (*see* **Note 17**).
6. Suspend the pellet in 30μL Molecular Grade Water and check the quality and quantity of DNA as in Subheading 3.1, **step 6** (Fig. 2a). Store the DNA at −20 °C.

3.2.2 Preparation of Cloning Vector

1. Set up a digestion reaction on pCC1BAC cloning vector in a 1.5-mL microcentrifuge by mixing the following components: 1μg pCC1BAC vector, 5μL 10× Restriction enzyme buffer, 1μL 20 U/μL *Hin*dIII (*see* **Note 15**), and top up the reaction to 50μL with Molecular Grade Water. Mix gently and incubate at 37 °C in a heat block for 1 h.
2. Dephosphorylate the digested pCC1BAC vector by adding 1μL 1 U/μL calf intestinal alkaline phosphatase (CIAP), 6μL 10× CIAP reaction buffer, and 3μL Molecular Grade Water to get 60μL the final volume (*see* **Note 18**). Incubate at 37 °C in a heat block for 30 min and add another 1μL 1 U/μL CIAP. Incubate at 37 °C in a heat block for another 30 min.
3. Purify the digested, dephosphorylated pCC1BAC using a PCR purification kit.
4. Check the quality and quantity of DNA as in Subheading 3.1, **step 6**. Store the DNA at −20 °C.

3.2.3 Ligation and Desalting of Ligation Products

1. Set up a ligation reaction between the partially digested saliva DNA and digested pCC1BAC vector by mixing the following components in a 1.5-mL microcentrifuge tube: 200–1000 ng human saliva partially *Hin*dIII digested insert DNA, 1μL *Hin*dIII digested, 25 ng/μL dephosphorylated pCC1BAC vector, 10μL Fast-Link™ 10× ligation buffer, 1μL 10 mM ATP, 2μL Fast-Link™ DNA Ligase, and top up with Molecular Grade Water to a final reaction volume of 100μL. Mix gently and incubate at 16 °C overnight (*see* **Note 19**).
2. Prepare an agarose cone by adding 0.9 g glucose and 0.5 g agarose in 50 mL water (1.8% and 1% w/v, respectively) and heat the solution to dissolve. Transfer 600μL of the solution to a 1.5-mL microcentrifuge tube and put a 0.5-mL microcentrifuge tube in as a mold for the agarose cone. After the solution solidifies, remove the 0.5-mL microcentrifuge tube, forming an agarose cone in the 1.5-mL microcentrifuge tube.

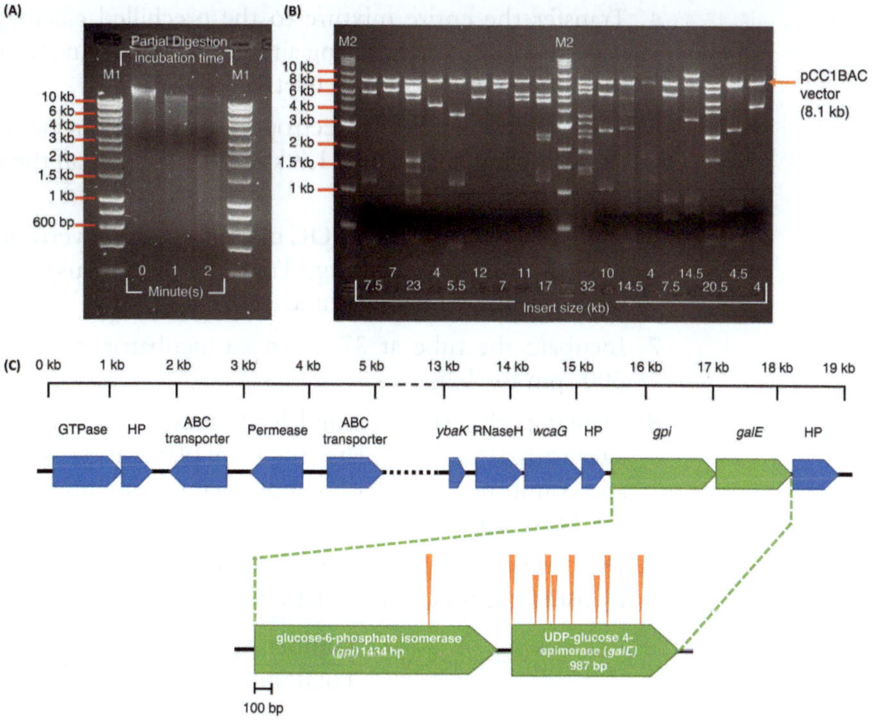

Fig. 2 Construction of a saliva metagenomic library and identification of resistance gene. (**a**) Agarose gel electrophoresis of partially digested metagenomic DNA shows a reduction of high molecular weight DNA with longer incubation time compared to undigested metagenomic DNA (0 min). Lane M1, 0.2–10 kb ladder. (**b**) Agarose gel electrophoresis of *Hin*dIII digested plasmids randomly extracted from the metagenomic library. The size of the insert DNA on each plasmid can be estimated by adding the size of each band other than the pCC1BAC vector band (8.1 kb). The average insert size from these 18 colonies is approximately 11.42 kb. Lane M2, 0.5–50 kb ladder. (**c**) Example of the identification of putative resistance genes on large insert DNA (19.1 kb) by transposon mutagenesis. The orange triangles indicate the positions where the Entrancepson inserts in each mutant based on the results from sequencing with SeqE and SeqW primers comparing to the sequencing data of the insert DNA. (Reproduced from Tansirichaiya et al. [19] and Reynolds [35])

3. Transfer the entire volume of ligated product to the agarose cone. Incubate on ice for 1 h, then transfer the ligated product to a new 1.5-mL microcentrifuge tube (*see* **Note 20**). Keep the ligated product on ice or store at 4 °C until use.

3.2.4 Electroporation of Ligation Product into E. coli *Surrogate Host*

1. Thaw TransforMax EPI300 Electrocompetent *E. coli* and pre-chill a 1.5-mL microcentrifuge tube and a 0.1 cm-gap electroporation cuvette on ice (*see* **Note 21**).
2. Aliquot 50μL of electrocompetent cells to the prechilled 1.5-mL microcentrifuge tube and keep on ice.
3. Add 2μL of the desalted ligation to the cells in the microcentrifuge tube.

4. Transfer the entire mixture to the prechilled electroporation cuvette without introducing air bubbles in the cuvette and wipe the condensation off the cuvette (*see* **Note 22**).
5. Place the cuvette in the electroporator and electroporate under the following conditions: 1.7 kV, 200 Ω, 25μF on the machine (*see* **Note 23**).
6. Immediately add 950μL SOC medium to the cuvette and mix it with the cells by pipetting. Transfer the cell suspension to a sterile 50-mL centrifuge tube (*see* **Note 24**).
7. Incubate the tube at 37 °C in an incubator with shaking at 200 rpm for 1 h.
8. Plate the cell suspension on LB–chloramphenicol IPTG/X-Gal agar with 100μL per plate. Incubate plates at 37 °C overnight and count the number of blue and white colonies on each plate (*see* **Note 25**).
9. Determine the ligation efficiency and transformation efficiency by using the following equations (*see* **Note 26**).

$$\text{Ligation efficiency} = \frac{\text{Number of white colonies}}{\text{Total number of colonies}} \times 100.$$

$$\text{Transformation efficiency} = \frac{\text{Total number of colonies}}{\mu\text{g of DNA transformed}}.$$

3.2.5 Library Size Determination

1. Subculture 10–50 random colonies into 10 mL of LB–chloramphenicol broth and extract the plasmids by using QIAprep Spin Miniprep Kit, following the protocol from the manufacturer.
2. Set up *Hin*dIII digestion on each plasmid as follows: 1μL Extracted plasmid, 1μL 10× Restriction enzyme buffer, 1μL 20 U/μL *Hin*dIII, and 7μL Molecular Grade Water. Mix gently and incubate at 37 °C in a heat block for 1 h. Add 5μL 6× loading buffer to digested products and run them on 1% agarose gel to determine the size of inserts in each clone by comparing it with the DNA Ladder (Fig. 2b).
3. The average insert size can be calculated by using the following equation.

$$\text{Average insert size (kb)} = \frac{\text{Total insert size in kb}}{\text{Number of extracted plasmid}}$$

4. If the average insert size is too short, repeat partial digestion on saliva DNA with either shorter incubation time or more diluted restriction enzyme (*see* **Note 14**). If the average insert size is appropriate already, the rest of the ligation product can be electroporated into *E. coli*.

5. The size of the metagenomic library can be calculated as follows (*see* **Note 27**):

$$\text{Library size (kb)} = \text{Average insert size (kb)} \times \text{Total number of white colonies}$$

*3.2.6 Preparation of Metagenomic Library Glycerol Stock for Optional Storage of Isolated Clones (See **Notes 28** and **29**)*

1. Prepare 96-well sterile microdilution plates by adding 100μL LB–chloramphenicol broth into each well.
2. Pick each white colony on the LB–chloramphenicol IPTG/X-Gal agar from Subheading 3.2.4 by using sterile toothpicks and inoculated into each well of the 96-well sterile microdilution plates. On each plate, it composes of 94 white colonies from the electroporation and two control wells: one inoculated with *E. coli* EPI300 containing empty pCC1BAC (wild-type) and another one with no inoculation (negative control).
3. Incubate the 96-well plates at 37 °C with shaking at 200 rpm overnight.
4. Add 100μL 40% glycerol solution to each well and store at −80 °C.

3.3 Screening of Metagenomic Library for Antimicrobial Resistance

*3.3.1 Determination of the Minimum Inhibitory Concentration (MIC) of Antimicrobial Compounds (See **Note 30**)*

1. Subculture *E. coli* EPI300 containing empty pCC1BAC (*E. coli* EPI300::pCC1BAC) in 5 mL LB–chloramphenicol broth and incubate at 37 °C with shaking at 200 rpm overnight.
2. Dilute the overnight culture with LB–chloramphenicol broth to an OD_{600} of 0.1.
3. Add 90μL LB broth supplemented with chloramphenicol (12.5μg/mL) and varying concentrations of antimicrobial compounds to a 96-well sterile microdilution plate.
4. Inoculate 10μL of the diluted culture to the 96-well plate and incubate overnight at 37 °C with shaking at 200 rpm.
5. Determine the growth at each concentration of the antimicrobials by measuring OD_{600}. The MIC is the lowest concentration that inhibits the growth of *E. coli* EPI300::pCC1BAC.

3.3.2 Screening of Metagenomic Library Antimicrobial Resistance Clone

1. Replicate each 96-well plate containing the saliva metagenomic glycerol stock by using a 96-well microplate replicator to inoculate another 96-well sterile microdilution plate filled with 100μL LB–chloramphenicol broth (*see* **Note 31**). Incubate overnight at 37 °C with shaking at 200 rpm.
2. Add 90μL LB broth supplemented with chloramphenicol (12.5μg/mL) and the screening antimicrobials at the MIC, determined in Subheading 3.3.1, to a sterile 96-well microdilution plate, and inoculate 10μL of the overnight culture to the screening plates.

3. Incubate overnight at 37 °C with shaking at 200 rpm and check for antimicrobial resistant clones by measuring OD_{600} (*see* **Note 32**).

3.4 Characterization of Genes Conferring Antimicrobial Resistance

3.4.1 Plasmid Extraction, Insert Size Determination, and Sequencing

1. All resistant clones are subcultured into 5 mL of LB–chloramphenicol broth and incubated overnight at 37 °C with shaking at 200 rpm.
2. Set up a copy number induction culture from each overnight culture by transferring 1 mL of overnight culture to 9 mL LB–chloramphenicol broth and 10μL of 1000× CopyControl Induction Solution. Incubate the tubes horizontally at 37 °C with shaking at 200 rpm for 4 h (*see* **Note 33**).
3. Extract plasmids from the copy number induction cultures and determine the insert size, *see* **steps 1** and **2** in Subheading 3.2.5.
4. Perform Sanger sequencing of the inserts with primer pairs targeting pCC1BAC vector: pCC1-F and pCC1-R. Additional primers are designed based on the sequencing results to extend the sequences of the inserts.

3.4.2 Identification of Genes Conferring Antimicrobial Resistance (See **Note 34**)

1. Transposon mutagenesis is performed to identify genes conferring antimicrobial resistance on the plasmids by using the Template Generation System II kit. Set up the reaction as follows: 60 fmol, extracted plasmid from Subheading 3.4.1, 1μL MuA Transposase (0.22μg/μL), 1μL Entranceposon (Kan^R-3) (20 ng/μL) (*see* **Note 35**), 4μL 5× reaction buffer, and top up to 20μL with Molecular Grade Water.

   Mix gently and incubate at 30 °C for 1 h in the heat block. Stop the reaction by incubating at 75 °C for 10 min.
2. Electroporate 5μL of the tenfold diluted mutagenesis reaction into *E. coli* EPI300 electrocompetent cells (50μL) as in Subheading 3.2.4. Plate the cell suspension on LB–chloramphenicol–kanamycin agar with 100μL per plate. Incubate plates at 37 °C overnight.
5. Pick each colony with sterile toothpick or pipette tip and subculture into the same well of two separate 96-well plates: containing LB–chloramphenicol–kanamycin broth and LB–chloramphenicol–kanamycin–antimicrobials broth. Incubate the 96-well plates at 37 °C with shaking at 200 rpm overnight.
3. Select for clones with a loss of resistance phenotype: Grow only in the plates with LB–chloramphenicol–kanamycin broth but not the LB–chloramphenicol–kanamycin–antimicrobials broth. Extract the plasmid as **steps 1–3** in Subheading 3.4.1.
4. Perform Sanger sequencing with primer pairs flanking the Entranceposon: SeqW and SeqE primers to determine the

location of Entranceposon within the plasmid by comparing with sequencing data from the experiments outlined in Subheading 3.4.1 (Fig. 2c).

3.4.3 Subcloning of Putative Antimicrobial Resistance Genes

1. Design DNA primers with added *Hin*dIII restriction sites to amplify putative antimicrobial resistance genes identified in Subheading 3.4.2 and set up a 30μL PCR reactions as follows (*see* **Note 36**):

   15μL 2× PCR master mix, 2μL 10μM forward primer, 2μL 10μM reverse primer, 1μL extracted plasmid from Subheading 3.4.1, and 10μL Molecular Grade Water.
2. Perform PCR with the following settings on the thermocycler: Initial denaturation at 95 °C for 3 min, 35 cycles of (95 °C 1 min, 55 °C 30 s, 72 °C 1 min) and a final extension at 72 °C for 5 min (*see* **Note 37**).
3. Check the size of the PCR product on 1% agarose gel, whether it matches the expected size before purifying the PCR product by using a PCR purification kit.
4. Digest the PCR product with *Hin*dIII restriction enzyme and clean the digested PCR product with a PCR purification kit.
5. Set up a ligation reaction between *Hin*dIII digested PCR amplicons and *Hin*dIII digested pCC1BAC (from Subheading 3.2.2) with the vector:insert molar ratio of 1:3. Incubate the ligation reaction at 16 °C overnight.
6. Desalt the ligation product and electroporate into *E. coli* EPI300 electrocompetent cells as in Subheading 3.2.4. Plate 100μL of cells on LB–chloramphenicol agars and incubate at 37 °C overnight.
7. Select white colonies and perform colony PCR to check their plasmids, whether they contain the amplified putative resistance genes.
8. Inoculate the colonies containing the putative resistance gene into LB–chloramphenicol–antimicrobial broth and check for the growth after overnight incubation to confirm that the identified resistance gene confers resistance phenotype.

4 Notes

1. The saliva collection tubes should have a suitable diameter to be convenient for the volunteers to expectorate and a line indicating a 2-mL volume should be marked on the tubes.
2. Other genomic DNA extraction protocols or kits can be used as an alternative, but it should include lysozyme to ensure the cell lysis of Gram-positive bacteria as they have a thick

peptidoglycan layer. Lysozyme can hydrolyze the linkages between *N*-acetylmuramic acid and *N*-acetyl-D-glucosamine residues in peptidoglycan.

3. pCC1BAC is maintained as a single copy per cell in *E. coli* EPI300 to ensure the stability of large-insert DNA and to allow the cloning and screening for toxin-producing genes.

4. The copy number of pCC1BAC vector can be induced to 10–20 copies per cell in TransforMax EPI300 Electrocompetent *E. coli*, to increase DNA concentrations for sequencing. A CopyControl Induction Solution can induce the expression of *trfA* in *E. coli* EPI300, which subsequently initiates the replication of pCC1BAC by a high-copy number origin of replication *oriV*.

5. The plasmids and bacterial surrogate hosts can be changed depending on the research purpose, which requires the users to modify the protocol carefully, such as selective marker, sequencing primers, restriction enzymes, and growth media. For example, a pHT01 shuttle vector has been used to construct soil metagenomic libraries in *E. coli* and *B. subtilis* for antimicrobial activity [30], and a small insert-high copy number pUC18 has been used to screen for antibiotic resistance genes in soil metagenomic DNA [31].

6. For the convenience in the preparation of media, it is advised to prepare the antimicrobial stocks with 1000× concentration of the working concentration. The stock solution should be filter sterilized if they are not prepared with organic solvents.

7. Alternative nucleic acid stains, which are less hazardous than ethidium bromide, can be used to stain an agarose gel as well such as GelRed Nucleic Acid Gel Stain (Biotium) and SYBR Green I Nucleic Acid Gel Stain (Thermo Fisher Scientific).

8. While we use Qubit fluorometer to measure DNA concentration, it cannot give information on DNA purity ($A_{260/280}$) that can be determined by Nanodrop spectrophotometer. Alternatively, DNA concentration can be estimated by comparing the intensity of DNA samples in an agarose gel to each band of the standard DNA ladder, where DNA fragments of specific lengths have defined different DNA concentrations indicated by the manufacturer.

9. A combined commercial IPTG/X-Gal solution could be used to add to the molten (50 °C) agar as well with the amount recommended by the manufacturer.

10. These criteria regarding the healthy volunteers are applied to make sure that the collected saliva samples would represent the oral microbiome under normal conditions as we want to look

at the background resistance genes in commensals in the oral cavity. The inclusion/exclusion criteria for each study should be based on the research questions.

11. Oral metagenomic DNA can also be extracted from other samples such as tongue swab, cheek swab, supragingival and subgingival plaque samples (e.g. [6]).
12. If the saliva samples cannot be processed immediately, it could be kept in −20 °C freezer. Alternatively, saliva DNA collection kits could be used as they contain buffer that can maintain and preserve DNA at ambient temperature for years, such as Norgen's saliva DNA Collection and Preservation Devices, which have been used to collect saliva samples for metagenomic studies previously [32]. DNA extraction from samples collected in these devices should be done as recommended by the manufacturer of the collection kits.
13. The DNA integrity can be checked on an agarose gel in which the extracted metagenomic DNA should have no or little degradation or smear on the gel.
14. Partial digestion on the saliva metagenomic DNA generates large DNA fragments for the construction of the metagenomic library (average insert size above 10 kb) because it will keep the large DNA operons intact and still functional in the screening. DNA flanking the target genes can also be used for taxonomic identification and to provide information on associated genes such as mobile gene elements. Optimization of partial digestion might require various incubation times and restriction enzyme concentrations to achieve the appropriate range of insert sizes. Complete digestion can also be performed to generate a library with small inserts, if the selective phenotype is conferred by a single gene, which can be overexpressed when they are cloned into a high-copy number vector, increasing the chance of selection in a screening assay.
15. Different restriction enzymes can be used as well, depending on available restriction sites in the cloning site on the vector. Vector and insert DNA should be digested with the same restriction enzyme or enzymes that leave compatible overhanging sequences for ligation. For example, metagenomic insert DNA can be partially digested with a 4-bp cutting enzyme *Sau*3AI, which has more restriction sites to generate more diverse fragments, and the vector can be digested with a 6-bp cutting enzyme *Bam*HI, which leaves the same GATC overhang compatible with that resulting from a *Sau*3AI digest.
16. Ethanol precipitation is performed to purify the partial digestion products to keep large DNA fragments after the purification since the spin columns from the PCR purification kit can purify DNA only up to a maximum size of 10 kb.

17. Do not overdry the DNA pellet as it could be difficult to redissolve the pellet.
18. Dephosphorylation is required to prevent self-ligation when the vector is digested with a single restriction enzyme. Alkaline phosphatase enzymes catalyze the removal of the 5′ phosphate groups, essential for ligation, from the digested DNA.
19. Low-temperature ligation (between 4 and 16 °C depending on the T4 DNA ligase) overnight is recommended, especially with large DNA fragments, as it will allow the sticky ends to keep annealed during the ligase works, maximizing the ligation efficiency.
20. Desalting of the ligation reaction with an agarose cone removes salt from the samples which can cause electric arcing during electroporation.
21. Keeping the cells, cuvettes, and tubes cold before the pulse is essential for high-efficiency electroporation.
22. Air bubbles and condensation on the cuvette can cause electrical arcing during electroporation, often accompanied by a loud and alarming popping sound.
23. The parameters for electroporation can be varied depending on the strains of bacteria.
24. Pre-warm the SOC medium to 37 °C before use.
25. Blue/white screening is a technique to differentiate between clones with and without insertion DNA in the cloning vector in the presence of IPTG/X-Gal. When insert DNA is ligated to the cloning site within the *lacZ* gene on the vector, it will disrupt the expression of *lacZ* and result in a white colony. Some of the blue colonies could sometimes contain a small insert DNA (false negative), which inserts in-frame with *lacZ*, so the read-through can result in a functional LacZ and blue colonies.
26. Ligation and transformation efficiencies are calculated to check whether the constructed metagenomic library has high quality to be used for the screening. If not, several factors can be optimized such as the ratio between insert and vector, the volume of ligation product, dephosphorylation of the vector, and more prolonged incubation in partial digestion.
27. As the saliva metagenomic DNA tends to contain DNA from both bacteria and human, the library size of bacterial DNA could be calculated by end sequencing each plasmid, determining whether each insert DNA is derived from bacteria or human using BlastN and calculating the library size as Library size of bacterial DNA (kb) = Average bacterial insert size (kb) × Total number of white colonies × Percentage of colonies containing bacterial DNA.

28. Preparing a metagenomic library as glycerol stocks in a 96-well format can be time-consuming, but it allows the library to be screened multiple times against multiple compounds at a time, and also in new unrelated assays when the opportunity presents itself. If long-term storage of the metagenomic library is not required, Subheadings 3.2.6, 3.3.1, and 3.3.2 can be skipped and immediate screening of the cells following electroporation can be achieved by plating directly onto selective media. Multiple selective screening could also be performed by using replica plating to transplant colonies from LB–chloramphenicol IPTG/X-Gal agar from the initial plating after electroporation onto multiple selective agar plates. Any positive clones can be characterized as described in Subheading 3.4.
29. Alternatively, the transformants could be subcultured into LB broth containing antibiotic corresponding to the selective marker on the vector and incubated overnight [33, 34]. The overnight culture is centrifuged, resuspended with LB with 20% glycerol, aliquoted into several tubes, and kept in −80 °C freezer. Functional screening can be performed on cells, subcultured from the glycerol stocks. However, clones containing rare DNA insert, large DNA insert, and toxin-producing genes are prone to be lost or outcompeted in each subculture, and could be missing out from the screening.
30. MIC determination is an essential step for the screening for antimicrobial resistance genes as it will be the minimum concentration used for the screening assay, which should be determined against the host strain containing the empty vector that is used in the library construction.
31. A 96-well pin replicator must be ethanol sterilized between the subculture of each 96-well glycerol plate to prevent cross contamination.
32. Check the growth in the control wells (wild-type *E. coli* and negative control) as well, where there should be no growth on both wells in the screening. If there is a growth in the control wells, the screening should be repeated as there could be contamination or incorrect concentration of the screening compounds.
33. Horizontally shaking of the culture increases the surface area and aeration of the culture to maximize the bacterial growth for the induction culture.
34. As the library is constructed with large DNA inserts, it could be difficult to identify the gene(s) conferring the phenotype and will be faster through transposon mutagenesis. It randomly inserts Entranceposon on the extracted plasmid, which will disrupt the expression of the gene in that location. If the insert

DNA is small or composed of a few genes, each gene could be subcloned to determine the gene responsible for the positive phenotype instead, as described in Subheading 3.4.3.

35. There are three different selective marker genes for Entranceposon provided by the manufacturer, including chloramphenicol, kanamycin, and tetracycline resistance genes. Other transposon-based mutagenesis systems are also commercially available.
36. Primers should be designed with space at least 50 bp upstream and downstream from the target gene. *Hin*dIII restriction site can be added at the 5′ end of both forward and reverse primers. An extra 4–6 bp nucleotide GC clamp should be added before the restriction site to ensure efficient DNA binding by the restriction enzyme. Therefore, the primers should have this structure: (5′-(4–6 bp GC clamp)-(Restriction site)-(18–22 bp target gene sequence)-3′).
37. The annealing temperature and elongation time could be different depending on the primers and expected amplicons size, respectively.

Acknowledgments

We would like to thank Chakraphan Hiranwongwira for artwork included in Fig. 1.

References

1. Puspita ID, Kamagata Y, Tanaka M, Asano K, Nakatsu CH (2012) Are uncultivated bacteria really uncultivable? Microbes Environ 27 (4):356–366. https://doi.org/10.1264/jsme2.me12092
2. Overmann J, Abt B, Sikorski J (2017) Present and future of culturing bacteria. Annu Rev Microbiol 71(1):711–730. https://doi.org/10.1146/annurev-micro-090816-093449
3. Dewhirst FE, Chen T, Izard J, Paster BJ, Tanner AC, Yu WH et al (2010) The human oral microbiome. J Bacteriol 192(19):5002–5017. https://doi.org/10.1128/jb.00542-10
4. Escapa IF, Chen T, Huang Y, Gajare P, Dewhirst FE, Lemon KP (2018) New insights into human nostril microbiome from the expanded Human Oral Microbiome Database (eHOMD): a resource for the microbiome of the human aerodigestive tract. mSystems 3(6). https://doi.org/10.1128/mSystems.00187-18
5. Wade W, Thompson H, Rybalka A, Vartoukian S (2016) Uncultured members of the oral microbiome. J Calif Dent Assoc 44 (7):447–456
6. Hall MW, Singh N, Ng KF, Lam DK, Goldberg MB, Tenenbaum HC et al (2017) Interpersonal diversity and temporal dynamics of dental, tongue, and salivary microbiota in the healthy oral cavity. npj Biofilms Microbiomes 3 (1):2. https://doi.org/10.1038/s41522-016-0011-0
7. Wade WG (2011) Has the use of molecular methods for the characterization of the human oral microbiome changed our understanding of the role of bacteria in the pathogenesis of periodontal disease? J Clin Periodontol 38(Suppl 11):7–16. https://doi.org/10.1111/j.1600-051X.2010.01679.x
8. Shaw L, Ribeiro ALR, Levine AP, Pontikos N, Balloux F, Segal AW et al (2017) The human salivary microbiome is shaped by shared environment rather than genetics: evidence from a large family of closely related individuals. mBio 8(5):e01237–e01217. https://doi.org/10.1128/mBio.01237-17

9. Tansirichaiya S, Mullany P, Roberts AP (2016) PCR-based detection of composite transposons and translocatable units from oral metagenomic DNA. FEMS Microbiol Lett 363 (18):fnw195. https://doi.org/10.1093/femsle/fnw195
10. Marathe NP, Janzon A, Kotsakis SD, Flach C-F, Razavi M, Berglund F et al (2018) Functional metagenomics reveals a novel carbapenem-hydrolyzing mobile beta-lactamase from Indian river sediments contaminated with antibiotic production waste. Environ Int 112:279–286. https://doi.org/10.1016/j.envint.2017.12.036
11. Hjort K, Presti I, Elväng A, Marinelli F, Sjöling S (2014) Bacterial chitinase with phytopathogen control capacity from suppressive soil revealed by functional metagenomics. Appl Microbiol Biotechnol 98(6):2819–2828. https://doi.org/10.1007/s00253-013-5287-x
12. Tasse L, Bercovici J, Pizzut-Serin S, Robe P, Tap J, Klopp C et al (2010) Functional metagenomics to mine the human gut microbiome for dietary fiber catabolic enzymes. Genome Res 20(11):1605–1612. https://doi.org/10.1101/gr.108332.110
13. Li Y, Wexler M, Richardson DJ, Bond PL, Johnston AWB (2005) Screening a wide host-range, waste-water metagenomic library in tryptophan auxotrophs of *Rhizobium leguminosarum* and of *Escherichia coli* reveals different classes of cloned trp genes. Environ Microbiol 7(12):1927–1936. https://doi.org/10.1111/j.1462-2920.2005.00853.x
14. Nasidze I, Li J, Quinque D, Tang K, Stoneking M (2009) Global diversity in the human salivary microbiome. Genome Res 19 (4):636–643. https://doi.org/10.1101/gr.084616.108
15. Vitorino R, Lobo MJC, Duarte JR, Ferrer-Correia AJ, Domingues PM, Amado FML (2005) The role of salivary peptides in dental caries. Biomed Chromatogr 19(3):214–222. https://doi.org/10.1002/bmc.438
16. Faveri M, Mayer MPA, Feres M, De Figueiredo LC, Dewhirst FE, Paster BJ (2008) Microbiological diversity of generalized aggressive periodontitis by *16S rRNA* clonal analysis. Oral Microbiol Immunol 23(2):112–118. https://doi.org/10.1111/j.1399-302X.2007.00397.x
17. Reynolds LJ, Roberts AP, Anjum MF (2016) Efflux in the oral metagenome: the discovery of a novel tetracycline and tigecycline ABC transporter. Front Microbiol 7(1923). https://doi.org/10.3389/fmicb.2016.01923
18. Diaz-Torres ML, McNab R, Spratt DA, Villedieu A, Hunt N, Wilson M, Mullany P (2003) Novel tetracycline resistance determinant from the oral metagenome. Antimicrob Agents Chemother 47(4):1430–1432. https://doi.org/10.1128/aac.47.4.1430-1432.2003
19. Tansirichaiya S, Reynolds LJ, Cristarella G, Wong LC, Rosendahl K, Roberts AP (2018) Reduced susceptibility to antiseptics is conferred by heterologous housekeeping genes. Microbial Drug Resist (Larchmont, NY) 24 (2):105–112. https://doi.org/10.1089/mdr.2017.0105
20. Card RM, Warburton PJ, MacLaren N, Mullany P, Allan E, Anjum MF (2014) Application of microarray and functional-based screening methods for the detection of antimicrobial resistance genes in the microbiomes of healthy humans. PLoS One 9(1):e86428. https://doi.org/10.1371/journal.pone.0086428
21. Sommer MOA, Dantas G, Church GM (2009) Functional characterization of the antibiotic resistance reservoir in the human microflora. Science (New York, NY) 325 (5944):1128–1131. https://doi.org/10.1126/science.1176950
22. Sukumar S, Roberts AP, Martin FE, Adler CJ (2016) Metagenomic insights into transferable antibiotic resistance in oral bacteria. J Dent Res 95(9):969–976. https://doi.org/10.1177/0022034516648944
23. Wild J, Szybalski W (2004) Copy-control pBAC/oriV vectors for genomic cloning. Methods Mol Biol (Clifton, NJ) 267:145–154. https://doi.org/10.1385/1-59259-774-2:145
24. Beloqui A, Nechitaylo TY, López-Cortés N, Ghazi A, Guazzaroni M-E, Polaina J et al (2010) Diversity of glycosyl hydrolases from cellulose-depleting communities enriched from casts of two earthworm species. Appl Environ Microbiol 76(17):5934–5946. https://doi.org/10.1128/AEM.00902-10
25. Ufarté L, Laville E, Duquesne S, Morgavi D, Robe P, Klopp C et al (2017) Discovery of carbamate degrading enzymes by functional metagenomics. PLoS One 12(12):e0189201. https://doi.org/10.1371/journal.pone.0189201
26. Cecchini DA, Laville E, Laguerre S, Robe P, Leclerc M, Doré J et al (2013) Functional metagenomics reveals novel pathways of prebiotic breakdown by human gut bacteria. PLoS One 8(9):e72766. https://doi.org/10.1371/journal.pone.0072766

27. Simon C, Herath J, Rockstroh S, Daniel R (2009) Rapid identification of genes encoding DNA polymerases by function-based screening of metagenomic libraries derived from glacial ice. Appl Environ Microbiol 75 (9):2964–2968. https://doi.org/10.1128/AEM.02644-08
28. Iqbal HA, Craig JW, Brady SF (2014) Antibacterial enzymes from the functional screening of metagenomic libraries hosted in Ralstonia metallidurans. FEMS Microbiol Lett 354 (1):19–26. https://doi.org/10.1111/1574-6968.12431
29. Meier MJ, Paterson ES, Lambert IB (2016) Use of substrate-induced gene expression in metagenomic analysis of an aromatic hydrocarbon-contaminated soil. Appl Environ Microbiol 82(3):897–909. https://doi.org/10.1128/aem.03306-15
30. Biver S, Steels S, Portetelle D, Vandenbol M (2013) *Bacillus subtilis* as a tool for screening soil metagenomic libraries for antimicrobial activities. J Microbiol Biotechnol 23 (6):850–855. https://doi.org/10.4014/jmb.1212.12008
31. McGarvey KM, Queitsch K, Fields S (2012) Wide variation in antibiotic resistance proteins identified by functional metagenomic screening of a soil DNA library. Appl Environ Microbiol 78(6):1708–1714. https://doi.org/10.1128/aem.06759-11
32. Tansirichaiya S, Rahman MA, Antepowicz A, Mullany P, Roberts AP (2016) Detection of novel integrons in the metagenome of human saliva. PLoS One 11(6):e0157605. https://doi.org/10.1371/journal.pone.0157605
33. Perron GG, Whyte L, Turnbaugh PJ, Goordial J, Hanage WP, Dantas G et al (2015) Functional characterization of bacteria isolated from ancient arctic soil exposes diverse resistance mechanisms to modern antibiotics. PLoS One 10(3):e0069533. https://doi.org/10.1371/journal.pone.0069533
34. Boolchandani M, Patel S, Dantas G (2017) Functional metagenomics to study antibiotic resistance. Methods Mol Biol (Clifton, NJ) 1520:307–329. https://doi.org/10.1007/978-1-4939-6634-9_19
35. Reynolds LJ (2017) The identification and characterisation of novel antimicrobial resistance genes from human and animal metagenomes. PhD Thesis, UCL (University College London)

Chapter 4

Isolation and Functional Characterization of *Fusobacterium nucleatum* Bacteriophage

Mwila Kabwe, Teagan Brown, Heng Ku, Stuart Dashper, and Joseph Tucci

Abstract

Bacteriophages are viruses that specifically lyse bacteria. They have demonstrated potential in applications as antibacterial agents in medicine, agriculture, and environmental remediation. Due to the complex and dynamic nature of the oral microbiome, antibiotic treatment of chronic, polymicrobial oral diseases may lead to dysbiosis. In these diseases, bacteriophages may provide targeted activity against oral bacteria without such disruption to the broader microbial community. In this chapter, we describe the methods for screening samples that may contain bacteriophages against oral pathogenic bacteria, and using the example of FNU1, the bacteriophage we isolated against *Fusobacterium nucleatum*, describe the process of bacteriophage purification and characterization.

Key words Bacteriophage, *Fusobacterium nucleatum*, Sample screening, Enrichment, Purification, One-step growth curve, Host range

1 Introduction

Fusobacterium nucleatum is a Gram-negative anaerobic pathobiont associated with a range of human diseases including but not limited to chronic polymicrobial oral diseases, gastrointestinal disorders, cancers, and adverse pregnancy outcomes [1]. *F. nucleatum* pathogenesis occurs mainly by adhesin-mediated binding to host cells and other bacterial species to form biofilms [2], invade host cells [3], and induce a host inflammatory response [1]. These mechanisms of *F. nucleatum* virulence facilitate their colonization of the host and evasion of the immune system [1]. Treatment with antibiotics is not usually effective in these circumstances as many of them fail to penetrate biofilms [4] or to reach therapeutic concentrations to control intracellular infections [5]. In addition, it has been recognized that broad-spectrum antibiotics have not been desirable in treatment of many human diseases due to their dysbiotic effects on the microbiome [6–10]. Antibiotics have also decreased efficacy due to increasing antimicrobial resistance

Guy R. Adami (ed.), *The Oral Microbiome: Methods and Protocols*, Methods in Molecular Biology, vol. 2327,
https://doi.org/10.1007/978-1-0716-1518-8_4,

by bacteria [11]. Bacteriophages are natural viral predators of bacteria. They have coevolved with bacteria and have emerged as alternatives and adjuncts to antibiotics. Their benefits are that they may provide a more precise means of targeting specific bacteria, they are not associated with the myriad side effects seen with antibiotics, they avoid microbiome disruption, and they are able to penetrate biofilms. While research in bacteriophages has continued for over 100 years in Eastern Europe, interest in the application of bacteriophage therapy has only increased in the last 20 years or so in Western countries, leading to large-scale clinical trials [12, 13]. Because bacteriophages are rapidly coevolving with bacteria [14], they hold a distinct advantage over antibiotics, whose design, testing, implementation, and redesign to improve efficacy consumes many years.

Since bacteriophages have coevolved with bacteria, they are usually found in locations where their hosts are present [15]. Bacteriophages against oral bacteria will mainly be found in saliva and around tissues such as the gingiva and tooth roots [16–18]. However, bacteria from the oral cavity may traverse the entirety of the gastrointestinal tract (GIT), as may their bacteriophages [19]. As such, wastewater is a potential source for bacteriophages of the GIT. More specifically, *F. nucleatum* has been associated with colon cancers, and uncharacterized bacteriophage against *F. nucleatum* have been isolated from the feces of murine colon cancer models [20]. Therefore, when selecting samples for *F. nucleatum* bacteriophage screening, it would be useful to explore samples of mouth wash, oral biopsies, wastewater, colon and colon cancer biopsies, or stools from such patients. Although oral samples may be obtained through dental clinics, it is worth noting that some patients report good oral hygiene, such as regular tooth brushing and application of mouth wash, which could contain antimicrobial agents [21]. These treatments may affect the oral microbiome [22, 23], and consequently the presence of bacteria and bacteriophage in dental samples collected. Furthermore, some preservatives that may be used in toothpastes and mouthwashes may have detrimental effects on the viability of bacteriophage [24]. A morning sample of saliva and/or dental swab before brushing teeth may be a preferred sample source for screening bacteriophages against oral bacteria, including *F. nucleatum.*

Despite increased interest in bacteriophages and their isolation, challenges in isolating some bacteriophages, especially those specific to oral pathogens, remain. For instance, there are many reports of bacteriophages isolated against skin [25], urogenital [26, 27], respiratory tract [28, 29], and lower GIT bacteria [30–33]. On the other hand, lytic bacteriophages specific for oral pathobionts, such as *Porphyromonas gingivalis*, *Treponema denticola*, *Prevotella intermedia*, *Streptococcus gallolyticus*, and *Streptococcus gordonii*, have remained elusive and are rarely reported [34]. These specialist oral bacteria may not be adapted to long-term starvation outside the oral cavity and may not survive as well as the *Enterobacteriaceae* under similar conditions [35]. As such, bacteriophages against them may not readily be found in environmental samples, unlike bacteriophages against bacteria infecting respiratory, lower GIT and skin tissue, which may be found in many natural and artificial environments. Bacterial resistance mechanisms, such as CRISPR defenses, may also contribute to the difficulties in isolating bacteriophage. For instance, genome sequencing analyses have revealed a range of CRISPR palindromic sequences in important oral pathogens, such as *P. gingivalis* [36] and *Streptococcus mutans* [37].

In this chapter, we provide a description of the methods used in the isolation and functional characterization of FNU1, a bacteriophage lytic against *F. nucleatum* [17].

2 Materials

2.1 Bacteriological Culture Media

1. Brain Heart Infusion (BHI) broth: Weigh 37 g of BHI broth powder and 0.5 g of cysteine. Add 500 mL Milli-Q water and 1 mL of 5 mg/mL Haemin stock solution. Make volume up to 1 L with Milli-Q water and autoclave for 30 min at 121 °C. Label appropriately and store at 4 °C.
2. BHI at varying agar concentrations: Weigh 37 g of BHI broth powder and 0.5 g of cysteine. Make volume close to 1 L with Milli-Q water and mix before adding desired amount of agar (add 15 g/L of bacteriological agar to make 1.5% agar, 10 g/L of bacteriological agar for 1% agar and 8 g/L of agar for 0.8% agar) and 1 mL of 5 mg/mL haemin. Make volume to 1 L with Milli-Q water and autoclave for 30 min, pour out into 15 mm culture plates in laminar flow hood. Label culture plates appropriately and store at 4 °C.

2.2 F. nucleatum Identification

1. 16S rRNA gene amplification via U27F: 5′ AGAGTTT GATCMTGGCTCAG- 3′ and U1492R: 5′- AAG GAGGTGWTCCARCC-3′ primers.
2. GoTaq Long range PCR Master Mix (2×) (Promega).
3. PCR purification kit to purify PCR products, conducted according to manufacturer's instructions (Qiagen).
4. Sanger sequencing using above primers. Our samples were sequenced at the Australian Genome Research Facility (AGRF), Brisbane, Australia.

2.3 Screening Samples for Bacteriophages

1. Once saliva or mouthwash is collected, add an equal volume of sterile BHI broth (*see* **Note 1**). Add BHI broth to make up 3 mL final volume for ease of handling and filtration. Centrifuge the mixture at 500 × *g* for 15 min. After centrifugation,

collect the supernatant and discard the pellet. Filter supernatant using a 0.2μm filter (*see* **Note 2**), to get rid of remnant bacterial cells.

2. Store filtrate in sealed glass container (*see* **Note 3**) at 4 °C or use immediately as described in Subheading 3 below.
3. Autoclaved cotton swabs.
4. 0.2μm filters and 5 mL syringes.

2.4 Bacteriophage Purification

1. Sodium chloride (NaCl).
2. Polyethylene Glycol (PEG_{8000}).
3. Triton X-100.

2.5 DNA Extraction

1. RNAse A solution, stored at 4 °C.
2. DNAse I solution, stored at −20 °C.
3. To prepare 2.5 M $MgCl_2$ (Magnesium chloride): Add 101.65 g $MgCl_2$ in 200 mL distilled water. Autoclave at 121 °C for 30 min to sterilize.
4. Prepare 0.5 M of EDTA, pH 8.0, [Ethylenediaminetetraacetic acid]: Add 186.1 g of EDTA-disodium salt to 800 mL of distilled water. Stir vigorously on a magnetic stirrer before adjusting pH to 8.0 by adding sodium hydroxide (NaOH). Adjust volume to 1 L with distilled water and sterilize by autoclaving.
5. Phosphate Buffered Saline (PBS); Make PBS to final concentrations of 137 mM sodium chloride (NaCl), 2.7 mM potassium chloride (KCl), 10 mM of hydrated disodium hydrogen phosphate ($Na_2HPO_4 \cdot H_2O$), and 2 mM potassium dihydrogen phosphate (KH_2PO_4): dissolve 8 g of NaCl, 0.2 g of KCl, 1.44 g of Na_2HPO_4, and 0.24 g of KH_2PO_4 in 800 mL of distilled water. Adjust pH to 7.4 with hydrochloric acid (HCl) and make up to 1 L with distilled water. Sterilize by autoclaving at 121 °C for 30 min.
6. 10% w/v Sodium dodecyl sulfate (SDS: $NaC_{12}H_{25}SO_4$).
7. Sodium Chloride (NaCl) powder.
8. 20 mg/mL Proteinase K.
9. Polyethylene glycol (PEG_{8000}) powder.
10. Absolute ethanol (prepare 70%).

2.6 Gel Electrophoresis

1. Prepare 1× TAE (Tris-acetate-EDTA) working stock; First make 50× TAE stock: Dissolve 242 g of Tris-base in 800 mL distilled water. Add 57.1 mL of glacial acetic acid to 100 mL of earlier prepared 0.5 M EDTA, pH 8.0. Adjust final volume to 1 L. Prepare 1× TAE working solution by dissolving 20 mL of 50× TAE into 1 L of distilled water.

2. 10 mg/mL ethidium bromide.
3. To make a 100 mL of 1% agarose gel: Add 1 g agarose to 100 mL of 1× TAE. Mix gently while heating until clear. Make sure not to boil. Cool to approximately 50 °C and add 20μg (2μL) of ethidium bromide. Pour agarose into a gel tray with appropriate comb and allow to cool until solidified (*see* **Note 4**).
4. Restriction enzyme: Use type II restriction enzymes. Our laboratory employs four to six base pairs cutters: *SalI*, *HincII*, *HindIII*, *EcoRI*, and *EcoRV.*
5. DNA gel loading dye (6×; NEB).
6. Lambda (λ) DNA-*HindIII* (Range 125 bp to 23,130 bp and concentration 500μg/mL) and NEB 1 kb plus ladder (Range 500 bp to 10 kb and concentration 500μg/mL).

2.7 Electron Microscopy

1. JEOL JEM-2100 transmission electron microscope.
2. Carbon-coated copper 400-mesh grids.
3. Gatan Orius SC200D 1 wide-angle camera.
4. Geneious software Version 11.0.5.
5. ImageJ software version 1.8.0_112.

3 Methods

3.1 Bacteria Identification

3.1.1 Bacterial DNA Extraction

1. Inoculate bacteria (*F. nucleatum*) by streaking onto a 1.5% agar BHI petri dish to give single colonies. Incubate for 48 h at 37 °C in anaerobic conditions. Visually inspect colonies for characteristic morphology. Take a loopful of colonies and resuspend in 50μL of nuclease-free water in a 1.5 mL microcentrifuge tube. Briefly vortex to mix.
2. Add 2μL of 0.5 M EDTA, pH 8.0, 2.5μL of 10% (w/v) SDS and 2.5μL of proteinase K (20 mg/mL) stock. Mix gently by inverting and pulse centrifuge. Incubate at 55 °C for 1 h.
3. Allow to cool to room temperature and then add 150μL of nuclease-free water. Add 200μL (equal volume) of phenol: chloroform:isoamyl alcohol (29:28:1). Mix by inverting 10 times and vortex until cloudy. Centrifuge the cloudy suspension at 12,000 × *g* for 10 min. This separates the mixture into two layers. If the layers are not distinct, centrifuge for a further 5 min at 12,000 × *g*.
4. Carefully pipette out the clear top aqueous layer (without disrupting the white turbid interface) into a fresh 1.5 mL microcentrifuge tube containing equal volume of isopropanol. Ensure that neither the interface nor the lower layer is pipetted

into the isopropanol but are discarded. Allow the DNA to precipitate overnight at −20 °C.

5. Centrifuge for 10 min at 12,000 × *g*. Pipette off supernatant and discard appropriately. Add 200μL 70% ethanol to the DNA pellet. Centrifuge for 5 min at 12,000 × *g*, then carefully pipette off and discard the ethanol. Allow the DNA pellet to air dry before resuspending in 50μL of nuclease-free water. Store at −20 °C or use immediately.

3.1.2 PCR for 16S rRNA Gene Amplification via U27F and U1492R Primers

1. Reconstitute and dilute primers to 100μM concentration in nuclease-free water. Add 1μL of the forward and 1μL reverse primer. Then add 12.5μL of 2× GoTaq Long range PCR Master Mix, 9.5μL of nuclease-free water, and 1μL (~10 ng) of the *F. nucleatum* DNA.
2. Thermocycling conditions: 95 °C for 3 min, 32 cycles of 95 °C for 30 s, 60 °C for 30 s, and 72 °C for 90 s, with a final extension at 72 °C for 10 min.
3. Add 1μL of DNA loading dye to 5μL of PCR product. Mix well by pipetting up and down before pulse centrifuge. Add the DNA loading dye and PCR product mix to a well on the agarose gel in TAE and run electrophoresis at 100 V for approximately 60 min. Visualize the 16S rRNA PCR product using 1% agarose gel electrophoresis to confirm the 1465 bp PCR product.
4. Once PCR product size is confirmed, clean the remaining PCR product to remove primers and enzymes using a PCR cleanup kit as per manufacturer's instructions. Perform gel electrophoresis on 2μL of the cleaned product as previously, to ensure purity after PCR cleanup.
5. Submit cleaned 16S rRNA product for Sanger sequencing (for our experiments we submitted to the AGRF).

3.2 Bacteriophage Isolation Using the Enrichment Method

Once the 16S rRNA sequence confirms *F. nucleatum*, screening for the *F. nucleatum* bacteriophages can commence (*see* **Note 5**).

1. Inoculate *F. nucleatum* on a petri dish with BHI and 1.5% agar by streaking to produce single colonies. Incubate under anaerobic conditions for 48 h. Pick a single colony of *F. nucleatum* and inoculate into 5 mL sterile BHI broth in a glass Universal or McCartney bottle. Incubate broth culture under anaerobic conditions for 18–24 h (Fig. 1b).
2. From the overnight *F. nucleatum* broth culture, take 100μL and add to 20 mL of fresh BHI media within a McCartney bottle. Incubate anaerobically until OD_{600} is 0.6. This allows for the bacteria to reach exponential growth phase. Standard anaerobic incubation without shaking is done.

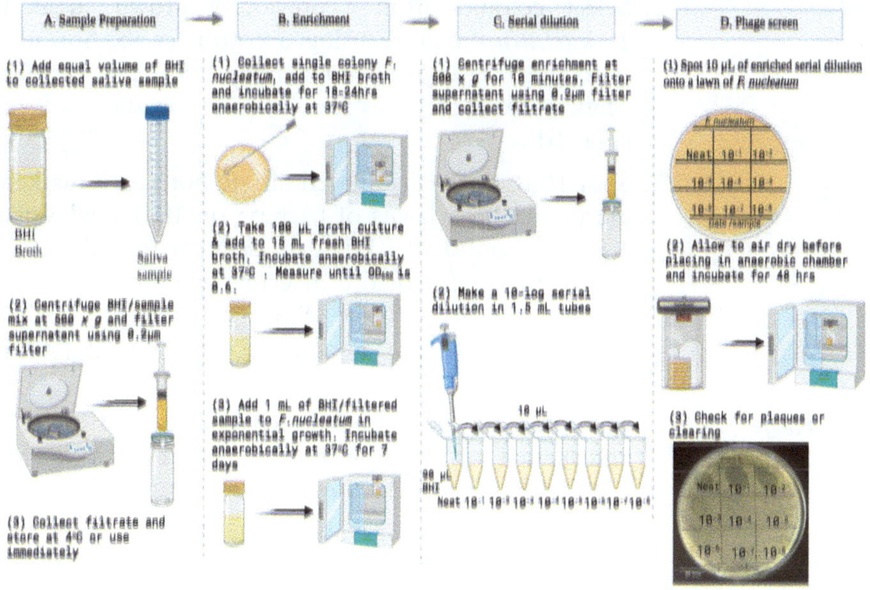

Fig. 1 Method for screening bacteriophage, from sample preparation to plaque visualization. (**a**) Initial preparation of sample to use in screening. (**b**) Sample is enriched in broth culture of *F. nucleatum* in exponential growth. (**c**) After enrichment, bacteria is filtered out such that bacteriophages remain in suspension, which is then serially diluted. (**d**) Dilutions are spotted onto divided segments of fresh lawn of *F. nucleatum*, incubated for 48 h anaerobically and visualized for plaque formation

3. Add 1 mL of filtered sample to the McCartney bottle and incubate the enrichment for 7 days anaerobically at 37 °C (*see* Fig. 1b) (*see* **Notes 6** and **7**).
4. Transfer enrichment into a 15 mL falcon tube and centrifuge at 500 × *g* for 10 min. Collect supernatant and filter the supernatant using a syringe and a 0.2μm filter (Fig. 1c).

3.3 Screening Enrichment

Once the enrichment is filtered, there are two plate screening methods used to determine the presence of bacteriophage. These methods are referred to as overlay method and spot method [38, 39] (*see* **Note 8**). We used the spot method to screen for the FNU1 bacteriophage (*see* Fig. 1).

1. Prepare nine sterile 1.5 mL microcentrifuge tubes and label them starting from the first as "neat" (highest concentration) to 10^{-8} (highest 10-log dilution). Add 90μL of BHI broth to sterile 1.5 mL microcentrifuge tubes labeled 10^{-1} to 10^{-8}. Add 100μL of filtered enrichment to the "neat" 1.5 mL tube. Take 10μL of the filtered enrichment from the "neat" and add to the tube marked 10^{-1}. Mix well by pipetting up and down. Then take 10μL of dilution from the 10^{-1} tube and add to the next tube (labeled 10^{-2}) and mix thoroughly by pipetting up and down. Repeat this for all subsequent tubes taking 10μL from the immediate previous dilution (*see* Fig. 1c).

2. Select colonies from a previously grown streak plate of *F. nucleatum* culture using a cotton swab. Spread colonies onto fresh BHI plates containing 1.5%, 1%, and 0.8% agar. Mark bottom of plates for each dilution spot as 9 rectangles (Fig. 1d).
3. Spot 10μL of serially diluted filtrate onto each marked rectangle of the fresh lawns of bacteria on BHI media plates with 1.5%, 1%, and 0.8% BHI agar (*see* **Note 9**). Allow for spotted filtrate to air dry before inverting your plates and incubate anaerobically at 37 °C for 48 h.
4. Visually inspect for plaque formation and clearing (*see* Fig. 1d).

3.4 Bacteriophage Purification

1. Locate individual plaques which can be used for purification and excised with the agar using the blunt end of the pipette tip (Fig. 2), as described below.
2. Prepare eight 1.5 mL tubes for a dilution series as detailed above. Additionally, prepare a 1.5 mL tube with 500μL of BHI broth and an empty tube: These will be for the resuspension of bacteriophages from the plaque.
3. Push the blunt end of the tip through the agar where the plaque is observed and gather the plaque together with the agar.
4. Place this agar plug in the tube with 500μL of broth. Briefly vortex so the broth becomes cloudy. Centrifuge the mixture at 12,000 × *g* for 5 min. Pipette the supernatant into the empty fresh 1.5 mL microcentrifuge tube (*see* Fig. 2).
5. Complete the 10-log serial dilution (as described in Fig. 1c) and spot 10μL of each dilution on fresh lawn of *F. nucleatum* bacteria. Allow the bacteriophage suspension to air dry. Incubate at 37 °C for 48 h under anaerobic conditions (*see* Fig. 2).
6. Check plaque morphology for consistency of size. Repeat **steps 1–5** to obtain bacteriophage with a consistent plaque morphology.
7. Proceed to the next section (Subheading 3.5) to increase the concentration of purified bacteriophage suspension.

3.5 Preparation of High Concentration Bacteriophage Stock

For further applications, it is often important to have a high concentration of bacteriophages. Several methods can be employed to increase bacteriophage numbers. A high titer stock for FNU1 was obtained by using the spread plate method.

1. Prepare cotton swab lawns of *F. nucleatum* on BHI media (with the agar concentration that allows bacteriophage plaque visualization—in the case of FNU1, 0.8% or 1% agar was appropriate).

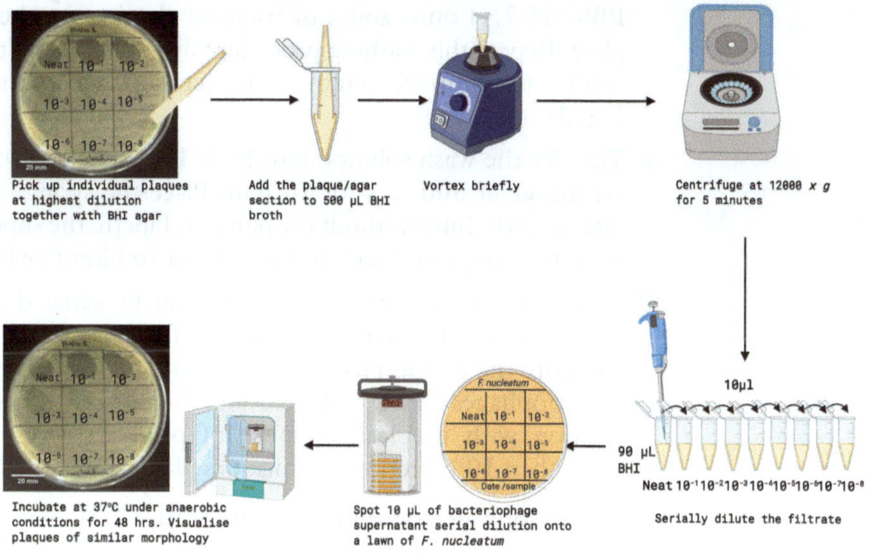

Fig. 2 Purification of bacteriophage plaques

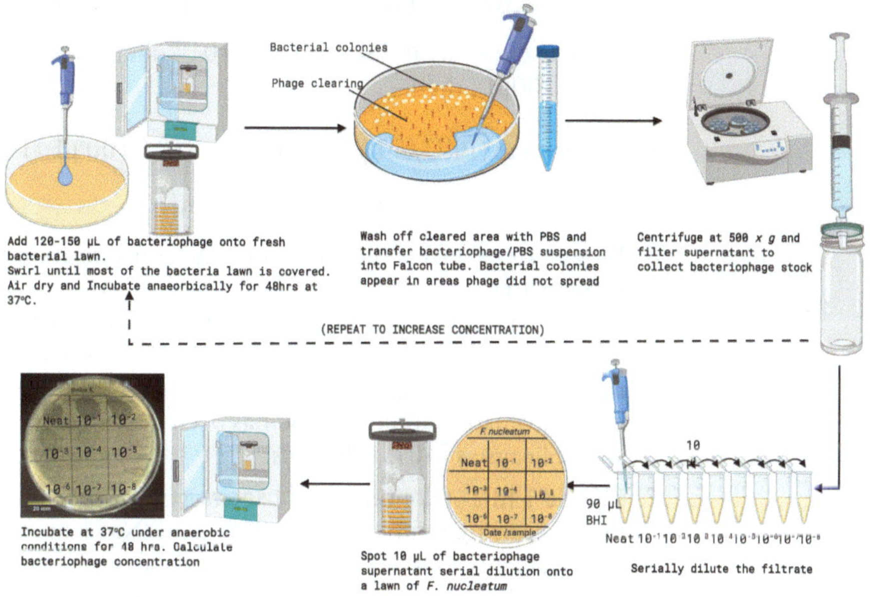

Fig. 3 Maximizing bacteriophage concentration and preparing stock suspension

2. 120–150μL of bacteriophage suspension is then dispensed onto these lawns. The plates are gently swirled to spread the suspension across the petri dish. The plates are allowed to dry within a laminar flow cabinet and incubated at 37 °C anaerobically for 48 h (Fig. 3).

3. Wash off bacteriophage particles by pipetting 2 mL of sterile BHI broth (or preferred bacteriophage storage medium, e.g.,

PBS pH 7.4) onto zones of bacterial clearing on the culture plate. Repeat this washing procedure five times, using the same media, to optimize removal of bacteriophage particles (*see* Fig. 3).

4. Transfer the wash solution into fresh 15 mL Falcon tubes and centrifuge at 500 × *g* for 10 min. Place the syringe into the filter unit (0.2μm) without the plunger. Pipette the supernatant into the syringe and replace the plunger to filter (*see* Fig. 3).
5. Determine bacteriophage concentration by serial dilution as shown in Fig. 3. Count the number of bacteriophage plaques and note the dilution factor. To calculate the concentration of bacteriophages (PFU/mL), locate the lowest concentrated dilution with the least number of plaques represented by more than 10 bacteriophage plaques. Then apply the formula:

$$\begin{aligned}&\text{Bacteriophage concentration (PFU/mL)}\\&=\text{Number of plaque forming units (PFU)}\\&\quad\times\text{Inverse of the dilution factor}\\&\quad/(\text{Amount placed on the plate [}\mu\text{L]}/1000\ \mu\text{L) mL}\end{aligned}$$

For example, if 25 plaques at 10^{-6} dilution were counted using 100μL, then

$$\begin{aligned}\text{Bacteriophage (PFU/mL)} &= 25\ \text{PFU}\times 10^{6}/(100\ \mu\text{L}/1000\ \mu\text{L})\\&= 2.5\times 10^{8}\ \text{PFU/mL}\end{aligned}$$

6. Repeat **steps 1–5** if the bacteriophage concentration needs to be increased.

   Depending on the final application for the bacteriophage, it may be useful to remove remnant bacterial endotoxin from the bacteriophage stock. A relatively simple non-chromatographic method has been described by Branston and colleagues [40], as follows:

7. Pipette 1 mL of concentrated bacteriophage stock into 1.5 mL microcentrifuge tube. Add 0.058 g of NaCl and 0.1 g of PEG_{8000}, mix well by gently inverting tube or place on a shaker until solubilized. Incubate the mix at 4 °C for 10 min. After incubation, add 20μL of Triton X-100 and mix using rotating mixer for 10 min. Incubate for 10 min at 4 °C.
8. After incubation, centrifuge at 12,000 × *g* for 15 min. Discard the supernatant and resuspend pellet in fresh 1 mL PBS, pH 7.4.
9. Repeat **steps** 7 and **8**, three times.

3.6 Testing Bacteriophage Lytic Capacity and Host Range

1. To determine the host range of the bacteriophage, inoculate other species or strains of bacteria on BHI agar plates in the same manner as the host *F. nucleatum* strain was grown.
2. Determine plaque formation by serial dilution of the concentrated bacteriophage stock. It is important to perform the serial dilutions and investigate for individual plaques, and not just test the concentrated stock (as explained above). Formation of plaques will be indicative of host susceptibility (*see* **Note 8**).

3.7 Bacteriophage FNU1 Replication Kinetics: One-Step Growth Curve

1. Inoculate individual colonies of *F. nucleatum* in 5 mL BHI broth and incubate in anaerobic conditions for 18–24 h.
2. Centrifuge the culture of *F. nucleatum* at 500 × *g* for 10 min and resuspend in fresh BHI broth media. Incubate anaerobically at 37 °C until OD_{600} is 0.6 to achieve bacterial concentration of 1×10^8 CFU/mL.
3. Dilute bacteriophage stock in BHI broth to make 1×10^7 PFU/mL.
4. Pipette 1 mL of 1×10^8 CFU/mL *F. nucleatum* into a 1.5 mL microcentrifuge. Centrifuge the bacteria at 12,000 × *g* for 10 min. Discard the supernatant and resuspend in 900μL of fresh BHI broth. Add 100μL of 1×10^7 PFU/mL bacteriophage and incubate bacteriophage and bacteria mixture for 15 min at 4 °C (*see* **Note 10**).
5. After incubation, centrifuge the bacteriophage/bacterial mixture at 12,000 × *g* at 4 °C for 10 min. Extract the supernatant and determine the bacteriophage count by serial dilution and spot technique (as described in Fig. 1c, d). This will give the number of bacteriophages unadsorbed to the bacteria during the adsorption phase. The difference between bacteriophage number added and unadsorbed will give the number of adsorbed bacteriophage (A) in the one-step analysis.
6. Resuspend the pellet in 50 mL fresh sterile BHI broth and place bacteria and adsorbed bacteriophage in suspension in a 37 °C anaerobic incubator.
7. Take 1 mL aliquots every 5 min, centrifuge at 12,000 × *g* at 4 °C for 5 min. Determine the concentration of bacteriophage (PFU/mL) in the supernatant by serial dilution and inoculating on a fresh lawn of *F. nucleatum*.
8. Plot graph of bacteriophage concentration (y-axis) against time (x-axis).
9. Calculate the burst of newly released bacteriophage (D) by subtracting the average concentration of the bacteriophage in the latent phase (B) from that of the plateau phase (C) that occurs after the rising phase. See formula for burst size below.

10. To calculate how much bacteriophage results from one viral particle after bacterial infection, divide the calculated burst of newly released bacteriophage by the number of adsorbed bacteriophages (PFU/bacteria), i.e.,

$$\text{Burst size} = \frac{\text{Burst of newly released bacteriophage } (D)}{\text{Number of adsorbed bacteriophage } (A)} \text{ [PFU per bacterial cell]}$$

3.8 Bacteriophage DNA Extraction

1. Pipette 5 mL of concentrated bacteriophage suspension (1×10^8 PFU/mL) into a 15 mL Falcon tube and add 10μL of 2.5 M $MgCl_2$ stock, 1μL RNase A, and 1μL DNase I. Mix gently and incubate at room temperature for 30 min.
2. After incubation, add 0.059 g NaCl per mL and mix well with gentle inversion or place on rotor for mixing until solubilized. Add 0.1 g of PEG_{8000} per mL, mix well by gently inverting tube or placing on a shaker until solubilized. Incubate the mixture overnight at 4 °C.
3. Decant 2 mL of the overnight mixture into a 2 mL microcentrifuge tube and centrifuge at 12,000 × *g* and 4 °C for 15 min. Discard supernatant and retain pellet. Decant another 2 mL of the overnight mixture into the same microcentrifuge tube with the retained pellet, and centrifuge again at 12,000 × *g* and 4 °C for 15 min. Discard supernatant and repeat until pellet from all the original 5 mL of the overnight mixture is collected into one 2 mL microcentrifuge tube. Resuspend the pellet in 50μL nuclease-free water.
4. Add 2μL of 0.5 M EDTA, pH 8.0, 2.5μL of 10% (w/v) SDS, and 2.5μL of 20 mg/mL proteinase K. Mix by inverting gently and incubate at 55 °C for 1 h.
5. Allow to cool to room temperature before adding nuclease-free water to bring the total volume to 200μL to which an equal volume (200μL) of phenol:chloroform:isoamyl alcohol (29:28:1) is added. Mix thoroughly and vortex until cloudy.
6. Immediately centrifuge the mixture at 12,000 × *g* for 10 min. The content separates into two layers often with a solid white interface of proteinaceous material. If two layers are not clear, centrifuge once more at 12,000 × *g* for 5 min and carefully pipette out the clear top (aqueous phase) layer into a fresh 1.5 mL microcentrifuge tube. This top layer will be approximately 200μL. Discard the white interface and the lower phenol layer appropriately, as referenced in your safety datasheet for discarding phenol waste.
7. To the aqueous phase, add 200μL isopropanol and mix well by inverting 10 times. Incubate this mixture overnight at −20 °C.
8. After incubation, centrifuge at 12,000 × *g* and 4 °C for 10 min. Taking care not to disturb the pellet, discard supernatant by

gently pipetting it out. Pulse centrifuge and remove all supernatant before adding 70% ethanol. Centrifuge for 5 min at 12,000 × *g*. Gently pipette out the ethanol and discard. Air dry the DNA pellet until all ethanol has evaporated and then resuspend in 25–50μL of nuclease-free water.

3.9 Restriction Enzyme Digestion of Bacteriophage DNA

1. To a 1.5 mL microcentrifuge tube, add 3–5μg of bacteriophage DNA and with nuclease-free water make up to 17μL final volume.
2. Add 2μL of 10× restriction enzyme buffer suitable for restriction enzyme used, then add a threefold excess of restriction enzyme (9–15 U) (*see* **Note 11**).
3. Pulse centrifuge and incubate for at least 4 h at the optimal temperature for the restriction enzyme. Pulse centrifuge every 30 min during incubation to allow condensation vapors to drop back into the reaction mix.
4. After incubation, visualize the restriction digestion pattern using 1% agarose gel electrophoresis. Pipette out 10μL of digested DNA into a fresh 1.5 mL microcentrifuge tube and add 2μL of DNA gel loading dye (6×). Mix loading dye and digested DNA by pipetting up and down and pulse centrifuge to settle the mixture to the bottom of the tube (*see* **Note 12**).
5. Place the agarose gel into the 1× TAE buffer in the electrophoresis tank. To the first well, add 10μL of 500μg/mL ladder, and to the subsequent wells, add 12μL of the mix of loading dye and digested DNA. Run the electrophoresis at 100 V for 1 h and visualize under ultra-violet light using specialized equipment (Fig. 4).

3.10 Whole Genome Sequencing (WGS) of FNU1

For FNU1, DNA extracted from the purified stock is subjected to WGS. We use the Illumina next generation sequencing platform (*see* **Note 13**). A DNA library is prepared by using a Nextera XT DNA sample preparation kit according to the manufacturer's instructions. The library is sequenced on an Illumina MiSeq using a MiSeq V2 reagent kit (300 cycles) with 150 bp paired end reads. For DNA sequence analysis, adaptor sequences are removed, and sequence reads are assembled de novo using Geneious software Version 11.0.5. Open reading frames (ORF) can be predicted within Geneious using genetic translation table 11 with a minimum nucleotide length of 50 bp. The ORFs are translated and analyzed by BLASTP (https://blast.ncbi.nlm.nih.gov/) to ascribe potential function.

3.11 Electron Microscopy

A JEOL JEM-2100 Transmission Electron Microscope with 400-mesh carbon-coated copper grids can be used to visualize purified FNU1 bacteriophage particles. This is achieved by allowing 30 s adsorbtion of the bacteriophage lysate onto the grids. After

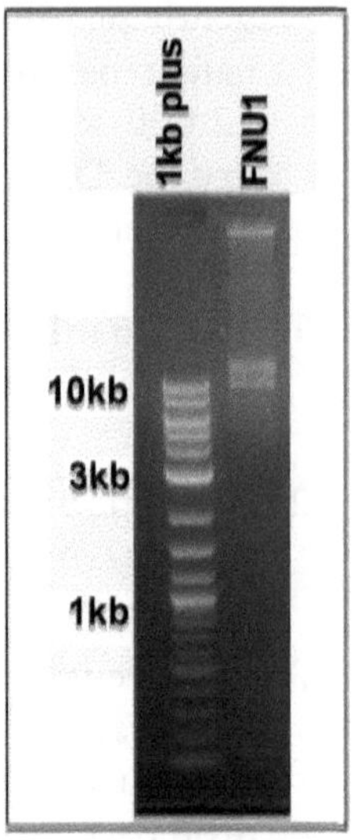

Fig. 4 Agarose gel electrophoresis of the *HincII* restriction digest of FNU1 DNA

adsorbtion, the grids are washed in Milli-Q water and negatively stained twice for 30 s with 2% [w/v] uranyl acetate. Filter paper is used to wipe off excess stain before air drying for 20 min. For our studies, a Gatan Orius SC200D 1 wide-angle camera is used at 200 kV to visualize the grids and capture images. The images are processed using ImageJ software version 1.8.0_112.

4 Notes

1. Sample is diluted in the sterile culture media used for growing target bacteria. This is done to ensure that when the sample is added to the bacteria for enrichment, the nutrients are not diluted down, which would occur if PBS or water were used for this process.
2. The choice of a 0.2μm or 0.45μm filter is a delicate balance between the risk of allowing bacteria to pass through the 0.45μm filter and the failure of large (jumbo) bacteriophages to pass through the 0.2μm filter.
3. The use of plastic microcentrifuge tubes such as Eppendorf tubes for long-term storage of bacteriophages may result in adherence of bacteriophages to tube wall, and loss of titer.
4. Do not boil agarose when dissolving. When agarose is clear, cool to approximately 50 °C before adding ethidium bromide; Make sure you have no bubbles in the agarose gel when it is poured into the electrophoresis tray. Ethidium bromide is carcinogenic and may not be available in every laboratory: Suitable alternative DNA stains may be used.

5. It is important to correctly identify bacteria used for the bacteriophage screening as bacteriophages are very specific and any errors in bacterial identification, or contamination, could lead to isolation of unwanted bacteriophages. In the case of contamination, target bacteria may be outcompeted during enrichment and bacteriophages against the intended host may not be detectable.
6. Filtration of the sample being screened is performed to ensure that the final sample added to the enrichment does not have any contaminating bacteria that would otherwise compete with target bacteria.
7. Enrichment can be incubated for longer than is required for optimal bacterial growth. In this instance, the enrichment was incubated for longer than the time taken for bacterial exponential growth. This may increase the possibility that bacteriophages will multiply to a detectable level. To ensure bacteria do not outgrow the bacteriophages, incubate the enrichment until bacterial growth is in decline phase, based on earlier experiments that define growth phases of the bacteria without bacteriophage present. The general idea is to have the bacteriophages present at the start of the early bacterial log growth to maximize opportunities for infection. To obtain complete bacterial lysis in the culture, bacteriophages must be virulent enough, and net growth of bacteria should not be more than what bacteriophages can kill [38]. Other authors have used chloroform to ensure all intracellular bacteriophages are released after enrichment; However, this is not favorable for chloroform-sensitive bacteriophages and can select for bacteriophages with lipid envelopes [41].
8. Bacteriophages can often be detected using the spot method. The overlay method may be preferred for ease of visualization of plaque morphology [38, 39]. Whichever method is used, media with the lower concentration of agar is more likely to show larger plaques, and plaque size is inversely proportional to the concentration of agar. Importantly, with both methods, there needs to be serial dilution of the enrichment to ensure that individual plaques are visible. This indicates the presence of virus particles. Areas of bacterial clearing that do not appear after serial dilution may be caused by other bactericidal constituents within the enrichment, or may be the result of incomplete bacteriophage infective processes such as abortive infection, the presence of prophages within the bacterial cell, or other bacterial resistance mechanisms [39, 42].
9. The *Fusobacterium* bacteriophage FNU1 did not cause noticeable plaques on the 1.5% agar plates. Plaques were visible on 1% and 0.8% agar plates only.

10. During the one-step growth analysis of FNU1, bacteriophages and bacteria were incubated at 4 °C to allow for adsorption. Centrifugation after adsorption will allow only bacteriophages bound to bacteria to be collected in the pellet and unabsorbed bacteriophages retained in supernatant.
11. It is often beneficial to use a combination of rare and common cutters for restriction enzyme digestion of bacteriophage DNA. Among the restriction enzymes we use, *SalI* (4 bp recognition sequence) is a more common cutter than *EcoRI*, *EcoRV*, and *HindIII* (6 bp recognition sequence). *HincII* has a six base-pair recognition sequence with four fixed bases and two degenerate bases within the recognition site, offering greater frequency of cutting than a six base pair cutter.
12. DNA methylation may affect the capacity of certain restriction enzymes to recognize specific sites in the genome, thus inhibiting their activity [43]. In some instances, fewer restriction fragments may be seen following gel electrophoresis, than those predicted *in-silico*.
13. Because of the novelty of the FNU1 genome, and the low level of genetic homology to other reported bacteriophages, the process of bacteriophage isolation from the enriched sample, DNA extraction and WGS was repeated three times to ensure accuracy.

References

1. Han YW (2015) Fusobacterium nucleatum: a commensal-turned pathogen. Curr Opin Microbiol 23:141–147. https://doi.org/10.1016/j.mib.2014.11.013
2. Rickard AH, Gilbert P, High NJ, Kolenbrander PE, Handley PS (2003) Bacterial coaggregation: an integral process in the development of multi-species biofilms. Trends Microbiol 11 (2):94–100. https://doi.org/10.1016/s0966-842x(02)00034-3
3. Han YW, Shi W, Huang GT, Kinder Haake S, Park NH, Kuramitsu H et al (2000) Interactions between periodontal bacteria and human oral epithelial cells: fusobacterium nucleatum adheres to and invades epithelial cells. Infect Immun 68(6):3140–3146. https://doi.org/10.1128/iai.68.6.3140-3146.2000
4. Høiby N, Bjarnsholt T, Givskov M, Molin S, Ciofu O (2010) Antibiotic resistance of bacterial biofilms. Int J Antimicrob Agents 35 (4):322–332. https://doi.org/10.1016/j.ijantimicag.2009.12.011
5. Imbuluzqueta E, Gamazo C, Ariza J, Blanco-Prieto MJ (2010) Drug delivery systems for potential treatment of intracellular bacterial infections. Front Biosci (Landmark Ed) 15:397–417. https://doi.org/10.2741/3627
6. Kilian M (2018) The oral microbiome - friend or foe? Eur J Oral Sci 126(Suppl 1):5–12. https://doi.org/10.1111/eos.12527
7. Dicks LMT, Mikkelsen LS, Brandsborg E, Marcotte H (2019) Clostridium difficile, the difficult “Kloster” fuelled by antibiotics. Curr Microbiol 76(6):774–782. https://doi.org/10.1007/s00284-018-1543-8
8. Theochari NA, Stefanopoulos A, Mylonas KS, Economopoulos KP (2018) Antibiotics exposure and risk of inflammatory bowel disease: a systematic review. Scand J Gastroenterol 53 (1):1–7. https://doi.org/10.1080/00365521.2017.1386711
9. Roubaud-Baudron C, Ruiz VE, Swan AM, Vallance BA, Ozkul C, Pei Z et al (2019) Long-term effects of early-life antibiotic exposure on resistance to subsequent bacterial infection. mBio 10(6). https://doi.org/10.1128/mBio.02820-19
10. Boursi B, Mamtani R, Haynes K, Yang Y (2015) Recurrent antibiotic exposure may promote cancer formation—another step in

understanding the role of the human microbiota? Eur J Cancer 51(17):2655–2664. https://doi.org/10.1016/j.ejca.2015.08.015
11. Dodds DR (2017) Antibiotic resistance: a current epilogue. Biochem Pharmacol 134:139–146. https://doi.org/10.1016/j.bcp.2016.12.005
12. Gordillo Altamirano FL, Barr JJ (2019) Phage therapy in the postantibiotic era. Clin Microbiol Rev 32(2). https://doi.org/10.1128/CMR.00066-18
13. Vandenheuvel D, Lavigne R, Brussow H (2015) Bacteriophage therapy: advances in formulation strategies and human clinical trials. Annu Rev Virol 2(1):599–618. https://doi.org/10.1146/annurev-virology-100114-054915
14. Shabbir MA, Hao H, Shabbir MZ, Wu Q, Sattar A, Yuan Z (2016) Bacteria vs. bacteriophages: parallel evolution of immune arsenals. Front Microbiol 7:1292. https://doi.org/10.3389/fmicb.2016.01292
15. Batinovic S, Wassef F, Knowler SA et al (2019) Bacteriophages in natural and artificial environments. Pathogens 8(3). https://doi.org/10.3390/pathogens8030100
16. Machuca P, Daille L, Vinés E, Berrocal L, Bittner M (2010) Isolation of a novel bacteriophage specific for the periodontal pathogen Fusobacterium nucleatum. Appl Environ Microbiol 76(21):7243–7250. https://doi.org/10.1128/AEM.01135-10
17. Kabwe M, Brown TL, Dashper S, Speirs L, Ku H, Petrovski S et al (2019) Genomic, morphological and functional characterisation of novel bacteriophage FNU1 capable of disrupting Fusobacterium nucleatum biofilms. Sci Rep 9(1):9107. https://doi.org/10.1038/s41598-019-45549-6
18. Khalifa L, Shlezinger M, Beyth S, Houri-Haddad Y, Coppenhagen-Glazer S, Beyth N et al (2016) Phage therapy against Enterococcus faecalis in dental root canals. J Oral Microbiol 8:32157. https://doi.org/10.3402/jom.v8.32157
19. Hillman ET, Lu H, Yao T, Nakatsu C (2017) Microbial ecology along the gastrointestinal tract. Microbes Environ 32(4):300–313. https://doi.org/10.1264/jsme2.ME17017
20. Zheng DW, Dong X, Pan P, Chen KW, Fan JX, Cheng SX et al (2019) Phage-guided modulation of the gut microbiota of mouse models of colorectal cancer augments their responses to chemotherapy. Nat Biomed Eng 3 (9):717–728. https://doi.org/10.1038/s41551-019-0423-2
21. Schneider C, Zemp E, Zitzmann NU (2019) Dental care behaviour in Switzerland. Swiss Dent J 129(6):466–478
22. Fernandez CE, Fontana M, Samarian D, Cury JA, Rickard AH, González-Cabezas C (2016) Effect of fluoride-containing toothpastes on enamel demineralization and Streptococcus mutans biofilm architecture. Caries Res 50 (2):151–158. https://doi.org/10.1159/000444888
23. Ardizzoni A, Pericolini E, Paulone S, Orsi CF, Castagnoli A, Oliva I et al (2018) In vitro effects of commercial mouthwashes on several virulence traits of Candida albicans, viridans streptococci and Enterococcus faecalis colonizing the oral cavity. PLoS One 13(11): e0207262. https://doi.org/10.1371/journal.pone.0207262
24. Brown TL, Ku H, Mnatzaganian G, Angove M, Petrovski S, Kabwe M et al (2020) The varying effects of a range of preservatives on Myoviridae and Siphoviridae bacteriophages formulated in a semi-solid cream preparation. Lett Appl Microbiol. https://doi.org/10.1111/lam.13299
25. Castillo DE, Nanda S, Keri JE (2018) Propionibacterium (Cutibacterium) acnes bacteriophage therapy in acne: current evidence and future perspectives. Dermatol Ther (Heidelb). https://doi.org/10.1007/s13555-018-0275-9
26. Garretto A, Miller-Ensminger T, Wolfe AJ, Putonti C (2019) Bacteriophages of the lower urinary tract. Nat Rev Urol 16(7):422–32. https://doi.org/10.1038/s41585-019-0192-4
27. Letkiewicz S, Miedzybrodzki R, Klak M, Jonczyk E, Weber-Dabrowska B, Górski A (2010) The perspectives of the application of phage therapy in chronic bacterial prostatitis. FEMS Immunol Med Microbiol 60 (2):99–112. https://doi.org/10.1111/j.1574-695X.2010.00723.x
28. Chang RYK, Wallin M, Lin Y, Leung SSY, Wang H, Morales S et al (2018) Phage therapy for respiratory infections. Adv Drug Deliv Rev 133:76–86. https://doi.org/10.1016/j.addr.2018.08.001
29. Vazquez R, Garcia E, Garcia P (2018) Phage lysins for fighting bacterial respiratory infections: a new generation of antimicrobials. Front Immunol 9:2252. https://doi.org/10.3389/fimmu.2018.02252
30. Lopetuso LR, Giorgio ME, Saviano A, Scaldaferri F, Gasbarrini A, Cammarota G (2019) Bacteriocins and bacteriophages: therapeutic weapons for gastrointestinal diseases? Int J Mol Sci 20(1). https://doi.org/10.3390/ijms20010183

31. Shkoporov AN, Hill C (2019) Bacteriophages of the human gut: the "known unknown" of the microbiome. Cell Host Microbe 25 (2):195–209. https://doi.org/10.1016/j.chom.2019.01.017

32. Shkoporov AN, Khokhlova EV, Fitzgerald CB, Stockdale SR, Draper LA, Ross RP et al (2018) PhiCrAss001 represents the most abundant bacteriophage family in the human gut and infects Bacteroides intestinalis. Nat Commun 9(1):4781. https://doi.org/10.1038/s41467-018-07225-7

33. Khan Mirzaei M, Khan MAA, Ghosh P, Taranu ZE, Taguer M, Ru J et al (2020) Bacteriophages isolated from stunted children can regulate gut bacterial communities in an age-specific manner. Cell Host Microbe 27 (2):199–212.e5. https://doi.org/10.1016/j.chom.2020.01.004

34. Mitchell HL, Dashper SG, Catmull DV, Paolini RA, Cleal SM, Slakeski N et al (2010) Treponema denticola biofilm-induced expression of a bacteriophage, toxin-antitoxin systems and transposases. Microbiology 156 (Pt 3):774–788. https://doi.org/10.1099/mic.0.033654-0

35. Baker JL, Hendrickson EL, Tang X, Lux R, He X, Edlund A et al (2019) Klebsiella and Providencia emerge as lone survivors following long-term starvation of oral microbiota. Proc Natl Acad Sci U S A 116(17):8499–8504. https://doi.org/10.1073/pnas.1820594116

36. Chen T, Olsen I (2019) Porphyromonas gingivalis and its CRISPR-Cas system. J Oral Microbiol 11(1):1638196. https://doi.org/10.1080/20002297.2019.1638196

37. van der Ploeg JR (2009) Analysis of CRISPR in Streptococcus mutans suggests frequent occurrence of acquired immunity against infection by M102-like bacteriophages. Microbiology 155(Pt 6):1966–1976. https://doi.org/10.1099/mic.0.027508-0

38. Hyman P, Abedon ST (2010) Bacteriophage host range and bacterial resistance. Adv Appl Microbiol 70:217–248. https://doi.org/10.1016/S0065-2164(10)70007-1

39. Hyman P (2019) Phages for phage therapy: isolation, characterization, and host range breadth. Pharmaceuticals (Basel) 12(1). https://doi.org/10.3390/ph12010035

40. Branston SD, Wright J, Keshavarz-Moore E (2015) A non-chromatographic method for the removal of endotoxins from bacteriophages. Biotechnol Bioeng 112 (8):1714–1719. https://doi.org/10.1002/bit.25571

41. Ackermann HW, Audurier A, Berthiaume L, Jones LA, Mayo JA, Vidaver AK (1978) Guidelines for bacteriophage characterization. Adv Virus Res 23:1–24. https://doi.org/10.1016/s0065-3527(08)60096-2

42. Kutter E (2009) Phage host range and efficiency of plating. Methods Mol Biol (Clifton, NJ) 501:141–149. https://doi.org/10.1007/978-1-60327-164-6_14

43. Nelson PS, Papas TS, Schweinfest CW (1993) Restriction endonuclease cleavage of 5-methyl-deoxycytosine hemimethylated DNA at high enzyme-to-substrate ratios. Nucleic Acids Res 21(3):681–686. https://doi.org/10.1093/nar/21.3.681

Chapter 5

Comparison of Microbiome in Stimulated Saliva in Edentulous and Dentate Subjects

Guy R. Adami, Michael J. Ang, and Elissa M. Kim

Abstract

Measurement of saliva microbes is promoted as a way to detect oral and systemic disease, yet there is a multitude of factors that affect the oral microbiome. The salivary microbiome is influenced by biofilm of shedding (epithelial) and non-shedding (tooth) surfaces. Methods for study of the salivary microbiome are by no means standardized, and differences in sample collection, storage, and processing can all affect results to some degree. Here we describe one method of saliva collection that has been validated for reproducibility. Standard 16S rRNA gene analysis is done using the Human Oral Microbiome Database library which results in analysis that is straightforward. Everything about this procedure except the library synthesis and DNA sequencing itself can easily be done in-house. To gauge the ability of salivary microbial analytics to distinguish between edentulous and dentate oral conditions, differences in the saliva microbiome of subjects with and without teeth were examined. Fifty-two dentate and 49 edentulous subjects provided stimulated saliva samples. 16S rRNA gene sequencing, QIIME-based data processing, and statistical analysis were done using several different analytical approaches to detect differences in the salivary microbiome between the two groups. Bacteria diversity was lower in the edentulous group. Remarkably, all 31 of the most significant differences in taxa were deficits that occurred in the edentulous group. As one might expect, many of these taxa are attributed to dental plaque and gingival sulcus-associated bacteria verifying that the measurement of 16S rRNA genes in the bacteria of the saliva can be used to reproducibly measure expected differences in the oral microbiome that occur with edentulism or other conditions and diseases.

Key words 16S rRNA gene, QIIME 2, MicrobiomeAnalyst, STAMP, Saliva

1 Introduction

The human oral microbiome consists of the microbes that live on the surfaces of the teeth and mucosa of the floor of the mouth, palate, gingival, buccal mucosa, tongue, and oral pharynx [1, 2]. The saliva serves as a pool of microbes shed from these surfaces and possibly others, including pharynx, tonsils, etc. Microbes from these oral surfaces have long been known to change with oral diseases such as caries and periodontal disease, but more recently these same or other oral microorganisms have been linked to a number of systemic diseases and certain cancers, cardiovascular

Guy R. Adami (ed.), *The Oral Microbiome: Methods and Protocols*, Methods in Molecular Biology, vol. 2327, https://doi.org/10.1007/978-1-0716-1518-8_5,

disease, stroke, diabetes, pneumonia, as well as oral, pancreatic, and gastrointestinal cancers [3, 4].

Saliva is secreted from salivary glands within the oral cavity. It contains a wide range of potential disease biomarkers including microbes, metabolites, proteins, nucleic acids, and hormones [5, 6]. Saliva contains between 10^7 and 10^9 bacteria per mL with an average value of 1.4×10^8 bacteria/mL, with a higher abundance of anaerobes than aerobes [7]. Since the average daily flow rate is nearly 1 L per day, it is easy to use saliva as a means of bacteria collection from oral surfaces, where bacteria live and proliferate. While the majority of oral bacteria is attached to exfoliated oral mucosal epithelial cells, about 1/3 are free [8]. Presumably, the teeth and mucosa shed cells and/or microbes that end up transiently in the saliva. The ease of collection of saliva makes it an ideal source for samples of the oral microbiome.

Past studies using 16S rRNA and saliva provide information of what bacteria constitute the oral microbiome in health [9]. Two groups have cataloged the bacteria that appear in the oral cavity of human beings [1, 10, 11]. These curated libraries, the Human Oral Microbiome Database and the core oral microbiome, or CORE, include 16S rRNA sequences for all taxa included. A key part of taxa identification based on 16S rRNA sequencing requires usage of a reference library of curated 16S rRNA sequences to allow conversion of 16S rRNA reads to bacteria taxa. Rather than using huge SILVA, RDB databases which include environment bacteria that are seldom if ever found in the oral cavity except at trace levels, a much smaller library of reference sequences from the oral microbiome is used [1, 12, 13]. This increases the accuracy of identification and greatly reduces the demand on computer storage space for taxonomic assignments, and simplifies and speeds the assignment process. In this protocol, we will explain an approach to this task which is practical for someone with minimal experience with command line computing.

When studying the oral microbiome, a choice must be made on what site to sample. Logically, the best place to sample is the site of interest. If you are studying occlusal caries development, then sample the occlusal surface of the teeth. If you are interested in oral cancer development in tobacco users, then sample the site of frequent tobacco-associated cancers, such as the lateral border of the tongue or the floor of the mouth [14]. However, there are other concerns, such as ease of sample collection. For example, if one is interested in the gingiva, then one must be careful not to mistakenly sample tooth surfaces with the swab or other collection device. This would limit self-sampling of gingiva by study subjects. An additional concern with sampling the gingiva is that, like the teeth, they are exposed to different levels of hygiene. We have found great variability in the taxa present at gingiva samples. It is possible that for some subjects who brush their teeth and parts of their gums, the identity of the taxa found at those sites depends on the last time they brushed their teeth. While subjects can be instructed not to brush their teeth 12–24 h prior to collection, this is not always practical. As an alternative, we collect saliva. It is representative of many different sites in the mouth, so it may reveal changes that can occur anywhere and it is less sensitive to variation at one site. It is also easier to collect, so it is amenable to self-collection [15–17]. It, of course, has the deficit that changes seen are an average of those that occur at many sites, so it may be less sensitive than the measurement of a single mucosal or hard surface site that makes up a single niche.

2 Materials

2.1 Sample Collection and Storage

1. Water in cup for subject to rinse mouth.
2. 50 mL sterile conical screw cap tube.
3. Hygienic white utility wax—square wax rope.
4. 1.5 mL Polypropylene microcentrifuge tubes.
5. Phosphate-buffered saline (PBS): 1×, pH 7.4. Prepare 800 mL of distilled water and add 0.2 M NaCl (11.6 g), 2.5 mM KCl (0.186 g), 8 mM Na_2HPO_4 (1.4 g), 1.5 mM KH_2PO_4 (0.2 g). Adjust the pH to 7.4 with HCl and add distilled water to prepare a 1 L solution of 1× PBS.
6. 1 mL and 200μL pipette with tips.
7. Vortex.
8. Biosafety hood.
9. −80 °C Freezer.

2.2 DNA Isolation

1. DNA isolation kit with bead-based lysis (*see* **Note 1**). The instructions as written here are for the Quick-DNA Fungal/ Bacterial Microprep Kit from Zymo Research, though other methods of microbe lysis and DNA isolation can be used.
2. Tube agitator; preferably rapid speed and multiple direction (*see* **Note 2**).
3. Vortex.
4. Microfuge.
5. NanoDrop spectrophotometer or similar (Thermo Fisher Scientific).

2.3 First and Second PCR Step in Amplicon Generation, in Preparation of Sequence Analysis

1. High-fidelity DNA polymerase.
2. iProof High-Fidelity polymerase (BioRad) or KAPA HiFi Hot-Start polymerase (Kapa Biosystems; recommended by Illumina).
3. Thermocycler.
4. U.V. emitting transilluminator to visualize U.V. absorbing DNA bands in gels.
5. 2% Agarose gel TAE buffer 1× and gel box.
6. Fluorometer, such as Qubit. And fluorescent dye kit to measure DNA.
7. Universal DNA primers to generate 16S V1–3 amplicons [18]

   CS1_27F: ACACTGACGACATGGTTCTACAAGAGTTTGATCCTGGCTCAG.

   CS2_534R: TACGGTAGCAGAGACTTGGTCTATTACCGCGGCTGCTGG.

8. PCR reaction components, prepared as below:
   Template DNA (test several volumes or ng—e.g., 2 ng).

Reaction Buffer	1×
dNTP	0.3 mM
Forward primer	0.3μM
Reverse primer	0.3μM
Polymerase	0.4 units per reaction KAPA HiFi Hotstart DNA Polymerase
ddH_2O (PCR-grade)	Up to 10μL total

2.4 Computational Analysis

All software is available for download from the indicated links. While the QIIME 2 program runs most quickly with 32 GB RAM and multi-core processor, this is not required. Because of the usage of a site-specific microbiome library for the oral cavity, which is much smaller than nonspecific 16S rRNA databases such as SILVA or RDP, we have successfully used a MacBook Air with 8 GB RAM. STAMP runs only on Windows computers as a stand-alone program. MicrobiomeAnalyst is run through a web browser.

2.4.1 Sequence Data Analysis

1. Personal computer; Preferably, a Macintosh is used.
2. QIIME 2 (https://docs.qiime2.org/2021.4/install/native) or the latest version.
3. Human Oral Microbiome Database reference databases and taxonomy files modified from QIIME (http://www.homd.org/?name=seqDownload&file&type=R).
4. 2021-02-09 eHOMD 16S rRNA RefSeq Version 15.22 Taxonomy File for QIIME and 2021-02-09 eHOMD 16S rRNA RefSeq Version 15.22 FASTA File.

5. STAMP (http://dparks.wikidot.com/stamp).
6. MicrobiomeAnalyst (https://www.microbiomeanalyst.ca/).

3 Methods

3.1 Experimental Design

Sample size for an oral microbiome study of course depends on the level of change expected and the variability. For something like comparison of edentulous and dentate subjects, we expect large differences and that is what is observed. In general, with saliva studies we find that 20–25 subjects in each group are sufficient to perform a pilot study. Optimally, sample size should be balanced in each group. This expands the statistical tools that are available to analyze the data. Various statistical tools such as PERMANOVA and ANOSIM to compare population diversity, and provide a statistical significance to the differences [18, 19], work best with balanced datasets [20]. Typically, a first question is whether there are any taxa in the two groups of subjects that are differentiated based on clinical state, lifestyle, diet, etc.? As shown some time ago, while beta diversity analysis has revealed large differences in specific taxa at different oral sites or niches [21, 22], comparison of the same site in different subjects shows much fewer differences. Similarly, when comparing saliva in different groups of people, taxa seen tend to be largely overlapping. Stability of the oral microbiome is an important concept when designing and interpreting salivary genomic studies. Importantly, it was suggested that there is variability in collection of oral microbe samples depending on food exposure, time of day, hyposalivation, precise method of collection, etc., making saliva an unreliable source for measurement of the oral microbiome [23–25]. However, several studies suggest that this is not the case [15, 26, 27]. Variability in oral microbiome measurement [28] based on 16S rRNA gene sequencing may be introduced due to computational error in operational taxon unit identification and assignment, especially for lower abundance taxa [29]. The exact method of sample collection can also cause differences [30]. As a first step to establish stimulated saliva collection as a reliable method to identify conditions or disease in individual subjects, 101 subjects with an obvious difference—tooth loss—were studied. It was determined that edentulous subjects, without their dentures, showed a different saliva microbiome than dentate subjects and these differences could be reproducibly calculated. In this chapter, methods used to accomplish those measurements of salivary taxa are discussed.

3.2 Identify and Measure Taxonomic Units in Saliva

The following method was applied to generate amplicon sequences from DNA purified from human saliva, and then used to identify bacterial taxa in the samples. Many methods of bacterial lysis and bacterial DNA purification are possible [17, 31, 32]. Many groups favor mechanical disruption using beads followed by silica purification based on speed, and ease of use. Region V1–V3 in the 16S rRNA gene is amplified and sequenced.

3.2.1 Sample Collection

1. Routine exclusion criteria for subjects are antibiotic usage in the last month, eating in the last hour, and usage of germicidal oral rinses, in the last 48 h.
2. Prior to saliva collection, the oral cavity of each subject is to be inspected for obvious signs of gross debris or existing prostheses. Edentulous patients need to remove their dentures.
3. Subjects are given a cup with tap water to rinse their mouths for 30 s.
4. A sterile 50 mL centrifuge tube and a 15-mm-long disinfected piece of periphery wax is given to each subject for saliva collection. The subject should be told to open the tube and begin chewing the wax to stimulate saliva flow. *Note*, unflavored wax is available in bulk or individual packs.
5. Subjects are timed using a digital timer for 5 min and asked to chew and spit in the tube as much as possible.
6. After 5 min, subjects are asked to screw the cap onto the tube and it is collected.

3.2.2 Sample Processing

1. Saliva samples are kept on ice for less than 2 h post-collection from subjects in 50 mL screw cap tubes. Samples are vortexed for 5 s, then 1 mL of saliva is transferred from a 50 mL tube to a 1.5 mL tube (*see* **Note 3**).
2. Saliva samples are centrifuged at 4500 × *g* for 5 min at 4 °C (*see* **Note 4**).
3. Supernatant is removed with pipette and discarded as biohazard (*see* **Note 5**).
4. 1 mL of cold phosphate-buffered saline (PBS) is added. Shake and/or vortex to resuspend.
5. Centrifuge at 5500 × *g* for 5 min at 4 °C.
6. Remove supernatant with 1 mL of PBS added to resuspend.
7. Repeat **steps 4–6**.
8. Store pellets in microcentrifuge tube at −80 °C.

3.2.3 DNA Isolation

1. Transfer to ZR BashingBead screw cap lysis tube with 0.1 and 0.5 mm high-density beads from the Zymo Research Quick-DNA Fungal/Bacterial Miniprep Kit.
2. Add 750μL BashingBead Buffer to the tube and close tightly.
3. Secure tubes in BioSpec Mini-beadbeater making sure that the tubes are well balanced, then shake for 1 min. Remove tube from apparatus and place on ice. After 4–5 min, repeat shaking

and again place the tube on ice. Samples should get warm but not hot while shaking to avoid degradation (*see* **Note 2**).

4. Remaining steps of the procedure are done according to the manufacturer's instruction until the final elution step from the silica column.
5. Remove the final column, then place a new lid on the tube containing the column eluate. Recentrifuge for 1 min, then collect the liquid, making sure not to disturb pellet in the tube made from column fines that may interfere with spectroscopic measurement of DNA concentrations but are largely inert.
6. Measure the DNA concentration of 1.5μL of the sample using the Nanodrop spectrophotometer.

3.2.4 Library Preparation and Amplicon Sequencing: Amplification of the V1–V3 Hypervariable Region of the 16S rRNA gene (See ***Note 6****)*

1. PCR reactions should be set up according to manufacturer's instructions in a final volume of 10μL. To facilitate the procedure, a mix is made for each set of 10 reactions or more. 1 and 0.1μL of each sample are amplified.
2. dNTP 10 mM 0.5μL + 5× KAPA HiFi Fidelity Buffer 2μL + 0.5μL of 10μM forward and 0.5μL of 10μM reverse primer, 3.5μL of sterile water, and 1μL of DNA. Primers hybridize to the 16S rRNA gene and produce about a 500 base amplicon.
3. The V1–V3 region amplified by heating to 95 °C for 3 min then cycling the reaction at 98 °C for 20 s, 65 °C for 15 s, and 72 °C for 20 s, followed by a 5 min 72 °C extension.
4. 27–28 cycles completed.
5. 2μL is added to 3μL loading buffer and separated on a 2% agarose Tris Acetate gel in an electrical field to verify a product at approximately 500 base pairs. If after ethidium bromide staining the expected band is visible with the UV transilluminator, then the sample is used for the next step. If both dilutions of a sample produce visible bands on the PCR, then the highest dilution is used.
6. Second PCR step with barcoding—this is routinely done by the DNA service facility where the next-generation sequencing is performed. 1μL of amplification product from the first stage is used as input to the second reaction. The primers for the second stage amplifications are the Access Array barcoding system primers containing Illumina sequencing adapters, sample-specific barcodes, and CS1 and CS2 linkers. Assay conditions are otherwise identical to that for the first step PCR, with fewer cycles.
7. Conditions—PCR conditions for the second reaction are: 5-min initial denaturation at 95 °C, followed by 8 cycles of 98 °C for 20 s, 65 °C for 15 s, and 72 °C for 20 s, followed by 72 °C for 5 min.

Samples were pooled in equimolar ratio after being quantified using Qubit 2.0 fluorometer. Sequencing is performed on an Illumina MiSeq sequencer using standard V3 chemistry with paired-end, 300 base reads. Fluidigm sequencing primers, targeting the CS1 and CS2 linker regions, are used to initiate sequencing (*see* **Note 7**). Demultiplexing of reads is performed on the sequencer.

3.2.5 Bioinformatics Analysis

1. Demultiplexed sequence, with nonbiological sequence, primer, etc. removed, each sample's two files of forward (R1) and reverse (R2) reads are downloaded from BaseSpace https://login.illumina.com/platform-services-manager/ in the form of FastQC files, which carry sequence quality information on each nucleotide.
2. For this analysis, only the reverse (R2) reads were used (*see* **Note 8**).
3. The QIIME 2 suite of tools is used to analyze data using command line interface. On a Mac computer, this is done by using the Terminal Utility, which allows text-based usage of Unix type commands. It is found within the Utilities Folder (*see* **Note 9**).
4. Move all FASTQ files that will be used, in this case all reverse (R2) reads, into a single folder/directory, which for our purposes will be called "16s-experiment." They should be stored as single files in the directory/folder (*see* **Note 10**). This can be done in the command line using Unix commands or after a directory is created in the command line, one can switch back to the Mac Desktop mode and transfer FASTQ files by drag and drop, after locating the "16s-experiment" folder.
5. Activate QIIME 2 using the command: conda activate [version name, ex. qiime2-2021.4]. Use the "change directory or cd" command to move to the directory that contains the "16s-experiment" folder with all the FASTQ files (*see* **Note 11**). Instructions for installation of the Conda environment and of QIIME 2 are on the QIIME 2 website. The version of QIIME 2 can be determined by using conda info --envs. It is important to move to the correct file path by typing cd /pathname. An easy way to do this is to drag the directory containing the "16s-experiment" folder into the Terminal after typing in cd. For example, if the "16s-experiment" folder is in the "QIIME 2" directory, one should drag and drop the QIIME 2 directory to the Terminal after typing in cd (*see* **Note 9**) (Fig. 1).
6. Reformatting sequence (single-end demultiplex). This first step imports the FASTQ files into the QIIME 2 environment, converting them to the right format, qza files.
7. The term "import" is not accurate here, as the files are already in the right folder, "16s-experiment." However, the

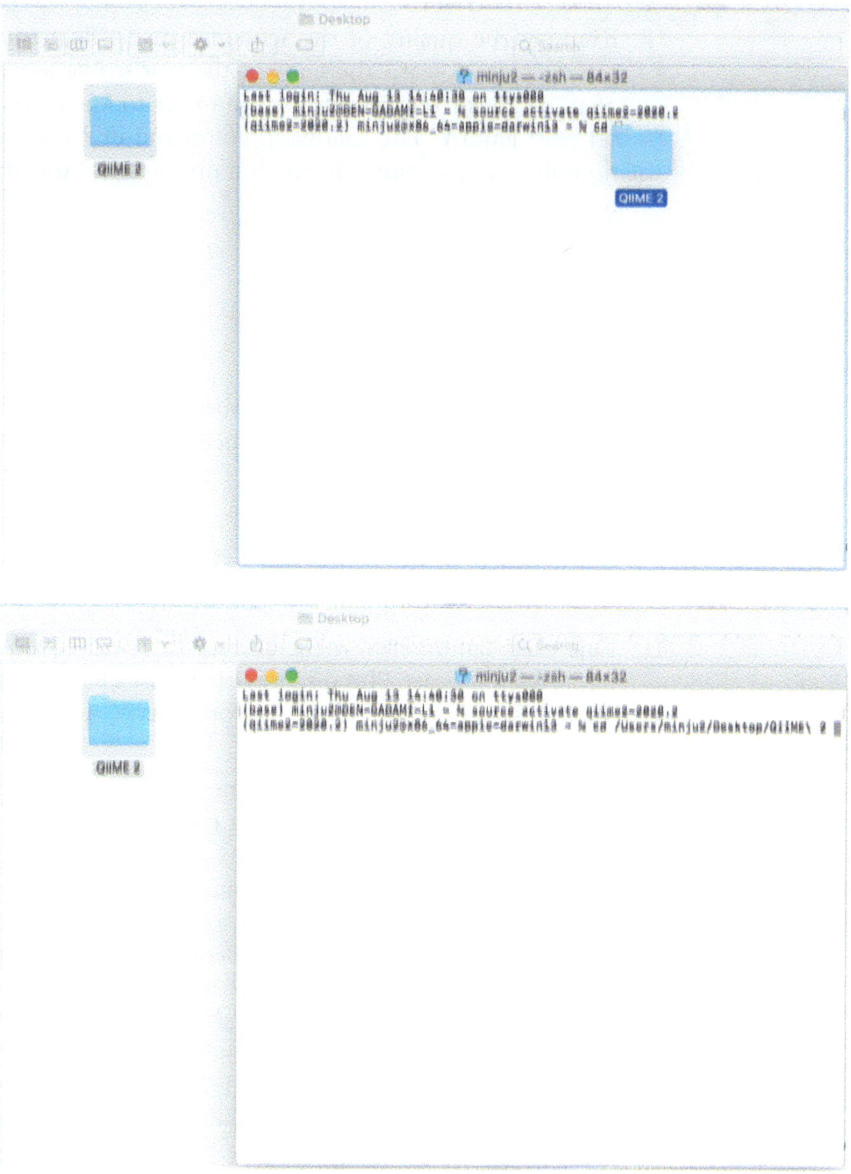

Fig. 1 Example of moving to a file path by drag and drop on the Terminal

conversion in format of the file that the import command performs is needed to run the program.

```
qiime tools import \
--type 'SampleData[SequencesWithQuality]' \
--input-path 16s-experiment \
--input-format CasavaOneEightSingleLanePerSampleDirFmt \
--output-path demux-single-end.qza
```

8. Examine the quality of the sequence reads. Sequence quality can decrease toward the ends of the reads. Average sequence quality needs to be determined in order to know when to cut off read length. The following command delivers a visualization file that contains information on sequence quality.

```
qiime demux summarize \
--i-data demux-single-end.qza \
--o-visualization demux-single-end-[date].qzv
```

9. An additional command allows visualization of this file, and all . qzv files, so that evaluation of sequence quality can be done (*see* **Note 11**):

```
qiime tools view demux-single-end-[date].qzv
```

10. One then can determine the average rates of sequence errors. Typical choices are Phred scores of at least 20 (one error every 10^2 bases) or 30 (one error every 10^3 bases) for most sequences.
11. Several years ago, DADA2 became available to do the operational taxon unit creation [33]. It takes the sequence reads for the samples and groups them into an amplicon sequence variant table after denoising, which includes identification of likely sequence errors, followed by chimera removal. In the past, the denoising was prone to error and could result in inappropriate grouping of sequences. DADA2 and other similar methods have been optimized to *minimize* error when variant calling and when using Illumina and other sequence data. This step will deliver a table with the representative sequences present and a table of the quantity of the representative sequences for each sample. The example below removes no sequence from the start and truncates at 293 bases of 16S rRNA sequence.

```
qiime dada2 denoise-single \
--i-demultiplexed-seqs demux-single-end.qza \
--p-trim-left 0 \
--p-trunc-len 293 \
--o-representative-sequences rep-seqs-dada2.qza \
--o-table table-dada2.qza \
--o-denoising-stats stats-dada2.qza
```

12. The sequence reads have been grouped and counted in each sample but have not yet been assigned to bacteria taxa, which depends, of course, on a reference library. This next step

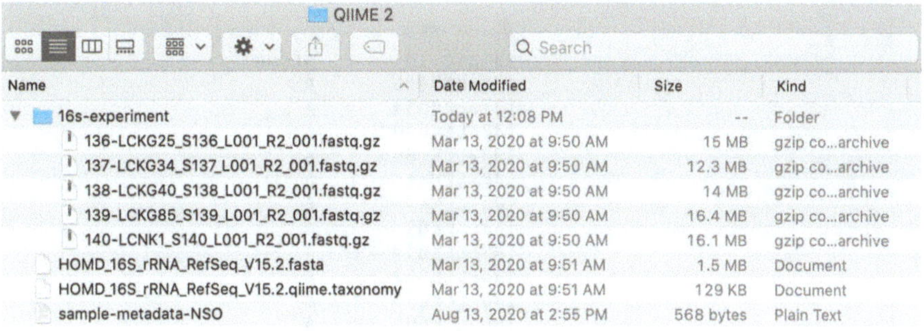

Fig. 2 HOMD files are placed in one directory above the FASTQ files

requires import of a FASTQ file containing the library of known sequence oral 16S rRNA gene sequence at the HOMD site and a file with the taxonomic tree of these sequences. At this point, it is possible to switch back to the Mac user interface. Go to the HOMD website (http://www.homd.org/?name=seqDownload&file&type=R) and download the latest annotated library of 16S rRNA sequence along with the taxonomy file. At the time of this writing, these are: HOMD_16S_rRNA_RefSeq_V15.22.fasta and HOMD_16S_rRNA_RefSeq_V15.22.qiime.taxonomy. These are downloaded and placed into one directory above the folder that contains the FASTQ files. For our purposes, these are placed in the "QIIME 2" directory, which is one above the "16s-experiment" folder containing all the FASTQ files (Fig. 2). Alternatively, these files can be directly accessed through the command line interface using Unix commands.

13. Return to the command line interface in QIIME 2. The following two commands convert the reference files, reformatting them so they become .qza files and are compatible with the QIIME 2 suite.

```
qiime tools import \
--type 'FeatureData[Sequence]' \
--input-path HOMD_16S_rRNA_RefSeq_V15.22.fasta \
--output-path homd-ref.qza

qiime tools import \
--type 'FeatureData[Taxonomy]' \
--input-format HeaderlessTSVTaxonomyFormat \
--input-path HOMD_16S_rRNA_RefSeq_V15.22.qiime.taxonomy \
--output-path homd-ref-taxonomy.qza
```

14. At this point, the HOMD 16S sequences, the taxonomy file, and the representative sequences from the samples are now all in the folder and it is possible to perform the sequence alignment necessary to assign denoised sequence reads to bacterial

sample-metadata-NSO

sample id	group
136-LCKG25	A
137-LCKG34	A
138-LCKG40	A
139-LCKG85	A
140-LCNK1	A
141-LCNK7	A
142-LCKG1	A
143-LCKG17	A
144-LCKG52	A
145-LCKG77	A
146-LCKG91	A
147-LCKG28	A
148-LCKG35	A
149-LCKG46	A
150-LCKG86	A
151-LCNK2	A
152-LCNK8	B
153-LCKG5	B
154-LCKG20	B
155-LCKG53	B
156-LCKG80	B
157-LCKG92	B
158-LCKG29	B
159-LCKG36	B
160-LCKG47	B
161-LCKG64	B
162-LCKG87	B
163-LCNK3	B
164-LCJS12	B
165-LCKG6	B
166-LCKG30	B
167-LCKG55	B
168-LCKG81	B
169-LCKG93	B
170-LCKG31	B

Fig. 3 Example of sample-metadata-NSO file

taxa. This is done using the alignment tool "BLAST" described below.

```
qiime feature-classifier classify-consensus-blast \
--i-query rep-seqs-dada2.qza \
--i-reference-reads homd-ref.qza \
--i-reference-taxonomy homd-ref-taxonomy.qza \
--p-perc-identity 0.98 \
--o-classification classification-reverse-NSO.qza \
--verbose
```

15. Next are a set of commands which take those sequences and taxa assignments and generate the figures to allow human examination of the data. To optimize comparison, import a sample metadata file that has the name and the group assignment for every sample (*see* **Note 12**, see example in Fig. 3).

```
qiime metadata tabulate \
--m-input-file classification-reverse-NSO.qza \
```

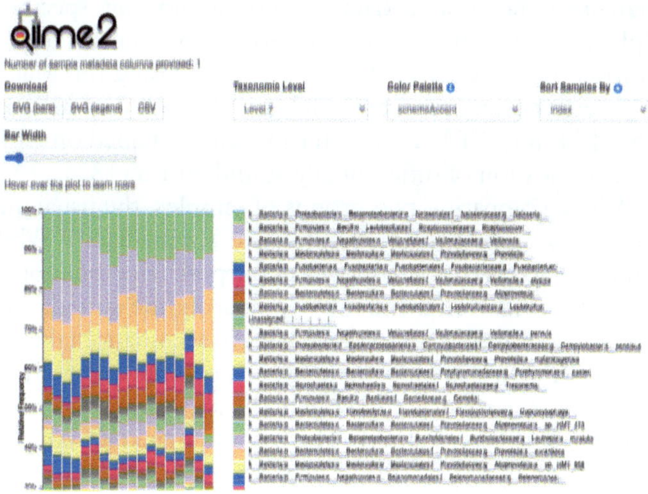

Fig. 4 Select "level 7" and "CSV" to download the abundance tables at the species level. Use "level 6" and "CSV" to generate an abundance table at the genus level

```
--o-visualization classification-reverse-NSO.qzv

qiime feature-table summarize \
--i-table table-dada2.qza \
--o-visualization table-reverse_NSO.qzv \
--m-sample-metadata-file sample-metadata-NSO.txt
```

16. The final command set generates the taxa barplot. It allows the production of the bacterial taxa abundance table which is necessary to identify relative amounts of each bacterial tax on in each sample. Select "Relabel X" and download tables corresponding to each phylogenetic level with 1 being phyla and 7 being species as shown in Fig. 4. These taxa abundance tables can be visualized with any spreadsheet program.

```
qiime taxa barplot \
--i-table table-dada2.qza \
--i-taxonomy classification-reverse-NSO.qza \
--m-metadata-file sample-metadata-NSO.txt \
--o-visualization taxa-bar-plots-reverse-NSO.qzv

qiime tools view taxa-bar-plots-reverse-NSO.qzv
```

3.2.6 Taxonomic Analysis: Identification of Differentially Abundant Samples

Analysis of taxonomic data, including plotting species richness (alpha diversity) and relative proportions of taxa and beta diversity, is carried out in QIIME 2 or can be done using the Macintosh or PC interface using programs such as PRIMER 7 or MicrobiomeAnalyst [34–36]. STAMP, which runs self-contained on a PC, allows for identification of differentially abundant taxa [37].

When comparing two groups of samples, the usual question is "do the taxa present differ in relative abundances?" While determining beta diversity can serve this purpose, an even more sensitive method is to detect changes by comparing the abundance of individual taxa. For this, we use STAMP. An abundance table created in QIIME 2, with a single sample in each column and each taxa, such as species, on rows as described (https://beikolab.cs.dal.ca/software/STAMP) is required. The sample metadata table is similar to that used in the QIIME 2 program (*see* **step 15** in Subheading 3.2.5). The first column is identical to the top row of the abundance table and contains each sample's name. The second column can be headed by the word "group" and in each cell is the type of sample. Here, the comparison is samples from dentate subjects versus those from edentulous subjects. Prior to importing the abundance table into the program, we used the "COUNTIF" command in Excel to identify taxa rows where >90% of the cells are zeroes. These are eliminated to make statistical comparison more robust.

Many sample groups can be designated in the metadata file. While using the program, the user has the ability to select and compare any two groups in the sample population against each other. This is a major asset of the STAMP program as it allows the use of a single abundance table and metadata table to run multiple comparisons. Additionally, the PCA component easily identifies outlier samples that are radically different from most samples based on several taxa and more closely resemble saliva-free negative controls. If these have low counts, they may be eliminated as artifactual [38, 39]. The Welch's *t*-test does not require an equal number of samples in each group and allows Benjamini-Hochberg or Storey-based correction for multiple testing to establish a False Discovery Rate for each differentially abundant taxon in the two chosen samples groups.

Microbiome analyst, like Primer7, allows determination of alpha diversity, beta diversity, and generation of bar plots for data comparison with a user-friendly interface. Microbiome Analyst additionally allows identification of taxa at different abundances, along with statistical analysis of the differences and correction for multiple testing, using programs such as DESeq2, univariate analysis, and edgeR.

4 Notes

1. Different approaches for DNA isolation are available. Bead-based or enzymatic/chemical-based cell wall and membrane lysis purification protocols are available in several kits, which are followed up by silica column-based purification. Other post-lysis purification methods include chemical purification or phenol extraction. With large projects, make sure to pick a method that you will be comfortable using over a long time, as the method (bead-based lysis versus chemical lysis) does contribute to bacterial abundance measurements and are not freely interchangeable [17, 31, 32].
2. We use the BioSpec Mini-bead beater. But any agitator that rapidly shakes the tube in three-dimensional space in multi-directions should be usable. Tubes should get warm, not hot to touch, when shaking. If they get hot, then it may be necessary to shake at shorter but more frequent intervals and cool samples on ice after each interval.
3. There are options to collect samples in preservative so that they are stable at room temperature. Several, such as RNAprotect cell reagent (Qiagen), when added at 2 or more volumes to saliva samples, allow preservation of bacteria structure, such that samples can be stabilized on collection and centrifuged on return to the laboratory without loss of DNA. Freezing in the reagent seems to be well tolerated but is not validated for all species. Usage of chaotropic agents, like RNA/DNA Shield, (Zymo Research) or others, has the advantage of inactivation of pathogens, but does not allow the removal of saliva supernatant, which can contain DNA and PCR inhibitors. Direct freezing of saliva followed by centrifugation to pellet the bacteria and remove aqueous phase should be done with caution, as there is the possibility of loss of DNA from some bacterial species to the supernatant while thawing.
4. Various forces for pelleting bacteria are noted in the literature. $4500 \times g$ is chosen as it is fast enough to pellet bacteria from most viscous samples, but slow enough to minimize damage to bacteria. It is not clear how sensitive oral bacteria species are to rapid centrifugation, which, in theory, could result in lysis of some taxa and loss of DNA.
5. Some samples are viscous and do not form stable pellets. In that case, after the first centrifugation, remove only the top of the supernatant that is clear. Add 1 mL PBS, then mix well and repeat centrifugation at $4500 \times g$. An extra PBS wash may be necessary, for 3 total, to ensure that a bacterial pellet free of excess liquid is generated for storage.

6. In our experience and that of others, the V1–V3 regions of 16S rRNA gene contain sequence variation that allows the differentiation of a maximal number of streptococcal species which are common in the oral cavity, so it is a good choice for 16S rRNA based oral microbiome studies [40].
7. The primers for the second PCR contain sequence that can hybridize to the first PCR generated amplicons via the CS1 or CS2 regions, and also contain unique barcode sequence for each sample, and sequence hybridizable to the sequence primers used for the next step.
8. There is an option to analyze the sequence reads as single-end reads (all from read 1 or read 2 of the DNA sequencing) or merged paired-end reads where each individual read is merged with the paired read from the opposite end of the amplicon. Because the merging of the sequence read pairs depends on there being high-quality overlapping sequence between the two reads, which is not always present in all reads, or even most reads, the method described for taxa identification uses single-end reads. Forward (R1) or reverse (R2) reads can be used.
9. The protocol as written requires minimal familiarity with command line interface which is required for QIIME 2 usage. When there is an option to use the Macintosh interface to proceed, it is provided first.
10. The term *folder* to denote a structure to hold files and other folders is the equivalent to *directory* which is the term used in Unix commands.
11. Alternatively, one can go to the Mac desktop and drag and drop these .qzv files into the QIIME 2 View tool at (https://view.qiime2.org/) when not in the command line interface.
12. The sample metadata file is generated in any spreadsheet program such as Excel. The first column can have the sample name in the same format as when entered as FASTQ files in the beginning of the program without the appended information. The next column is the group, in this case, dentate (A) or edentulous (B). This file is saved in tab-delimited text format and then is dragged and dropped in the directory used for this analysis. It is used directly in the Terminal in this format.

References

1. Dewhirst FE, Chen T, Izard J, Paster BJ, Tanner AC, Yu WH et al (2010) The human oral microbiome. J Bacteriol 192(19):5002–5017. https://doi.org/10.1128/JB.00542-10
2. Paster BJ, Boches SK, Galvin JL, Ericson RE, Lau CN, Levanos VA et al (2001) Bacterial diversity in human subgingival plaque. J Bacteriol 183(12):3770–3783. https://doi.org/10.1128/JB.183.12.3770-3783.2001
3. Kumar PS (2013) Oral microbiota and systemic disease. Anaerobe 24:90–93. https://doi.org/10.1016/j.anaerobe.2013.09.010
4. Seymour GJ, Ford PJ, Cullinan MP, Leishman S, Yamazaki K (2007) Relationship

between periodontal infections and systemic disease. Clin Microbiol Infect 13(Suppl 4):3–10. https://doi.org/10.1111/j.1469-0691.2007.01798.x

5. Kodukula K, Faller DV, Harpp DN, Kanara I, Pernokas J, Pernokas M et al (2017) Gut microbiota and salivary diagnostics: the mouth is salivating to tell us something. Biores Open Access 6(1):123–132. https://doi.org/10.1089/biores.2017.0020
6. Zhang CZ, Cheng XQ, Li JY, Zhang P, Yi P, Xu X et al (2016) Saliva in the diagnosis of diseases. Int J Oral Sci 8(3):133–137. https://doi.org/10.1038/ijos.2016.38
7. Lazarevic V, Whiteson K, Gaia N, Gizard Y, Hernandez D, Farinelli L et al (2012) Analysis of the salivary microbiome using culture-independent techniques. J Clin Bioinforma 2:4. https://doi.org/10.1186/2043-9113-2-4
8. Dawes C (2003) Estimates, from salivary analyses, of the turnover time of the oral mucosal epithelium in humans and the number of bacteria in an edentulous mouth. Arch Oral Biol 48(5):329–336
9. Krishnan K, Chen T, Paster BJ (2017) A practical guide to the oral microbiome and its relation to health and disease. Oral Dis 23 (3):276–286. https://doi.org/10.1111/odi.12509
10. Griffen AL, Beall CJ, Firestone ND, Gross EL, Difranco JM, Hardman JH et al (2011) CORE: a phylogenetically-curated 16S rDNA database of the core oral microbiome. PLoS One 6(4):e19051. https://doi.org/10.1371/journal.pone.0019051
11. Wade WG, Prosdocimi EM (2020) Profiling of oral bacterial communities. J Dent Res 99 (6):621–629. https://doi.org/10.1177/0022034520914594
12. Cole JR, Wang Q, Fish JA, Chai B, McGarrell DM, Sun Y et al (2014) Ribosomal Database Project: data and tools for high throughput rRNA analysis. Nucleic Acids Res 42(Database issue):D633–D642. https://doi.org/10.1093/nar/gkt1244
13. Quast C, Pruesse E, Yilmaz P, Gerken J, Schweer T, Yarza P et al (2013) The SILVA ribosomal RNA gene database project: improved data processing and web-based tools. Nucleic Acids Res 41(Database issue): D590–D596. https://doi.org/10.1093/nar/gks1219
14. Perry BJ, Zammit AP, Lewandowski AW, Bashford JJ, Dragovic AS, Perry EJ et al (2015) Sites of origin of oral cavity cancer in nonsmokers vs smokers: possible evidence of dental trauma carcinogenesis and its importance compared with human papillomavirus. JAMA Otolaryngol Head Neck Surg 141(1):5–11. https://doi.org/10.1001/jamaoto.2014.2620
15. Belstrom D, Holmstrup P, Bardow A, Kokaras A, Fiehn NE, Paster BJ (2016) Temporal stability of the salivary microbiota in oral health. PLoS One 11(1):e0147472. https://doi.org/10.1371/journal.pone.0147472
16. Vogtmann E, Chen J, Kibriya MG, Amir A, Shi J, Chen Y et al (2019) Comparison of oral collection methods for studies of microbiota. Cancer Epidemiol Biomark Prev 28 (1):137–143. https://doi.org/10.1158/1055-9965.EPI-18-0312
17. Lim Y, Totsika M, Morrison M, Punyadeera C (2017) The saliva microbiome profiles are minimally affected by collection method or DNA extraction protocols. Sci Rep 7(1):8523. https://doi.org/10.1038/s41598-017-07885-3
18. Ionescu D, Overholt WA, Lynch MD, Neufled JD, Naqib A, Green SJ (2016) Microbial community analysis using high-throughput amplicon sequencing. In: Yates MV, Cindy HN, Miller RV, Pillai SD (eds) Manual of environmental microbiology. Wiley, Hoboken, NJ. https://doi.org/10.1128/9781555818821.ch2.4.2
19. Kelly BJ, Gross R, Bittinger K, Sherrill-Mix S, Lewis JD, Collman RG et al (2015) Power and sample-size estimation for microbiome studies using pairwise distances and PERMANOVA. Bioinformatics 31(15):2461–2468. https://doi.org/10.1093/bioinformatics/btv183
20. Anderson MJ, Walsh DCI (2013) PERMANOVA, ANOSIM, and the Mantel test in the face of heterogeneous dispersions: what null hypothesis are you testing? Ecol Monogr 83 (4):8
21. Mark Welch JL, Dewhirst FE, Borisy GG (2019) Biogeography of the oral microbiome: the site-specialist hypothesis. Annu Rev Microbiol 73:335–358. https://doi.org/10.1146/annurev-micro-090817-062503
22. Segata N, Haake SK, Mannon P, Lemon KP, Waldron L, Gevers D et al (2012) Composition of the adult digestive tract bacterial microbiome based on seven mouth surfaces, tonsils, throat and stool samples. Genome Biol 13(6): R42. https://doi.org/10.1186/gb-2012-13-6-r42
23. Hall MW, Singh N, Ng KF, Lam DK, Goldberg MB, Tenenbaum HC et al (2017) Interpersonal diversity and temporal dynamics of dental, tongue, and salivary microbiota in the healthy oral cavity. NPJ Biofilms Microbiomes

3:2. https://doi.org/10.1038/s41522-016-0011-0

24. Lazarevic V, Whiteson K, Hernandez D, Francois P, Schrenzel J (2010) Study of inter- and intra-individual variations in the salivary microbiota. BMC Genomics 11:523. https://doi.org/10.1186/1471-2164-11-523
25. Rasiah IA, Wong L, Anderson SA, Sissons CH (2005) Variation in bacterial DGGE patterns from human saliva: over time, between individuals and in corresponding dental plaque microcosms. Arch Oral Biol 50(9):779–787. https://doi.org/10.1016/j.archoralbio.2005.02.001
26. Belstrom D, Holmstrup P, Fiehn NE, Rosing K, Bardow A, Paster BJ et al (2016) Bacterial composition in whole saliva from patients with severe hyposalivation—a case-control study. Oral Dis 22(4):330–337. https://doi.org/10.1111/odi.12452
27. Nasidze I, Li J, Quinque D, Tang K, Stoneking M (2009) Global diversity in the human salivary microbiome. Genome Res 19 (4):636–643. https://doi.org/10.1101/gr.084616.108
28. Hall M, Beiko RG (2018) 16S rRNA gene analysis with QIIME2. Methods Mol Biol 1849:113–129. https://doi.org/10.1007/978-1-4939-8728-3_8
29. Edgar RC (2017) Accuracy of microbial community diversity estimated by closed- and open-reference OTUs. PeerJ 5:e3889. https://doi.org/10.7717/peerj.3889
30. Gomar-Vercher S, Simon-Soro A, Montiel-Company JM, Almerich-Silla JM, Mira A (2018) Stimulated and unstimulated saliva samples have significantly different bacterial profiles. PLoS One 13(6):e0198021. https://doi.org/10.1371/journal.pone.0198021
31. Rosenbaum J, Usyk M, Chen Z, Zolnik CP, Jones HE, Waldron L et al (2019) Evaluation of oral cavity DNA extraction methods on bacterial and fungal microbiota. Sci Rep 9 (1):1531. https://doi.org/10.1038/s41598-018-38049-6
32. Vesty A, Biswas K, Taylor MW, Gear K, Douglas RG (2017) Evaluating the impact of DNA extraction method on the representation of human oral bacterial and fungal communities. PLoS One 12(1):e0169877. https://doi.org/10.1371/journal.pone.0169877
33. Callahan BJ, McMurdie PJ, Rosen MJ, Han AW, Johnson AJ, Holmes SP (2016) DADA2: high-resolution sample inference from Illumina amplicon data. Nat Methods 13 (7):581–583. https://doi.org/10.1038/nmeth.3869
34. Bolyen E, Rideout JR, Dillon MR, Bokulich NA, Abnet CC, Al-Ghalith GA et al (2019) Reproducible, interactive, scalable and extensible microbiome data science using QIIME 2. Nat Biotechnol 37(8):852–857. https://doi.org/10.1038/s41587-019-0209-9
35. Chong J, Liu P, Zhou G, Xia J (2020) Using MicrobiomeAnalyst for comprehensive statistical, functional, and meta-analysis of microbiome data. Nat Protoc 15(3):799–821. https://doi.org/10.1038/s41596-019-0264-1
36. Clarke KR, Gorley RN, Somerfield PJ, Warwick RM (2014) Change in marine communities: an approach to statistical analysis and interpretation. Primer-E, Plymouth, Devon
37. Parks DH, Tyson GW, Hugenholtz P, Beiko RG (2014) STAMP: statistical analysis of taxonomic and functional profiles. Bioinformatics 30(21):3123–3124. https://doi.org/10.1093/bioinformatics/btu494
38. Davis NM, Proctor DM, Holmes SP, Relman DA, Callahan BJ (2018) Simple statistical identification and removal of contaminant sequences in marker-gene and metagenomics data. Microbiome 6(1):226. https://doi.org/10.1186/s40168-018-0605-2
39. Salter SJ, Cox MJ, Turek EM, Calus ST, Cookson WO, Moffatt MF et al (2014) Reagent and laboratory contamination can critically impact sequence-based microbiome analyses. BMC Biol 12:87. https://doi.org/10.1186/s12915-014-0087-z
40. Cabral DJ, Wurster JI, Flokas ME, Alevizakos M, Zabat M, Korry BJ et al (2017) The salivary microbiome is consistent between subjects and resistant to impacts of short-term hospitalization. Sci Rep 7(1):11040. https://doi.org/10.1038/s41598-017-11427-2

Chapter 6

Host DNA Depletion in Saliva Samples for Improved Shotgun Metagenomics

Clarisse Marotz, Cristal Zuniga, Livia Zaramela, Rob Knight, and Karsten Zengler

Abstract

Host DNA makes up the majority of DNA in a saliva sample. Therefore, shotgun metagenomics can be an inefficient way to evaluate the microbial populations of saliva since often <10% of the sequencing reads are microbial. In this chapter, we describe a method to deplete human DNA from fresh or frozen saliva samples, allowing for more efficient shotgun metagenomic sequencing of the salivary microbial community.

Key words Metagenomics, Propidium monoazide (PMA), Oral, Bacteria, Microbial DNA enrichment

1 Introduction

16S rRNA gene amplicon (16S) sequencing is the most common way to assess microbial community composition via next-generation sequencing. While this has greatly expanded the field of microbiology, there are several disadvantages to this approach including limited taxonomic resolution, primer bias, and inability to evaluate full genetic diversity. Accordingly, shotgun metagenomic sequencing has become an increasingly common way to assess microbial community composition and functional potential. As opposed to 16S sequencing, which requires primer-based amplification, shotgun sequencing works by randomly fragmenting DNA strands into short fragments that are simultaneously tagged with an oligo ("tagmentation") that can then be used to create sequencing libraries containing all of the DNA in a given sample. Because the human genome is roughly 1000× larger than the average bacterial genome, human DNA can quickly drown out microbial signal when applying shotgun metagenomics sequencing to host-derived samples. In saliva samples, the percentage of sequencing reads aligning to the human genome is typically ~90%

Guy R. Adami (ed.), *The Oral Microbiome: Methods and Protocols*, Methods in Molecular Biology, vol. 2327, https://doi.org/10.1007/978-1-0716-1518-8_6,

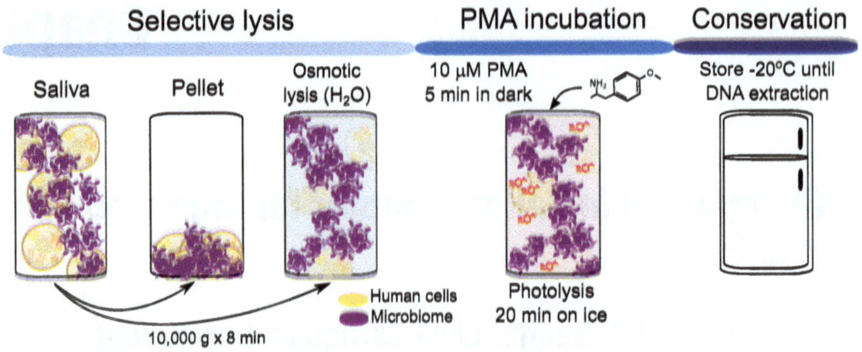

Fig. 1 Schematic diagram of the protocol. Microbes and host cells are separated from saliva by centrifugation. Osmotic lysis in H_2O followed by PMA addition allow selective degradation of host cell DNA

[1]. To improve the efficiency of shotgun sequencing metagenomics from saliva, we validated a simple benchtop method to selectively deplete host DNA.

This method works by first selectively lysing human cells, which are more vulnerable to osmotic pressure than microbial cells (Fig. 1). Once human DNA is exposed, extracellular DNA signal is removed using the chemical propidium monoazide (PMA) [2]. PMA intercalates into DNA that is not protected by a cell membrane, and when exposed to visible light forms a covalent bond with the DNA, breaking it into small fragments that are excluded from downstream analysis [3]. Traditionally, this method has been used to differentiate live from dead cell signal, but by performing this reaction on selectively lysed human cells, we can remove dead cell signal and dramatically improve the percentage of microbial sequencing reads specific to the DNA from bacteria and fungi that are not lysed.

This method can be compared to commercially available kits that also perform host DNA depletion prior to DNA extraction (pre-extraction). Pre-extraction approaches generally follow a two-step procedure. The first step is to selectively lyse mammalian cells, which takes advantage of the fragility of the mammalian cell membrane that lyses much more readily than that of the vast majority of microbial membranes/cell walls. The second step removes exposed DNA enzymatically, leaving only the intact microbial cells for downstream analysis [4]. The current protocol does not require enzyme and eliminates some of the washing steps, and thus reduces the hands-on time for sample processing and increases reproducibility. It also is a method that has been optimized specifically for saliva samples. Another approach is to eliminate host DNA post-extraction by affinity-based subtraction of host methylated DNA [5]. This approach, like others, has various limitations (Fig. 2).

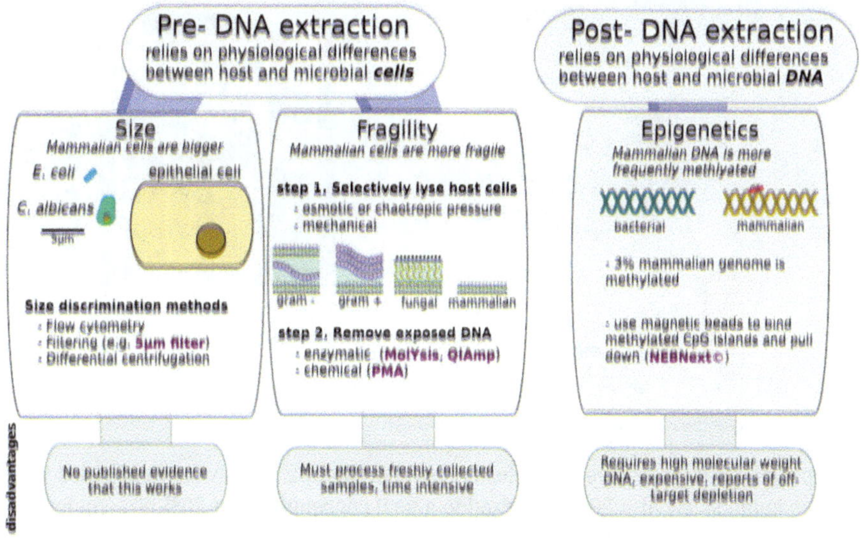

Fig. 2 Summary of various physical ways to deplete host cells and DNA from mixed populations of host (mammalian) and bacteria cells

2 Materials

2.1 Chemical Reagents

1. Nuclease-free H_2O.
2. Propidium monoazide (PMA) at 0.2 mM stock concentration (*see* **Note 1**).

2.2 Equipment

1. Light source (*see* **Note 2**).
2. Vortex.
3. Centrifuge for microcentrifuge tubes.
4. Standard semitransparent polypropylene tubes.

3 Methods

3.1 Saliva Collection and Cryopreservation

1. Ask individuals to provide at least 2 mL saliva in a sterile container (*see* **Note 3**).
2. If saliva samples need to be stored before processing, sterile glycerol can be added to a final concentration of 10–20%, thoroughly mixed, and the sample stored at −80 °C (*see* **Note 4** and Fig. 3).

3.2 Selective Lysis

1. Thaw cryo-preserved saliva samples.
2. Vortex well and aliquot 1 mL into a sterile microcentrifuge tube.

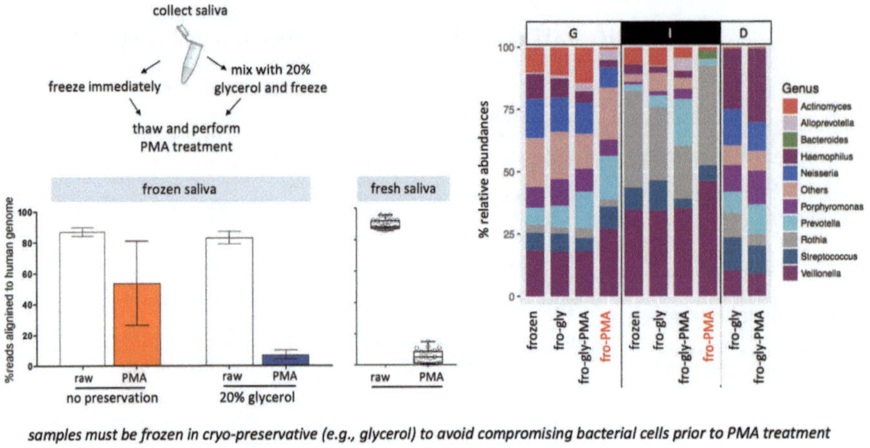

Fig. 3 Host depletion for cryopreserved samples is an important option. Raw saliva samples were aliquoted and either frozen immediately at −20 °C or mixed to a final concentration of 20% glycerol for cryopreservation. The percentage of human reads was assessed by Bowtie2, and the top 15 most abundant genera were assessed by MetaPhlAn2. As expected, samples frozen without 20% glycerol then PMA treated (Fro-PMA) show taxa distribution in the bar plot distinct from Fro-Gly-PMA samples and Fro-PMA samples, indicating usage of glycerol during freezing is necessary when PMA enrichment is to be done

3. Centrifuge saliva aliquot at 10,000 × *g* for 8 min to pellet cells at room temperature.
4. Remove supernatant with a pipette, careful not to dislodge the cell pellet.
5. Resuspend cell pellet in 200 μL nuclease-free H_2O by pipetting up and down and brief vortexing.
6. Leave at room temperature for 5 min to allow for osmotic lysis of human cells.

3.3 Host DNA Depletion

1. Add 10 μL of 0.2 mM propidium monoazide, to a final concentration of 10 μM, and vortex to mix.
2. Incubate at room temperature protected from light for 5 min to allow for intercalation.
3. Place samples horizontally on ice, <15 cm from a light source containing 488 nm wavelength light.
4. Expose to light for >20 min, vortexing and rotating periodically to ensure full light penetration.
5. Samples can now either be processed immediately for DNA extraction or stored at ≤−20 °C for later DNA extraction (*see* **Note 5**).

3.4 DNA Extraction and Shotgun Metagenomic Sequencing

1. DNA extraction and shotgun metagenomic library preparation can be performed per the users' typical pipeline (*see* **Note 6**).

4 Notes

1. PMA is available as 20 mM stock or can be purchased as powder and dissolved in H_2O. Concentrated stock is stored at −20 °C. All stocks and aliquots should be stored in opaque vials protected from light.
2. PMA is maximally activated at 488 nm wavelength light. Specialty tube holders with blue light sources are available through the manufacturer of PMA. Alternatively, fluorescent light bulbs also contain wavelengths in the excitable range for PMA and can be used in this protocol by setting up an ice bucket to lay the samples on <15 cm from the light source.
3. The amount of microbial cells to expect in a saliva sample can vary depending on many factors, including salivary flow rate. If you are collecting passive saliva, especially from individuals with high salivary flow rate, you may need to process up to 2 mL of saliva in order to obtain sufficient DNA for sequencing.
4. For saliva sample cryo-preservation, 7% DMSO may be a suitable alternative to glycerol.
5. There is no need to pellet the microbes at this point because any mammalian DNA is nearly totally degraded and excess PMA will react with water and become inert.
6. To develop this methodology, we used Qiagen's PowerSoil DNA kit for DNA purification and the Kapa Hyper Plus kit for library generation.

Acknowledgments

This work was supported in part through the Center for Microbiome Innovation at UC San Diego, and mentored by Drs. Karsten Zengler and Rob Knight.

References

1. Marotz CA, Sanders JG, Zuniga C, Zaramela LS, Knight R, Zengler K (2018) Improving saliva shotgun metagenomics by chemical host DNA depletion. Microbiome 6(1):1–9
2. Nocker A, Sossa-Fernandez P, Burr MD, Camper AK (2007) Use of propidium monoazide for live/dead distinction in microbial ecology. Appl Environ Microbiol 73 (16):5111–5117
3. Soejima T, Iida K, Qin T, Taniai H, Seki M, Takade A et al (2007) Photoactivated ethidium monoazide directly cleaves bacterial DNA and is

applied to PCR for discrimination of live and dead bacteria. Microbiol Immunol 51 (8):763–775

4. Hansen WLJ, Bruggeman CA, Wolffs PFG (2013) Pre-analytical sample treatment and DNA extraction protocols for the detection of bacterial pathogens from whole blood. Methods Mol Biol 943:81–90

5. Feehery GR, Yigit E, Oyola SO, Langhorst BW, Schmidt VT, Stewart FJ et al (2013) A method for selectively enriching microbial DNA from contaminating vertebrate host DNA. PLoS One 8(10):e76096

Chapter 7

A Standardized Approach for Shotgun Metagenomic Analysis of Ancient Dental Calculus

Nicole E. Moore and Laura S. Weyrich

Abstract

Ancient dental calculus provides a challenging, yet unparalleled, opportunity to reconstruct ancient oral microbial communities and trace the origins of modern microbiota-associated diseases. Metagenomic analysis of ancient dental calculus using high-throughput DNA sequencing has proven itself as an effective method to accurately reconstruct microorganisms that once lived in the mouths of ancient humans. Here, we provide the strategy, methodologies, and approaches used to establish an ancient dental calculus project, from project conception, community engagement, sampling, extracting DNA, and preparing shotgun metagenomic DNA libraries for sequencing on an Illumina platform. We also discuss techniques to minimize background or contaminant DNA by monitoring and reducing contamination in calculus data sets, utilizing appropriate protective gear, and employing the use of sample decontamination strategies. In this methodology chapter, we hope to promote transparency in the ancient dental calculus research field and encourage collaboration across the ancient DNA research community.

Key words Ancient DNA, Microbiome, Bacteria, Dental calculus, Metagenomics, Oral microbiota

1 Introduction

Ancient DNA (aDNA) research has revolutionized our understanding of the evolutionary history of ancient humans and other hominins. For example, we now have a detailed picture of the Neanderthal genome [1] because of techniques adapted and harnessed to sequence highly degraded, fragmented pieces of DNA, otherwise known as ancient DNA [2]. While we are limited to sequencing DNA no more than 1.5 million years old in the best conditions [3], novel aDNA applications and techniques are paving the way to examine the once unimaginable details about the origins of human life and disease.

In particular, recent novel applications of aDNA techniques have sparked a flurry of interest from medical and dental researchers. One application in particular uses aDNA techniques to examine ancient human microbiota—a collection of microorganisms that

Guy R. Adami (ed.), *The Oral Microbiome: Methods and Protocols*, Methods in Molecular Biology, vol. 2327,
https://doi.org/10.1007/978-1-0716-1518-8_7,

includes bacteria, archaea, viruses, fungi, and parasites. Reconstructing ancient microbiota, along with their environmental and genetic context (i.e., microbiome), can provide insights into the origin and spread of both infectious and chronic diseases in humans today, providing unique opportunities to develop novel prevention and treatment strategies in the future. These ancient microbiota techniques have been adapted not only to investigate the human microbiota but also reveal insight on past climate change [4], identify key ecological shifts [5, 6], and explore ancient animal health [7].

Most ancient human microbiome research conducted to date has focused on a single biological specimen—calcified dental plaque or dental calculus. This cement-like matrix forms on the enamel surface of the tooth and encases and protects microorganisms present in this oral biofilm for millennia; in fact, DNA has been examined in a dental calculus specimen that is ~48,000 years old [8]. Dental calculus is now appreciated as the only reliable source of ancient human microbiota, as it forms during the life of an individual and is less susceptible to postmortem damage or alterations than other ancient sample types. While ancient feces, otherwise known as coprolites, can also preserve some ancient human microbiota, coprolites are more susceptible to degradation and environmental contamination over time [9, 10]. Despite its recent integration into ancient DNA research, dental calculus has already revealed much about the past, including information on the evolution of specific oral pathogens in the mouth, including the rise of specific oral pathogens linked to periodontal disease and caries, such as *Porphyromonas gingivalis* or *Streptococcus mutans* [11, 12]; the origin of unique oral microbiota in discrete human groups [8]; the link between long-term dietary changes and microbiota shifts [11]; and the potential microbiota contributions to ancient noncommunicable diseases [13].

Reconstructing ancient oral microbial communities from dental calculus is no easy task. Any aDNA study of dental calculus will be fraught with limitations and considerations that must be taken into account before embarking on such a study. For example, some ancient dental calculus samples will not have any biological or "endogenous" DNA signal, as the DNA could have been destroyed over time from heat, extreme drying, or other chemical processes [3]. One must authenticate and verify the presence of ancient DNA in any dental calculus specimen, typically using a handful of different bioinformatic analyses [14, 15]. This means that only a proportion of samples, or perhaps none of the samples, from a given site will yield any results. Second, DNA from ancient oral microbiota can be swamped out by modern DNA contaminants from the technician, laboratory reagents, or its storage environment [16, 17]. Incredible care must be taken to monitor DNA contamination throughout the process, including sampling, documenting, extracting, and sequencing. Third, ancient dental calculus samples are inherently contaminated with soil and environmental microorganisms, as many samples are obtained from postmortem individuals buried underground or were stored in secondary environments after their discovery [18]. Therefore, researchers must acknowledge, measure, and account for non-oral signals within their calculus data sets. Lastly, ancient dental calculus samples are rare, as the preservation of ancient individuals is rare. Ancient calculus specimens must be sampled and utilized with the utmost consideration with sample management and curation, taking special care to not deplete a single resource and be intentional in their study questions and design to limit irreversible sample destruction [19]. Despite these limitations, dental calculus research has the opportunity to reveal the origins of modern disease and provide foundational information to reverse microbiota-associated diseases today (Fig. 1).

In this chapter, we describe the strategies, methodologies, and approaches used to generate metagenomic DNA sequencing libraries from ancient human dental calculus. We provide information on minimizing contamination and improving biological signal during project design, sample collection and documentation, DNA extraction, and DNA library preparation. Bioinformatic approaches to accurately examine ancient dental calculus are not included in this chapter but should also be closely examined prior to the start of any ancient DNA study. We hope this work will provide transparency in the aDNA research field and encourage collaboration between research groups and other communities.

2 Materials

Volumes and quantities listed here are for one calculus sample.

2.1 Protective Gear for Sample Collection

1. Gloves.
2. Surgical face masks.
3. Tyvek suits with hoods and foot coverings (e.g., has booties attached) or lab coat.

2.2 Protective Gear for Lab Work (Fig. 2)

1. Gloves.
2. Surgical face masks.
3. Tyvek suits with hoods and foot coverings (e.g., has booties attached).
4. Face shield.
5. Optional: Hair net and shoes dedicated to the laboratory (e.g., boots or foam shoes).

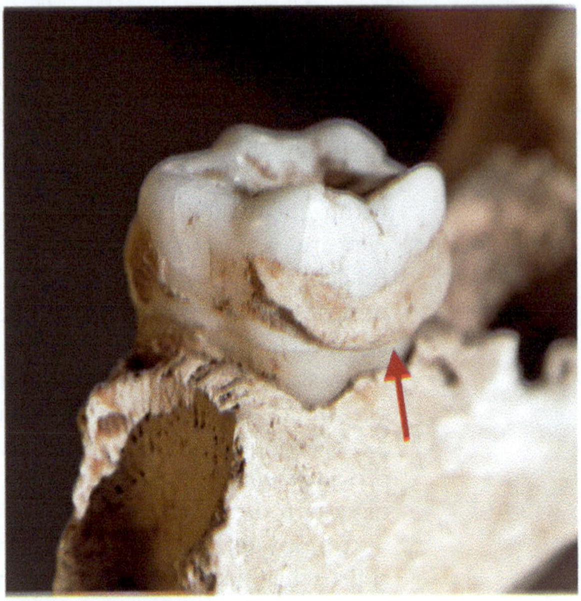

Fig. 1 Calcified dental plaque shown on the buccal surface of a molar tooth. Downward pressure should be applied in the direction of the red arrow to remove calculus from the tooth surface

Fig. 2 An example of proper protective gear worn in an ancient DNA lab or during sampling, which includes a full body Tyvek suit, a face mask, a plastic face shield, hair net, and two pairs of sterile gloves

2.3 Labware and Decontamination Solutions

1. Swabs for control collection: Sterile 6-in. foam swabs (e.g., Puritan) in sterile dry transport tubes are recommended for collection (*see* **Note 1**).

2. Sterile resealable bags or plastic tubes for sample and control collection.
3. UltraPure molecular grade water (Invitrogen; for preparing bleach and ethanol, below).
4. 5% Sodium hypochlorite (bleach) solution for decontamination of dental calculus: For 5 mL, make a master mix of 12 mL molecular grade water + 15 mL household bleach in a 50 mL falcon tube and use 5 mL from this; these values will depend on the concentration of bleach being used.
5. 80% Ethanol for ethanol solution for decontamination of dental calculus: For 5 mL per sample, mix 1 mL molecular grade water + 4 mL absolute ethanol.
6. Weigh boat (1 per sample).
7. Sterile plastic petri dishes (2 per sample).

2.4 Reagents

2.4.1 Reagents for PB Extraction

1. 2 mL 80% Ethanol: 0.4 mL Molecular grade water + 1.6 mL absolute ethanol.
2. 100 μL Silica solution. Prepare a bulk preparation of silica the day before, as follows: Add 8 g silica to 50 mL molecular grade water and vortex. Leave to settle for 1 h. Pipette off ~40 mL of suspension into a new 50 mL tube. Leave overnight. Pipette or pour off supernatant to leave 8–10 mL of silica suspension at the bottom, which represents medium size silica particles. Aliquot into tubes and store in the refrigerator for up to 1 month.
3. 200μL TLE buffer: Make a larger mix of 500μL Tris–HCl (1 M), 10μL EDTA (0.5 M), 50 mL molecular grade water, and aliquot 200μL for use. Aliquot into smaller tubes and store at −20 °C until needed.
4. 980μL Digestion buffer: 900μL EDTA (0.5 M) and 80μL molecular grade water.
5. 20μL Proteinase K (20 mg/mL).
6. 12.6 mL Modified PB binding buffer: Qiagen PB buffer 12.2 mL, 7μL Tween-20, 378μL NaOAc (3 M).

2.4.2 Reagents for 16S ribosomal RNA Amplification

1. PCR mastermix. For each 25μL reaction, create a mastermix: 18.05μL molecular grade water, 2.5μL Platinum Taq DNA polymerase High Fidelity buffer (10×), 1μL $MgCl_2$ (50 mM), 0.2μL dNTPs (25 mM), 0.25μL Platinum Taq High Fidelity DNA polymerase, 1μL Forward primer (10μM) [20], 1μL Reverse primer (10μM) [20]; this will be added to 1μL DNA extract.

2.4.3 Reagents for Library Preparation for Metagenomic Shotgun Sequencing

1. Repair mastermix. For each 20μL reaction, create a mastermix: 4μL Tango buffer/10× NEB2 buffer, 0.4μL each dNTP (25 mM), 4μL ATP (10 mM), 8.1μL molecular grade water, 2μL T4 PNK (10 U/μL), 1.5μL T4 DNA Polymerase (3 U/μL).
2. Ligation mastermix. For each 18μL reaction, create a mastermix: 4μL 10× T4 Ligase Buffer, 4μL PEG-4000 (50% solution), 9μL molecular grade water, 1μL T4 DNA Ligase.
3. Bst fill-in mastermix. For each 20μL reaction, create a mastermix: 4μL 10× Thermopol buffer, 1μL dNTPs (25 mM each—100 mM total), 13.5μL molecular grade water, 1.5μL Bst DNA Polymerase (8 U/μL).
4. Amplification mastermix 1—prepared in the ancient DNA lab. For each 20μL reaction, create a mastermix: 13.25μL molecular grade water, 2.5μL 10× Platinum Taq DNA polymerase High Fidelity buffer, 1.25μL $MgCl_2$ (50 mM), 0.25μL dNTPs (25 mM each—100 mM total), 1.25μL IS7 adapter [21], 1.25μL IS8 adapter [21], 0.25μL Platinum Taq High Fidelity DNA polymerase.
5. Amplification mastermix 2—prepared in modern molecular biology lab. For each 21μL reaction, create a mastermix: 12.75μL molecular grade water, 2.5μL 10× Platinum Taq DNA polymerase High Fidelity buffer, 2.5μL $MgCl_2$ (25 mM), 0.625μL dNTPs (10 mM), 1.25μL IS4 adapter [21], 1.25μL GAII-X adapter [21], 0.25μL Platinum Taq High Fidelity DNA polymerase.
6. Qiagen MinElute Reaction Cleanup kit.
7. Agencourt AMPure XP beads or Axygen AxyPrep Mag PCR Clean-Up kit.

2.4.4 Reagents for Library Quantification

1. Qubit—dsDNA BR Assay kit (Cat. Number Q32850) or equivalent.
2. KAPA qPCR kit with Illumina Standards (Kit number KK4824).
3. Agilent High Sensitivity DNA Tapestation D1000 (ScreenTape, Ladder and Reagents).

2.5 Equipment

2.5.1 Sample Documentation

1. High-resolution camera.
2. Millimeter scaled ruler.
3. Scale.
4. Forceps.

2.5.2 Decontamination of Samples

1. Still air hood with internal UV lights.
2. Forceps.
3. UV Crosslinker. UV crosslinkers should be available to decontaminate equipment and racks should be separate to those that decontaminate samples.
4. Timer.

2.5.3 DNA Extraction

1. Still air hood with internal UV lights.
2. Forceps.
3. Pipettes.
4. Heat block.
5. Rotary mixer.
6. Standard microbial incubator—no CO_2 required.
7. Centrifuge for 15 mL tubes (Eppendorf 5910R or similar, with swinging bucket rotor).
8. Microcentrifuge (Eppendorf 5424 R or similar, with fixed rotor for 1.5/2.0 mL tubes)
9. Mini centrifuge, for example, https://us.vwr.com/store/product/24130031/vwr-mini-centrifuges.
10. Vortex.
11. −20 °C Freezer.
12. 4 °C Refrigerator.
13. Timer.
14. Sterile delicate task wipes, such as Kimwipes (Kimtech Science).
15. Screw-top 15 mL tubes.
16. Sterile Screw-top 2.0 mL tubes.
17. Eppendorf DNA LoBind 1.5 mL tubes.

2.5.4 16S rRNA Amplification

1. Still air hood with internal UV lights.
2. Pipettes.
3. Mini centrifuge.
4. Thermal cycler.
5. Gel electrophoresis equipment.
6. −20 °C Freezer.
7. 4 °C Refrigerator.
8. Strip tubes with attached flat caps (Thermo Fisher Scientific EasyStrip Plus Tube Strip with attached flat caps—AB2000 or similar).
9. Eppendorf DNA LoBind 1.5 mL tubes.

2.5.5 Library Preparation for Shotgun Sequencing

1. Still air hood with internal UV lights.
2. Pipettes.
3. Mini centrifuge.
4. Thermal cycler.
5. Microcentrifuge, such as Eppendorf 5424 R or similar, with fixed rotor for 1.5/2.0 mL tubes.
6. Magnetic rack.
7. Heatblock.
8. Ice block/rack.
9. −20 °C Freezer.
10. 4 °C Refrigerator.
11. Timer.
12. Eppendorf DNA LoBind 1.5 mL tubes.
13. Strip tubes with attached flat caps (Thermo Scientific EasyStrip Plus Tube Strip with attached flat caps—AB2000 or similar).

2.5.6 Preparing Libraries for Sequencing

Not all equipment is required, but an assortment may be needed to provide the correct information.

1. Pipettes.
2. Qubit.
3. Agilent Tapestation, Bioanalyzer, or Fragment Analyzer.
4. Real time thermal cycler.

3 Methods

3.1 Create a Project

1. When establishing an ancient DNA project using dental calculus, it is critical to create a sustainable and transparent project plan (Fig. 3). A project plan needs to be developed prior to sampling, so the correct stakeholders (such as local communities, ancestors, or descendants of ancient individuals) can be integrated; the research questions can be appropriately formulated and discussed; contamination risks can be assessed and mitigated; adjustments made to selecting the best laboratory procedures; and the appropriate controls and tools can be included in the study (*see* **Note 2**).

3.2 Develop a Sampling Strategy for Collecting the Dental Calculus

1. Sampling of dental calculus from the surface of a tooth may occur in a variety of settings, such as museums, private collections, archaeological dig sites, or ancient DNA facilities. An appropriate sampling strategy should be designed prior to sample collection occurring to ensure enough sample is collected, contamination is minimized, controls are included, metadata for each sample is collected, and samples are stored appropriately when it is possible (*see* **Notes 3** and **4**).

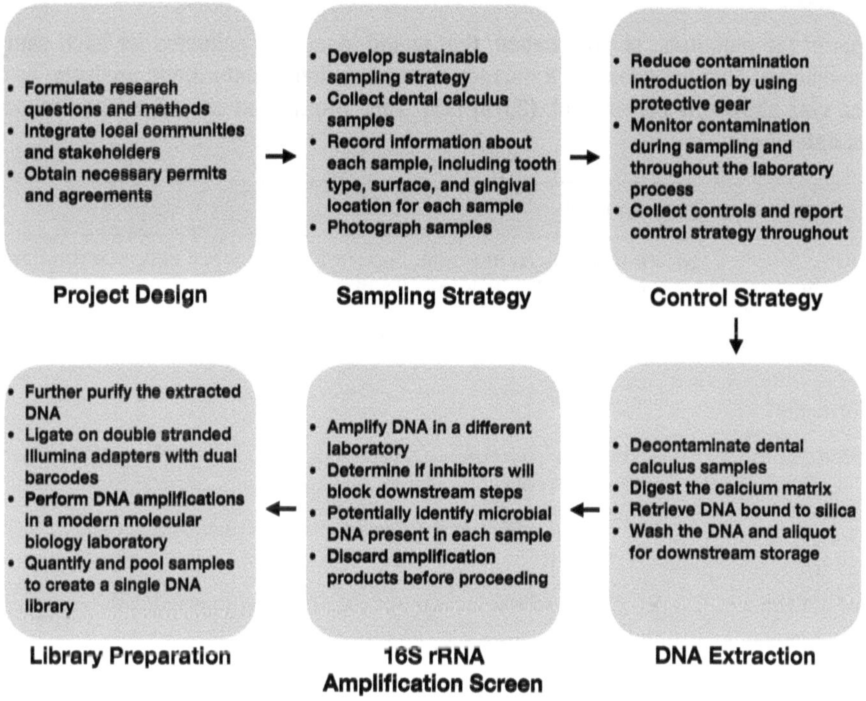

Fig. 3 An overview of the workflow from beginning a project to having DNA prepared for metagenomic shotgun sequencing

3.3 Sample Documentation During the Sampling Procedure

1. Samples should be photographed using a high-resolution camera with millimeter-scaled ruler, measured according to length, width, and height, and weighed. This will provide a record for any visual aspects of the sample before destructive analysis begins, which can be critical in downstream interpretations; for example, the maturation or size of calculus may be linked to certain microbial communities. Additional written notes accompanying photographs, measurements, and weights should also be recorded before sampling occurs during destructive analysis.
2. Researchers should collect as much background information on the samples as possible (Table 1). Critical information includes the location (e.g., tooth type and surface) in the oral cavity where the calculus was collected.

3.4 Sample Collection

Calculus samples should be collected from teeth as described previously and updated below [10].

1. Where possible, all sampling areas and tools need to be decontaminated to reduce DNA contamination. In between all samples, the area and tools should be wiped down again to limit cross contamination. Specifically, all tools, e.g., dental picks and tweezers, and areas should be decontaminated with 3–5%

Table 1
An example of the metadata, or information, that would ideally be collected for each sampled calculus specimen. This provides more detailed context during bioinformatic analysis, as the importance was shown by Farrer et al. (2018) [13]. This information should not be limited to the categories listed here

Metadata type	Description
Collection information	Location of collection origin, current location when collected if different, number of individuals in collection
Collector's information	Name, location, occupation, date, and time of collection
Host information	
Sample or Museum ID	An ID from the collection that can be used to link samples and information with their original collection if they are given new numbers when being
Archaeological age estimate	To determine if differences observed may be connected with age e.g child vs. adult
Gender of individual	To determine if differences observed may be connected with estimated gender of individual
Tooth	Type of tooth, tooth surface, sub-gingival or supra gingival, location in mouth
Number teeth sampled	Number of teeth sampled from an individual and which teeth were sampled
Pathologies and abnormalities	Dental or skeletal pathologies/abnormalities. These may be important in determining health status of individuals such as oral heath status and systemic health status
Region	Local, regional, or national geographic information corresponding to the original location of the sample
Cultural information	Contextual information regarding the cultural or historical affiliation of an individual
Environmental information	Information regarding the environmental context where the sample was initially located

sodium hypochlorite (bleach) and rinsed with >70% ethanol and DNA-free water. Tools can also be UV treated for at least 30 min to help decontaminate them further. Multiple tools may be decontaminated together to speed up the process if sampling larger numbers of dental calculus, if enough are available.

2. To begin the sampling procedure, place the skeletal material over a piece of clean, unused aluminum foil to catch all fragments of calculus. Further, foil can be wrapped around the skull or bone to capture as much calculus as possible when removing it from the tooth surface (Fig. 1).
3. Using a sturdy dental pick, the researcher should apply lateral force or pressure to the most pronounced edge of the dental

calculus (e.g., where the calculus once sat on the gum line) to dislodge the calculus from the tooth surface. It is generally recommended not to use large instruments (e.g., chisel) as this may damage the tooth or tooth surface. We recommend sampling calculus from a single surface of a single tooth at a time, as the oral microbiota have been shown to be different in these ecological niches in ancient and modern specimens [13]. Further, researchers should also be aware of differences in expected oral microbiota from sampling supra- vs. subgingival calculus.

4. The calculus sample is likely to fragment during the removal process, but all fragments should be caught on the foil beneath or surrounding the sampling areas. These fragments on the foil should be poured or placed using decontaminated tweezers in a small, labeled sterile plastic bag. In general, small sterile screw-top plastic tubes can also be used, but it may be difficult to remove the sample later due to static electricity.
5. Each sample should be uniquely and appropriately documented (e.g., photographed, measured, and weighed) and labeled to provide sample context, including information such as a sample or museum ID; the tooth type, surface, and gingival relationship of the calculus; sample collection location, personnel, timing; and other general metadata (e.g., archaeological age and site location, sex, oral health status; Table 1).
6. Before moving onto another calculus sample or individual, surfaces and tools should be cleaned again with bleach and ethanol. Clean, unused foil should be used for each tooth or tooth surface that is sampled. Any used tools need to be decontaminated with bleach, followed by ethanol and UV, if available, before they are used again.
7. Gloves should be replaced or wiped with 3–5% bleach between specimens; other protective gear can be retained between individuals if it remains clean and unsoiled.
8. Control samples should be collected at the time of sampling. This can be done by swabbing the area, including the bench, storage box, sampling tools, etc., or by collecting air samples, environmental samples (e.g., soil potentially associated with the skeleton or site), or empty sterile bags/tube (*see* Subheading 3.5 for more information).
9. Store samples as soon as possible at 4 °C, preferably in a temperature and humidity-controlled environment to limit growth of contaminants and aid in DNA preservation. If transportation is required, *see* **Note 5** for further information.

3.5 Control Strategy

This is the process of monitoring and limiting contamination throughout the procedure. Including controls and systematically and continually monitoring contamination can be used to help account for contaminant species in the results and better interpret the data. Two sample types (sampling and laboratory controls) can be included in each ancient calculus study. In addition to monitoring contamination, it can also be minimized, although never completely removed, using specific methodologies (e.g., by working in dedicated clean facilities and wearing protective gear; *see* **Note 6**).

1. Sampling controls are necessary to survey contamination present in the environment where ancient samples were found, collected, or stored. Sampling controls may include dry swabs of surfaces where ancient samples were collected, such as benchtops (*see* **Note 7**). To collect these samples, medium pressure should be applied to a foam swab for 60 s in a defined area (such as a 10″ × 10″ square). Be sure to duplicate controls and store them at −20 °C as soon as you are able.
2. Soil or other physical samples can also be collected to help assess microbes present during sampling and storage. Other sampling controls may include swabs or soil from an archaeological dig, the skeleton, or a coffin or box where the skeleton is placed. Tooth roots can also be taken as a sampling control for the contamination of biological material from an individual; however, additional legal, social, ethical, and methodological concerns need to be addressed before these are introduced into the project.
3. Once collected, all sampling controls should be treated the same throughout the DNA extraction, library preparation, and sequencing processes as ancient samples. Including controls throughout the project is critical to monitor background levels of DNA in your samples and unintended new or cross contamination, collectively called "contamination" [16].
4. Laboratory controls are included to detect contamination that may be present in the laboratory environment, researchers, reagents, or labware and to detect cross-contamination between calculus samples during lab work. Laboratory controls are added during the processing of samples, typically including both DNA extractions and amplifications/library preparations. Extraction blank controls are included in each batch of extractions, in addition to any sampling controls that may be present. For example, an empty sterile tube could be opened inside a clean still air hood for 10–20 s prior to the decontamination of samples but before the beginning of the extraction process (*see* **Note 8**). This will detect possible contamination introduced into samples during the DNA extraction process, as well as contaminants added during amplification steps. These extraction blank controls should be treated as if they are biological samples and be prepared for DNA sequencing accordingly.

5. The number of extraction controls included in an extraction depends on the total number of samples being extracted. It is generally assumed there should be at least 1 extraction control per 12 samples [16], but more controls may be added if samples are older, have poor preservation, or are processed using robotics.
6. Other laboratory controls that should be included are an amplification or library negative controls. These are blank tubes introduced at the beginning of the amplification or library preparation step, again without any biological material added to the tube. These controls are used to detect contamination of reagents specifically used during the amplification and library preparation procedures. These should be included at a rate of at least 1 negative control per 20 samples; more or less may be needed if other methodologies are applied.
7. If no amplification or DNA products are observed in these samples after the amplification or library preparation procedure, carrier RNA or other known aDNA can be added to these controls to improve amplification efficiency in the control samples [16]. Many extraction blank controls contain incredibly little DNA, so adding carrier RNA can help the efficiency of the reaction by increasing the total nucleic acids present in the sample.

3.6 DNA Extraction: Collection of Nucleic Acids from Ancient Samples

A commonly used protocol for extracting aDNA was developed by Dabney et al. [22]. We have modified their method for small calculus samples, as described below. This method takes 2 days to complete. All volumes are provided for a single sample and should be increased accordingly when processing multiple samples.

3.6.1 Day 1

1. In a room with no biological specimens present, prepare the reagents to prevent cross-contamination prior to commencing any work with samples (*see* Subheading 2.4.1). In a still air hood, prepare the following reagents needed for Day 1 of extraction, by first decontaminating the outside of any tube with 3–5% bleach before entering the hood: Invitrogen UltraPure molecular grade water aliquots from 500 mL bottle, freezing any that will not be used; 2 mL of 80% ethanol per sample; 980μL Digestion buffer (but do not yet add 20μL Proteinase K per sample). Remember to also begin the silica solution preparation at this point for use on Day 2.
2. In a sample processing room using sterilized forceps (as completed during sampling), place a single piece of dental calculus into a weigh boat and label with the sample ID. Forceps can be decontaminated with bleach and ethanol between samples, or new sterile forceps can be used for each new sample or procedure.

3. To decontaminate the samples, this method utilizes ultraviolet (UV) radiation and bleach treatment. Place the weigh boat with the dental calculus into the UV crosslinker for 15 min (*see* **Note 9**).
4. After 15 min, flip the dental calculus with sterile forceps.
5. Replace the weigh boat with the calculus carefully back into the UV crosslinker and treat the sample for a further 15 min.
6. Inside a clean still air hood, place 5 mL of freshly prepared bleach (*see* Subheading 2.3) into a sterile petri dish. Remove the calculus from the weight boat with sterile forceps and submerge it in the bleach for 3–5 min. Be sure to include an extraction blank prior to the start of this process by opening up a new sterile tube for 10–20 s inside the hood and replacing the lid.
7. Inside a clean still air hood, place 5 mL of freshly prepared ethanol solution (*see* Subheading 2.3) into a separate sterile petri dish. Using forceps, pick up the calculus from the bleach solution and rinse the calculus to remove residual bleach by submerging it in the ethanol for 3 min.
8. Using the forceps, remove the calculus and place it on a sterile delicate task wipe (e.g., Kimwipes) placed on a fresh, sterile petri dish for 3 min to let it dry.
9. Place the dried calculus into a sterile 2.0 mL screw-top tube and replace the lid. Ensure the tube is labeled appropriately with the sample information.
10. Discard the used bleach, ethanol, and delicate task wipe. Sterilize the surfaces with 3–5% bleach between samples, and decontaminate the forceps between samples, as above. Change or bleach the outer gloves between working with each sample.
11. The next step is to crush the ancient dental calculus sample, so it can easily be digested during the first step of the extraction process. Sterilization of forceps and surfaces will need to be done as before and between each sample. Decontaminated samples should be moved directly into the first steps of DNA extraction and not stored for future use.
12. Decontaminate the outer surface of the tube containing the decontaminated calculus sample and place it in the hood.
13. Using sterilized forceps, open the tube and apply pressure downward onto the calculus specimen with the forceps. Continue to apply pressure repeatedly and twist the forceps around, pushing the calculus against the tube until the calculus is ground up into a course powder. Take care to keep the calculus sample within the tube.

14. Place the cap back on the tube and remove the sample tube from the hood.
15. Decontaminate the workspace and the forceps with 3–5% bleach between samples, as above. Change or bleach your outer pair of gloves between samples.
16. After all samples have been ground up, introduce another extraction blank control into the sample set by opening up a new sterile tube for 10–20 s inside the hood and replacing the lid.
17. The chemical and enzymatic digestion can now begin. Take all tubes, both calculus and control samples, into a working room and perform the next steps in a still air hood. If suits or facemasks were soiled in earlier steps, please replace these before moving forward.
18. First, add 20 μL proteinase K (20 mg/mL) to 980μL digestion buffer (*see* **Note 10**).
19. Add the 1 mL of digestion buffer with proteinase K to each sample or control. Be careful not to create bubbles, and work as quickly as possible to ensure that the digestion buffer is mixed well due to the warm temperature.
20. Screw the lids on tight and wrap each tube lid in parafilm to prevent any leakages.
21. Place the tubes, evenly spaced, on the rotary mixer inside an incubator set at 55 °C.
22. Rotate the samples to incubate at 55 °C until the next day (~20–24 h).

3.6.2 Day 2

1. Prepare the reagents for the day in the clean room in a clean still air hood prior to work with samples beginning (*see* **Note 11**): 100μL silica solution (On Day 1, as shown in **step 1** of Subheading 3.6.1, be sure to finish preparing the stock silica solution and aliquot into 2 mL tubes); 12.6 mL modified PB binding buffer; 200μL TLE buffer; 1.8 mL 80% ethanol in molecular grade water (*see* Subheading 2.4.1).
2. Take the buffers into the workroom and decontaminate the tubes. Complete the following inside a clean still air hood. Be sure to decontaminate any reagents, tubes, or samples with bleach prior to placing them inside the hood.
3. Decontaminate and place 1 × 15 mL conical tube per calculus sample in the still air hood.
4. In each of the clean 15 mL tubes, add 12.6 mL of the modified PB binding buffer and 100μL silica.
5. Turn off the incubator and rotator, and remove the calculus and control samples.

6. Centrifuge all samples, from Subheading 3.6.1, **step 22**, for 3 min at 19,500 × *g* in microcentrifuge.
7. Transfer the supernatant from the each of the samples into each of the sterile 15 mL tubes prepared earlier. Do not transfer the pellet, and store it in the freezer at −20 °C for potential downstream use.
8. Tighten the lids and parafilm the seal of the lid to prevent any leaks.
9. Place the 15 mL tubes on a rotary mixer for 1 h at room temperature.
10. Set the heat block at 37 °C.
11. Remove all the tubes from the rotary mixer, and centrifuge the tubes for 5 min at 4400 × *g*.
12. Pour off the supernatant into a waste bottle, and seal the bottle after its use. Use a pipette to remove as much supernatant off the pellet as possible without disturbing it.
13. Add 900μL 80% ethanol to each pellet, and resuspend the pellet using a long reach pipette tip by mixing up and down.
14. Prepare and label new sterile 1.5 mL DNA LoBind tubes.
15. Transfer the resuspended solution from the 15 mL tubes to the new sterile 1.5 mL DNA LoBind tubes.
16. Centrifuge all of the samples for 1 min at 14,000 rpm in microcentrifuge.
17. Pipette off the supernatant and pipette into the waste bottle used earlier. Seal the bottle after the last use.
18. Add 900μL 80% ethanol to each pellet, and resuspend the pellet using a vortex. If the pellet is recalcitrant to resuspension, use a pipette tip to pipette up and down to fully resuspend the pellet.
19. Centrifuge the samples for 1 min at 14,000 rpm in microcentrifuge.
20. Pipette off the supernatant, as above.
21. Place the samples to dry in the heat block for 15 min with the lids slightly ajar. Place a sterile delicate task wipe over the tubes to prevent particles from landing in the tubes. Ensure that the heat block is placed inside the still air hood. Prewarm the tube of TLE on the heat block at the same time.
22. After 15 min, inspect the pellet. Continue the incubation for more time if the pellet is not yet dry (e.g., does not have a glossy/shiny appearance but is not cracked or over dried).
23. Increase the temperature on the heat block to 50 °C.

24. Add 100 μL TLE to each dried pellet, and resuspend the pellet using a vortex.
25. Place the tubes back on the heat block for 10 min.
26. Centrifuge the samples for 1 min at 14,000 rpm in microcentrifuge.
27. Transfer the supernatant to a new sterile 1.5 mL tube.
28. Repeat **steps 24–27** to complete a double elution for a total of 200μL per extract.
29. 30μL DNA extract can be placed into a tube for downstream work, while the remainder of the extract can be preserved in a separate tube for later use, minimizing long-term risk of downstream contamination.
30. Store all extracts at −20 °C (*see* **Note 12**).
31. Following the completion of lab work in a work room for the day, the workspace should be cleaned down with 3–5% bleach (*see* **Note 13**).

3.7 16S ribosomal RNA Amplification Screen

An amplicon amplification is used to examine inhibitors and explore microbial DNA present in the DNA extracts prior to library preparation (*see* **Note 14**). We utilize the Caporaso et al. [20] method to amplify the 515 to 806 V4 region of the 16S ribosomal RNA (rRNA) encoding gene, resulting in DNA fragments that are approximately 300–350 bp in size when visualized on a gel.

1. The 16S rRNA PCR can be prepared as follows in a clean hood within the working room: Prepare the 16S rRNA mastermix inside a clean still air hood (*see* Subheading 2.4.2). Clean reagents down before placing them inside the hood.
2. Remove the DNA extracts from calculus and controls from the freezer and thaw. Clean the outer surface of the tubes as they are transferred into a decontaminated still air hood.
3. Place a decontaminated PCR microtube (often in strips) for each sample in the hood; label each accordingly.
4. In the PCR strip tubes, add 24μL PCR mastermix containing forward and reverse primer and polymerase enzyme to each tube.
5. Add 1μL DNA extract to each of the respective tubes.
6. Be sure to include an amplification negative control; do not add DNA to this tube.
7. Ensure the lids of the strip tubes are sealed properly.
8. Spin the strip tubes in a mini centrifuge quickly to ensure there is no liquid on the side of the tubes.
9. Remove the samples from the lab, and transfer to modern DNA molecular lab as quickly as possible.

10. In a modern molecular biology lab, place the tubes in a thermal cycler, and run the program: 95 °C (6 min); ×38 cycles of 95 °C (30 s), 50 °C (30 s), and 72 °C (1 min 30 s); 72 °C (10 min); and (optional) hold at 4 °C.
11. Remaining in the modern molecular biology lab following the completion of the PCR amplification, samples should be run on a 2% agarose gel to visualize the banding patterns and inspect the presence of DNA fragments. The absence of these DNA fragments after amplification could be indicative of DNA extraction failure for a sample, inhibitors present, or little microbial DNA within the sample and suggests that further assessment of the DNA extracts is needed before proceeding with shotgun library preparation. This approach also allows the researcher to crudely assess the levels of contamination in the extraction blank controls, providing insights into the cleanliness of the extraction; it is not expected the researchers will sequence the resulting fragments from this step.

3.8 Library Preparation

DNA extracts are prepared from ancient samples for shotgun metagenomic DNA sequencing on an Illumina machine, utilizing a common approach in the ancient DNA research field developed by Kircher et al. [25]. This process should ideally contain no UDG treatment or a partial UDG treatment, so DNA damage may be used to authenticate ancient microbes downstream during analysis. Please note that all values provided below are for preparing a single DNA extract and should be upscaled when completing multiple samples at once.

1. Prepare all reagents (elution buffer, aliquots from Qiagen MinElute kits of ERC, and PE buffers) needed from the clean room in a still air hood prior to beginning work (*see* Subheading 2.4.3).
2. Store any enzymes or temperature-sensitive reagents on a decontaminated freezer block until they are needed.
3. Prepare the Repair mastermix (*see* **Note 15**).
4. Add 20 μL Repair mastermix to sterile PCR strip tubes.
5. Add 20 μL DNA extract to the strip tubes containing the mastermix.
6. Gently mix by pipetting the solution up and down. Then briefly spin the strip tubes in a mini centrifuge.
7. Incubate on a thermal cycler for 15 min at 25 °C.
8. In sterile 1.5 mL DNA LoBind tubes, add 300 μL ERC buffer to them to prepare for Qiagen MinElute Reaction Cleanup for repaired DNA (*see* **Note 16**—details follow below).
9. Turn the heat block onto 50 °C.

10. Add DNA from the repair step to the ERC buffer, and pipette up and down to mix.
11. Transfer ~400μL to the MinElute column.
12. Let sit for 1 min.
13. Centrifuge for 1 min at 13,000 rpm in microcentrifuge.
14. Discard the flow through, and tap the top of the collection tube dry with paper towel.
15. Replace the spin column into the collection tube and add 700μL PE buffer.
16. Let sit for 1 min.
17. Centrifuge for 1 min at 13,000 rpm in microcentrifuge.
18. Place the elution buffer in the heat block.
19. Discard the flow through, and tap the top of the collection tube dry with paper towel.
20. Using a 10μL pipette, remove all the excess liquid from the purple O-ring inside the spin column. Be careful not to touch the silica membrane.
21. Centrifuge for 1 min at top speed in microcentrifuge.
22. Prepare 1.5 mL DNA LoBind tubes without a lid, and place the spin column into it. Discard the collection tube.
23. Add 20μL elution buffer directly to the membrane without touching it with the pipette tip.
24. Let sit for 1–2 min.
25. Centrifuge for 1 min at 13,000 rpm in microcentrifuge.
26. Transfer the flow through to a new sterile strip tube in preparation for the next step.
27. Ligate adapters onto to repaired purified DNA from prior step by first adding 1μL P7 adapter to the side of the strip tube.
28. Add 1μL P5 adapter to the opposite side of the strip tube.
29. Prepare the ligation mastermix (*see* Subheading 2.4.3).
30. Add 18μL ligation mastermix to each of the strip tubes.
31. Spin the strip tubes in a mini centrifuge briefly, and gently mix by pipetting the solution up and down.
32. Incubate on the thermal cycler for 60 min at 22 °C.
33. Following this reaction, place the strip tubes on ice to stop the reaction.
34. Purify the ligated DNA using the Qiagen MinElute method described in **steps 8–26** of this section.
35. Prepare the Bst fill-in mastermix (*see* Subheading 2.4.3).

36. Add 20μL of DNA purified in **step 34** of this section to each strip tube.
37. Incubate on the thermal cycler as follows: Preheat the lid to 95 °C, then incubate at 37 °C (30 min) and 80 °C (10 min).
38. Prepare the Amplification mastermix 1 (*see* Subheading 2.4.3).
39. Add 20μL of Amplification mastermix 1 to new sterile strip tubes, as these PCRs are performed with 5 reactions per library.
40. Add 1μL library to each Amplification mastermix 1 reaction in the strip tube.
41. Spin down briefly using a mini centrifuge, and relocate samples to a modern molecular biology laboratory as quickly as possible.
42. Perform the amplification on a thermal cycler using the following conditions. Preheat the lid to 95 °C; Incubate at 94 °C (12 min); ×13 cycles of 94 °C (30 s), 60 °C (30 s), 72 °C (45 s) with a 2 s increment per cycle; and 72 °C (10 min) (*see* **Note 17**).
43. Following this amplification, the 5 reactions from each biological sample are pooled, and 10μL is aliquoted as a backup.
44. The remainder of the amplified library is then purified using Ampure beads on magnetic racks as per manufacturer's instructions, resulting in DNA eluted into 30μL of preheated elution buffer.
45. Pipette the supernatant into new sterile 1.5 mL DNA LoBind tubes.
46. This can be stored at −20 °C until it is needed, or it can be used in a second amplification if the concentration of DNA is too low for sequencing after the first round of amplification.
47. If a second round of amplification is needed, prepare the second amplification mastermix in a modern molecular biology laboratory (*see* Subheading 2.4.3). Triplicate reactions should be prepared for each sample.
48. Add 23μL Amplification mastermix 2 to sterile strip tubes.
49. Add 2μL purified DNA from amplification 1 to the strip tubes.
50. Briefly spin down the liquid to the bottom of the tube using a mini centrifuge.
51. Perform the amplification on a thermal cycler using the following conditions: Preheat the lid to 95 °C; Incubate at 94 °C (12 min) 13 cycles of 94 °C (30 s), 60 °C (30 s), and 72 °C (45 s) with a 2 s increment per cycle; 72 °C (10 min).
52. Following amplification, purify the samples using Ampure beads by repeating **step 44**.

53. Store the DNA at −20 °C, or proceed to quantification. The calculus sample is now ready for quantification prior to shotgun sequencing (*see* **Note 18**).

4 Notes

1. All swabs will have their own microbial profile, so individuals need to be consistent with brand and type of swab for the duration of project.
2. Before the sampling of dental calculus occurs, projects should integrate the local community into their project design where possible, importantly including the ancient individual's descendants or ancestors [23]. This allows local communities and their own knowledge to be incorporated into the study. This approach also ensures that the local community is fully aware of what will take place, including how many individuals could be sampled, which individuals would be the best candidates and why, what the specific research questions are, and how the results may be used. This also allows for any questions or concerns from the community to be addressed by the researchers and for both parties to establish collaborative agreements about the future use of these samples, if any sample or data repatriation is required, the storage of any data generated, and who has long-term access to the data. Open and honest communication from the conception of a project with these communities should also include a plan for the researchers to return upon completion of the work to thoroughly explain the results and answer any further questions the community may have.
3. A detailed sampling plan also ensures excessive sampling does not occur and that the minimum number of individuals are destructively sampled to answer the project's specific research questions. This approach also ensures controls are collected at the time of sampling to understand the impacts of sample storage on the microbial profile. These controls can be processed alongside the samples and used to identify and remove contaminant species present during bioinformatic analyses, therefore gaining a more accurate picture of the microbes that may have been present in the individual when they were alive. Contamination from the researcher can also be minimized during the sampling process if an assessment is completed prior to sampling and protective gear is worn during the collection process. Different sampling strategies or collection of environmental controls may be required depending on where the calculus samples are collected, e.g., a museum setting versus an archaeological dig.

4. During sampling, protective gear should be worn when collecting samples to minimize the contamination of the sample by the individual collecting the specimens (Fig. 2) [26]. Typical protective gear includes gloves, surgical face masks, and a lab coat or disposable Tyvek suit that covers exposed skin in proximity to the sample. The specific protective gear worn needs to be detailed in the sampling strategy and in resulting publications, as this is often tailored to the sampling environment. For example, more protection might be worn when sampling at an archeological site, as samples have not yet been exposed to other individuals and previous contamination is limited to environmental microbes. In this scenario, protective gear may involve the use of a full body Tyvek suit, a face mask, a plastic face shield, hair net, and sterile gloves to limit the introduction of any human or microbial DNA during sampling. In other settings where secondary contamination from other individuals is very high, researchers may rely more on decontamination methods rather than aiming to minimize the risk of introducing new contamination.
5. If the samples require transport, it is important that samples are sent with padding to prevent further fragmentation or destruction during transportation.
6. Ancient DNA samples should always be processed in an ultra-clean laboratory to limit contamination from the outside environment and researchers processing them. These facilities ideally should be located separately from any laboratories working on modern or any amplified DNA, as quantities of amplified DNA can easily override the low concentrations of target DNA in some ancient samples. Ancient DNA laboratories have a number of systems in place to further protect samples, including temperature and humidity controls (e.g., relative humidity between 40% and 60% ± 5% per day and the temperature between 40 and 70 °F ± 3–5 °F); HEPA filtration air systems; positive air pressure; UV lights for decontamination; and specific usage and operational procedures throughout the lab. Operational procedures typically include the activation of UV for 2 h overnight and routine cleaning of surfaces and floors with 3–5% bleach to further decontaminate the laboratory.

   Clean laboratories should be built with particular workflow in mind, and contain areas designated for certain types of work. For example, researchers should enter and dress in protective gear in a room separate to any laboratory work. Sample processing (sample collection, decontamination, and pulverization; i.e., sample processing room) should also be located in another area separate to DNA extractions and library preparations (i.e., working room). If possible, a separate space to prepare reagents should also be included in a clean area where no biological samples are ever taken (i.e., clean room). Amplifications should also not be completed in this laboratory but in an entirely separate laboratory to minimize the risk of contaminating ancient samples with modern, amplified DNA products. Further, all sample work should occur with the utmost care inside still air hoods with UV lights. This provides a space that can be easily cleaned with bleach and further decontaminated using the internal UV lights, while limiting contamination from the researchers and environment. Anything taken inside the hood should be decontaminated first by wiping it with 3–5% bleach. During incubations or down time, the hood should remain closed, particularly if people are moving around the workspace or entering or leaving the room.

Researchers working within ancient DNA laboratories should also aim to minimize contamination of samples from the outside environment and themselves. Researchers should wear protective gear that includes shoe covers, Tyvek suits with hoods, sterile gloves, facemasks, and googles or face shields; the goal is to minimize skin and aerosol exposures to the laboratory. Users should take the utmost care to ensure the outside of the protective gear is not contaminated by the user upon entry into the lab; for example, users should not put on goggles or face masks with bare hands. We also recommend that users utilize two pairs of gloves, as the inner pair can be affixed to the suit and remain in place for the entire session, while the outer pair of gloves can be regularly changed or bleached throughout the procedure. As with sampling, this protective gear minimizes the contamination of ancient samples throughout the laboratory analysis process.

7. The dry swab method is preferred over swabs moistened with a solution, because the solution can add an additional microbial signal to the control.
8. An empty tube is used for this control, without adding additional reagents, such as water, as each reagent has its own microbial signal—even DNA-free reagents and molecular grade water [16].
9. Ancient dental calculus samples will have contamination from the environment where they are found, collected, or stored that can be partially mitigated prior to DNA extraction. While multiple decontamination methods are utilized in the field, several have proven to be effective in reducing the contaminants present (Farrer et al. [27]). In this chapter, we outline a method of calculus decontamination that utilizes ultraviolet (UV) radiation and bleach treatment to cross-link and destroy contaminant DNA on the outer surface of samples.

10. Another option is not to perform the protein digestions during decalcification. Proteinase K can be added separately to each tube after digestion, as recently published by Fagernas et al. [24], to ensure that DNA and proteins can be extracted from the same sample.
11. Buffer preparation is done in a clean room in a clean still air hood. Be sure to decontaminate any reagents, tubes, or samples with bleach prior to placing them inside the hood.
12. These DNA extracts are now ready for quantification, amplification, or library preparation.
13. Remove all items from within the still air hood and clean down all surfaces with bleach, followed by a rinse of 70% ethanol or molecular grade water (depending on the hood material, as ethanol can cause some deterioration of plastics over time). All pipettes, pipette tip boxes, etc. that were present in the hood should be wiped with bleach, followed by ethanol, and replaced inside the hood, before the door is closed and UV turned on for at least 1 h. All racks should be cleaned down with bleach, followed by ethanol and molecular grade water if sensitive to ethanol. Ideally these should be UV treated for decontamination for at least 30 min. All equipment that was used should be wiped down with bleach, followed by ethanol. All benches, light switches, and door handles should be cleaned with bleach and ethanol. Garbage should be emptied, and floors vacuumed with HEPA vacuum, followed by cleaning with bleach and a wet vacuum to remove residual bleach from the floor. Ideally UV decontamination will occur for 1–2 h overnight.
14. As ancient DNA samples often have low concentrations of target DNA, it may be difficult to quantify the amount of extracted DNA, even with the most sensitive quantification methods. Therefore, it can be hard to determine whether or not the DNA extraction has been successful for a sample and should be utilized for library preparation. One fast and inexpensive way of quality checking ancient DNA extracts is by completing amplification of a section of the 16S rRNA encoding gene. This is particularly useful for old or poorly preserved samples or those that have high levels of inhibitors, which would limit the success of shotgun library preparations. While this method has been applied to explore microbial diversity of ancient calculus specimens, it is not recommended for that purpose, due to biases resulting from the length of ancient DNA fragments [8, 24]. However, this single PCR result provides a cost-effective and preliminary way to screen DNA extracts to assess the probability of a successful shotgun library preparation.

15. The repair mastermix should be prepared in a workroom, in a clean still air hood (*see* Subheading 2.4.3). Any items should be wiped down with bleach before placing them inside the hood, including all reagents, tubes, boxes, and tubes of DNA extracts.
16. The repaired DNA is purified using a modified version of the Qiagen MinElute Reaction Cleanup kit protocol.
17. The PCR amplification should be performed in a thermal cycler in a modern molecular biology laboratory, separate to the ancient DNA lab. This is to avoid the high concentration of amplified material contaminating the lab and near other ancient samples.
18. For the preparation of libraries for shotgun sequencing, all libraries require quantification prior to DNA sequencing. Quantification is used to determine how much of each library should be loaded onto the sequencing device, as well as check library preparation success, dimer presence, and the average fragment size. A Qubit can be used to determine the concentration of DNA present in individual libraries, while a TapeStation, Bioanalyzer, or Fragment Analyzer can be used to estimate both the concentration and the average fragment size present in libraries. Similarly, quantitative PCR can also provide information on the molarity of DNA present in the sample. These quantification methods may also help you to determine if further purification is needed, due to the presence of dimer, and provide information for accurate pooling (e.g., mixing together multiple samples) in a single sequencing run.

References

1. Green RE, Krause J, Briggs AW, Maricic T, Stenzel U, Kircher M et al (2010) A draft sequence of the Neandertal genome. Science 328:710–722. https://doi.org/10.1126/science.1188021
2. Pääbo S, Poinar H, Serre D, Jaenicke-Despres V, Hebler J, Rohland N et al (2004) Genetic analyses from ancient DNA. Annu Rev Genet 38:645–679. https://doi.org/10.1146/annurev.genet.37.110801.143214
3. Allentoft ME, Collins M, Harker D, Haile J, Oskam CL, Hale ML et al (2012) The half-life of DNA in bone: measuring decay kinetics in 158 dated fossils. Proc Biol Sci 279:4724–4733. https://doi.org/10.1098/rspb.2012.1745
4. Turney CSM, Fogwill CJ, Golledge NR, McKay NP, van Sebille E, Jones RT et al (2020) Early last interglacial ocean warming drove substantial ice mass loss from Antarctica. Proc Natl Acad Sci U S A 117:3996. https://doi.org/10.1073/pnas.1902469117
5. Frisia S, Weyrich LS, Hellstrom J, Borsato A, Golledge NR, Anesio AM et al (2017) The influence of Antarctic subglacial volcanism on the global iron cycle during the last glacial Mmaximum. Nat Commun 8:15425. https://doi.org/10.1038/ncomms15425
6. Zobel M, Davison J, Edwards ME, Brochmann C, Coissac E, Taberlet P et al (2018) Ancient environmental DNA reveals shifts in dominant mutualisms during the late quaternary. Nat Commun 9:139. https://doi.org/10.1038/s41467-017-02421-3
7. Boast AP, Weyrich LS, Wood JR, Metcalf JL, Knight R, Cooper A (2018) Coprolites reveal ecological interactions lost with the extinction of New Zealand birds. Proc Natl Acad Sci U S A 115:1546. https://doi.org/10.1073/pnas.1712337115

8. Weyrich LS, Duchene S, Soubrier J, Arriola L, Llamas B, Breen J et al (2017) Neanderthal behaviour, diet, and disease inferred from ancient DNA in dental calculus. Nature 544:357–361. https://doi.org/10.1038/nature21674
9. Warinner C, Speller C, Collins MJ, Lewis CM (2015) Ancient human microbiomes. J Hum Evol 79:125–136. https://doi.org/10.1016/j.jhevol.2014.10.016
10. Weyrich LS, Dobney K, Cooper A (2015) Ancient DNA analysis of dental calculus. J Hum Evol 79:119–124. https://doi.org/10.1016/j.jhevol.2014.06.018
11. Adler CJ, Dobney K, Weyrich LS, Kaidonis J, Walker AW, Haak W et al (2013) Sequencing ancient calcified dental plaque shows changes in oral microbiota with dietary shifts of the Neolithic and industrial revolutions. Nat Genet 45:450–455. https://doi.org/10.1038/ng.2536
12. Warinner C, Rodrigues JFM, Vyas R, Trachsel C, Shved N, Grossmann J et al (2014) Pathogens and host immunity in the ancient human oral cavity. Nat Genet 46:336–344. https://doi.org/10.1038/ng.2906
13. Farrer AG, Bekvalac J, Redfern R, Gully N, Dobney K, Cooper A et al (2018) Biological and cultural drivers of oral microbiota in Medieval and Post-Medieval London, UK. bioRxiv 343889. https://doi.org/10.1101/343889
14. Jónsson H, Ginolhac A, Schubert M, Johnson PLF, Orlando L (2013) mapDamage2.0: fast approximate Bayesian estimates of ancient DNA damage parameters. Bioinformatics 29:1682–1684. https://doi.org/10.1093/bioinformatics/btt193
15. Warinner C, Herbig A, Mann A, Fellows Yates JA, Weiß CL, Burbano HA et al (2017) A robust framework for microbial archaeology. Annu Rev Genomics Hum Genet 18:321–356. https://doi.org/10.1146/annurev-genom-091416-035526
16. Eisenhofer R, Minich JJ, Marotz C, Cooper A, Knight R, Weyrich LS (2019) Contamination in low microbial biomass microbiome studies: issues and recommendations. Trends Microbiol 27:105–117. https://doi.org/10.1016/j.tim.2018.11.003
17. Salter SJ, Cox MJ, Turek EM, Calus ST, Cookson WO, Moffatt MF et al (2014) Reagent and laboratory contamination can critically impact sequence-based microbiome analyses. BMC Biol 12:87. https://doi.org/10.1186/s12915-014-0087-z
18. Tito RY, Knights D, Metcalf J, Obregon-Tito AJ, Cleeland L, Najar F et al (2012) Insights from characterizing extinct human gut microbiomes. PLoS One 7:e51146. https://doi.org/10.1371/journal.pone.0051146
19. Austin RM, Sholts SB, Williams L, Kistler L, Hofman CA (2019) Opinion: to curate the molecular past, museums need a carefully considered set of best practices. Proc Natl Acad Sci U S A 116:1471. https://doi.org/10.1073/pnas.1822038116
20. Caporaso JG, Lauber CL, Walters WA, Berg-Lyons D, Huntley J, Fierer N et al (2012) Ultra-high-throughput microbial community analysis on the Illumina HiSeq and MiSeq platforms. ISME J 6:1621–1624. https://doi.org/10.1038/ismej.2012.8
21. Ziesemer KA, Mann AE, Sankaranarayanan K, Schroeder H, Ozga AT, Brandt BW et al (2015) Intrinsic challenges in ancient microbiome reconstruction using 16S rRNA gene amplification. Sci Rep 5:16498. https://doi.org/10.1038/srep16498
22. Dabney J, Knapp M, Glocke I, Gansauge M-T, Weihmann A, Nickel B et al (2013) Complete mitochondrial genome sequence of a Middle Pleistocene cave bear reconstructed from ultrashort DNA fragments. Proc Natl Acad Sci U S A 110:15758. https://doi.org/10.1073/pnas.1314445110
23. Wagner JK, Colwell C, Claw KG, Stone AC, Bolnick DA, Hawks J et al (2020) Fostering responsible research on ancient DNA. Am J Hum Genet 107:183–195. https://doi.org/10.1016/j.ajhg.2020.06.017
24. Fagernäs Z, García-Collado MI, Hendy J, Hofman CA, Speller C, Velsko I et al (2020) A unified protocol for simultaneous extraction of DNA and proteins from archaeological dental calculus. J Archaeol Sci 118:105135. https://doi.org/10.1016/j.jas.2020.105135
25. Kircher M, Sawyer S, Meyer M (2012) Double indexing overcomes inaccuracies in multiplex sequencing on the Illumina platform. Nucleic Acids Res 40:e3. https://doi.org/10.1093/nar/gkr771
26. Llamas B, Valverde G, Fehren-Schmitz L, Weyrich LS, Cooper A, Haak W et al (2016) From the field to the laboratory: controlling DNA contamination in human ancient DNA research in the high-throughput sequencing era. Sci Technol Archeol Res 3:1–14. https://doi.org/10.1080/20548923.2016.1258824
27. Farrer AG, Wright SL, Skelly F et al (2021) Effectiveness of decontamination protocols when analyzing ancient DNA preserved in dental calculus. Sci Rep 11, 7456. https://doi.org/10.1038/s41598-021-86100-w

Chapter 8

Whole-Genome Sequencing of Pathogens in Saliva: A Target-Enrichment Approach for SARS-CoV-2

David J. Speicher, Jalees A. Nasir, Peng Zhou, and Danielle E. Anderson

Abstract

Outbreak analysis and transmission surveillance of viruses can be performed via whole-genome sequencing after viral isolation. Such techniques have recently been applied to characterize and monitor SARS-CoV-2, the etiological agent of the COVID-19 pandemic. However, the isolation and culture of SARS-CoV-2 is time consuming and requires biosafety level 3 containment, which is not ideal for many resource-constrained settings. An alternate method, bait capture allows target enrichment and sequencing of the entire SARS-CoV-2 genome eliminating the need for viral culture. This method uses a set of hybridization probes known as "baits" that span the genome and provide sensitive, accurate, and minimal off-target hybridization. Baits can be designed to detect any known virus or bacteria in a wide variety of specimen types, including oral secretions. The bait capture method presented herein allows the whole genome of SARS-CoV-2 in saliva to be sequenced without the need to culture and provides an outline of bait design and bioinformatic analysis to guide a bioinformatician.

Key words Salivary diagnostics, COVID-19, SARS-CoV-2, Enrichment, Whole-generation sequencing, Bait capture

1 Introduction

Saliva has considerable diagnostic potential: It is abundant and noninvasive to collect and is representative of oral and systemic health. Saliva contains a range of molecular and serological biomarkers useful for diagnostics and surveillance of infectious pathogens including Human Herpesviruses, Zika virus, and Severe Acute Respiratory Syndrome Coronavirus 2 (SARS-CoV-2), which are readily transmitted via saliva [1–3]. SARS-CoV-2, the etiological agent of coronavirus disease (COVID-19), is a spherical, enveloped virus, with a non-segmented, positive-sense, RNA genome ~30 kb in length [4]. Infectious SARS-CoV-2 virion can be transmitted in salivary droplets by infected people breathing, talking, coughing, or sneezing in close contact and infecting another nearby person through the mouth, nose, or eyes [5]. SARS-CoV-2 infects

Guy R. Adami (ed.), *The Oral Microbiome: Methods and Protocols*, Methods in Molecular Biology, vol. 2327, https://doi.org/10.1007/978-1-0716-1518-8_8,

human epithelial cells through the host cell receptor angiotensin-converting enzyme II (ACE2), which is expressed on cells lining the lungs, oral buccal, and gingiva [6]. While the gold standard to detect SARS-CoV-2 is by reverse transcriptase polymerase chain reaction (RT-PCR) from nasopharyngeal swabs (NPS), the viral titers in NPS are comparable to those in saliva during the first week of symptoms and decrease over time [7]. Saliva can be positive for more than 20 days post-symptom onset, even after NPS becomes negative [8]. The salivary viral loads correlate with symptom severity and degree of tissue damage [9]. While RT-PCR is often used for diagnostics and monitoring transmission, it cannot be used to assess detailed phylogenetic analysis, which requires whole-genome sequencing (WGS) and mutation analysis [10].

In a viral outbreak, high transmission rates and increased viral replication across populations can introduce mutations and create divergent variants [11]. Consequently, surveillance during an outbreak is crucial. However, detection of multiple variants using RT-PCR is difficult due to the high specificity and location of primers. Sequences divergent from the primers may suboptimally bind and result in false negatives. Additionally, the short amplicon length in a conserved region provides only partial information on genetic diversity [11]. In contrast, metagenomic analysis allows diversity to be observed within a viral species but is often expensive and time consuming due to high background noise as most sequences in biological samples are not of viral origin. Nucleic-acid amplification tests, like PCR, aim to increase the quantity of viral nucleic acids but can introduce mutations with subsequent amplification cycles due to lack of polymerase fidelity. These mutations hinder the ability to examine variants from a surveillance perspective. A further alternative, target-enrichment methods reduce the quantity of non-viral nucleic acids while increasing the overall abundance of viral nucleic acid within the sequencing reaction. Originally, target-enrichment procedures were developed for human genomic studies referred to as exome sequencing (i.e., sequencing of all protein coding genes) [12]. Bait capture methods (also called target-enrichment in-solution, or hybridization capture-based methods) use specially designed small, biotinylated RNA hybridization probes, 80–120 nucleotides long, called "baits" that act like molecular fishhooks to separate viral sequences from background human/animal material and allow construction of high-quality assemblies when sequenced in conjunction with WGS technologies [13, 14].

The entire in-solution bait-capture enrichment process takes 1.5 days, ~4 days from sample preparation to bioinformatic analysis (Fig. 1). In brief, the process starts with preparing cDNA from fragmented RNA, 3′ adenylating the cDNA, and then hybridizing and ligating library prep adapters to the ends. Blocking oligos are then added to the adapter ligated fragments to prepare single-

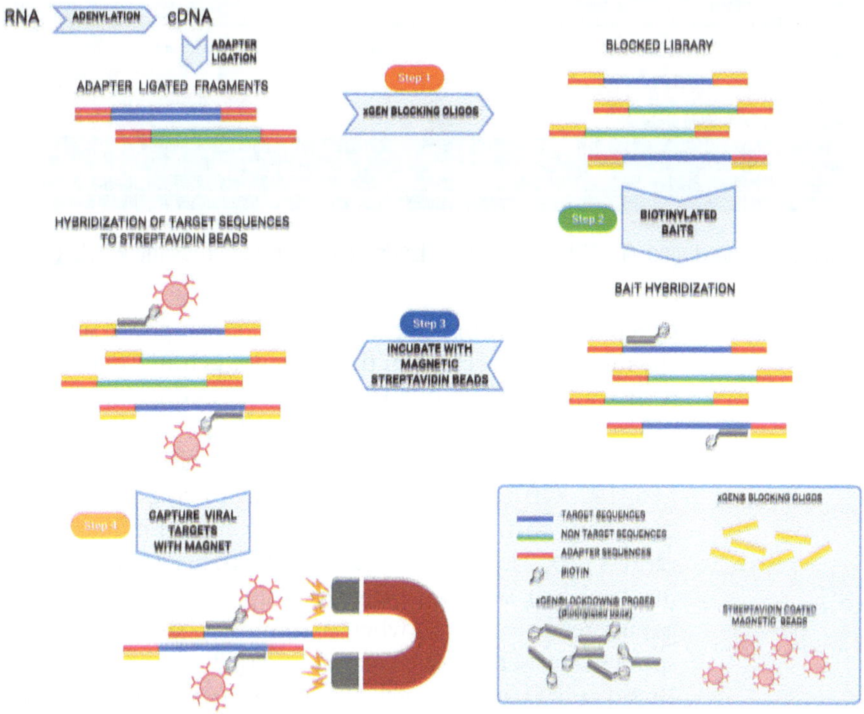

Fig. 1 The pre-sequencing target-enrichment workflow including blocking, hybridization, and isolation of viral sequences

stranded fragments prior to adding biotinylated baits. These blocking oligos are synthetic oligonucleotide sequences that hybridize to specific regions, including the adapters, to prevent cross-hybridization between library fragments. Biotinylated baits are then hybridized to fragments to be sequenced to facilitate their isolation with streptavidin-coated magnetic beads and the discard of non-target genetic material. The isolated fragments can then be sequenced using an Illumina MiSeq or other sequencing technology at the expertise of a sequencing facility for WGS and then analyzed bioinformatically.

For bait capture to be an effective protocol for viruses, hybridization probes must be designed such that they are able to anneal to the viral nucleotide sequences of known pathogens. Given one or more genome sequences of a known virus reference, software tools [15–17] can tile across a given sequence creating in silico genomic fragments (Fig. 2). Post-processing of these fragments is required as often the initial bait set is too large to affordably synthesize and can result in an excess of redundant baits due to overlapping tile sequences. The post-processing of these fragments into a complete bait set varies between software tools. PanArray [15] and CATCH (Compact Aggregation of Targets for Comprehensive Hybridization) [17] employ a greedy algorithm [18] that infers the removal

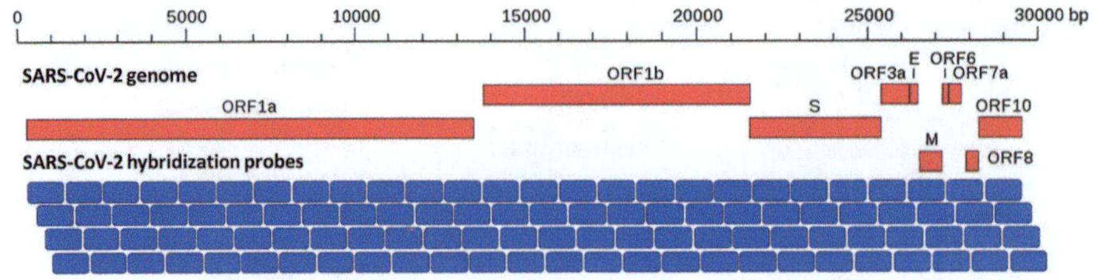

Fig. 2 Schematic representation of SARS-CoV-2 hybridization probes tiled across the whole genome

of redundant probes to produce an optimal probe set that provides the minimum coverage of the reference genome(s). A filtration step is then applied since not all fragments are equal and can have varied physical properties, such as melting temperature and GC content, which can affect the optimal binding of the baits to viral genomes [16, 19]. Off-target hybridization must also be considered to produce baits with high fidelity and low ambiguity to intended virus targets [20, 21]. Overall, there is no single design tool for generating viral bait sets and often re-evaluation of the post-processing parameters is required when designing a bait set for another viral species. Bait sets can be created to target one or more viral species. Broad range bait sets, such as VirCapSeq-VERT [22] and ViroFind [23], were designed to cover 207 viral taxa and 535 DNA/RNA viruses, respectively. VirCapSeq-VERT, a set of 1,993,176 baits 50–100 bp in length, allowed for 100- to 10,000-fold enrichment in viral content while reducing background human DNA [22]. ViroFind, a set of 165,433 baits 125 bp in length, was applied to five brain biopsy samples with progressive multifocal leukoencephalopathy, characterizing the genetic divergence of Human Polyomavirus 2 (JC virus) [23]. One caveat of broad bait sets is that the large number of baits substantially increases cost and restricts clinical utility due to affordability. More specific bait sets, such as Li et al.'s set of 4,303 baits, 120 bp in length, which target the majority of coronavirus species and were used to characterize novel bat-borne coronaviruses, are relatively cheaper due to the fewer number of baits needed [24, 25]. A species-specific SARS-CoV-2 virus bait set containing 1,310 baits 80 bp long was designed by Nasir et al. for surveillance of the emergent novel SARS-CoV-2 virus [21]. Many of these commercially available and custom-made bait sets can be ordered through Arbor Bioscience (myBaits) or Integrated DNA Technologies (xGen Lockdown baits) for use in viral research and surveillance studies.

In this chapter, we provide the methodology for enrichment and sequencing of the SARS-CoV-2 genome using the SARS-CoV-2 bait set designed by Nasir et al. [21], Integrated DNA Technologies xGen Lockdown baits, or the Arbor Biosciences myBaits Expert Virus SARS-CoV-2 panel. Using this method, the viral surveillance and outbreak analysis can be performed without the need to culture or purify the virus. As these bait sets are designed solely for SARS-CoV-2, this method can be used to sequence the virus from any sample type once the RNA is appropriately extracted. Alternatively, this method can be modified using any bait set specifically designed for any individual or panel of viruses or bacteria of interest and used on any extracted sample type. Bait design and detailed bioinformatics is beyond the scope of this chapter, but information is provided to give an understanding of the steps needed. We encourage anyone attempting this protocol to collaborate with a bioinformatics team to aid in proper design and analysis.

2 Materials

2.1 Sample Collection and Extraction

1. Sterile 50 mL conical tube or urine container.
2. Ice packs, crushed ice or dry ice.
3. Trizol (Thermo Fisher Scientific) or Extraction kit designed for viral RNA or total nucleic acid, e.g., QIAamp Viral RNA Mini Kit (Qiagen), High Pure Viral RNA Kit (Roche Diagnostics), or the NucliSENS easyMag (bioMérieux) (*see* **Note 1**).
4. Refrigerated centrifuge able to accommodate 50 mL conical tubes.
5. −80 °C Freezer.
6. Elution buffer (10 mM Tris–HCl, pH 8.0) or TE Buffer (10 mM Tris–HCl, pH 7.5, 1 mM EDTA).
7. NanoDrop or Qubit Fluorometer (Thermo Fisher Scientific).

2.2 Library Preparation for Illumina

1. TruSeq stranded mRNA library preparation kit (Illumina).
2. Agencourt AMPure XP beads (Beckman Coulter Genomics).
3. 80% (v/v) Ethanol (*see* **Note 2**).
4. Agilent 2100 Bioanalyzer system (Agilent Technologies).
5. Bioanalyzer high-sensitivity DNA Kit (Agilent Technologies).
6. Magnetic rack.
7. Thermocycler or heating block for 94 °C.
8. SuperScript II One-Step RT-PCR System with Platinum Taq DNA Polymerase (Thermo Fisher Scientific).
9. Low binding 1.7 mL microcentrifuge tubes (sterile and RNase/DNase-free).
10. Low-binding 0.2 mL PCR tubes (sterile and RNase/DNase-free).
11. Centrifuge.

12. Microcentrifuge with adaptors for 0.2 and 1.7 mL tubes.
13. Vortex.

2.3 Bait Capture Enrichment and Sequencing

1. IDTE (10 mM Tris–HCl, 0.1 mM EDTA) buffer, pH 8.0.
2. Biotinylated xGen Lockdown baits (Integrated DNA Technologies) or SARS-CoV-2-specific baits (myBaits; Arbor Bioscience) (*see* **Note 3**).
3. xGen Universal Blocker-TS Mix (Integrated DNA Technologies).
4. Human Cot-1 DNA (Thermo Fisher Scientific).
5. NimbleGen 2× hybridization buffer and NimbleGen 2× hybridization solution (Roche Diagnostics).
6. Dynabeads M-270 Streptavidin magnetic beads (Thermo Fisher Scientific).
7. ThermoMixer C shaker (Eppendorf).
8. SeqCap EZ Hybridization and Wash kit (Roche Diagnostics).
9. NGS library primers (Illumina).
10. Agencourt AMPure XP beads (Beckman Coulter Genomics) or other DNA cleanup kit.
11. Agilent 2100 Bioanalyzer system (Agilent Technologies).
12. Bioanalyzer High-Sensitivity DNA Kit (Agilent Technologies).
13. Freeze Dryer (Heto PowerDry LL3000, Thermo Fisher Scientific).
14. Nuclease-free molecular grade water.
15. Magnetic rack.
16. MiSeq v2 Nano Reagent Kit (Illumina).
17. PhiX Control v3 Library (Illumina).

2.4 Bioinformatic Analysis

1. Trimmomatic.
2. Galaxy analysis platform.
3. ORFfinder, BLASTN and BLASTX (NCBI).
4. CLC Genomics Workbench version 12.0 (Qiagen).
5. Circos (version 0.69.8).
6. Prism (GraphPad Prism 7).

3 Methods

3.1 Sample Collection and Extraction

1. Request that the patient refrain from eating, drinking, smoking, or oral hygiene procedures for at least 1 h prior to collection.

2. Request that the patient rinse their mouth with distilled drinking water, which can then be swallowed or expectorated.
3. Collect 2–3 mL unstimulated whole saliva by having the patient expectorate (*see* **Notes 4** and **5**) in a sterile container (*see* **Note 6**).
4. Immediately following collection, cool the sample on ice packs or dry ice, or put in the fridge or aliquot and freeze (*see* **Note 7**).
5. Extract samples as soon as possible. Extraction can be done directly on 200μL nonviscous saliva (*see* **Note 8**).
6. Extract viral RNA with either Trizol or a commercial extraction kit following manufacturer's instructions (*see* **Note 1**).
7. Store samples at −80 °C if processing is delayed (*see* **Note 9**).
8. Determine the quantity and quality of RNA via NanoDrop or Qubit Fluorometer (*see* **Note 10**).

3.2 Library Preparation

The method uses the TruSeq® Stranded mRNA Sample Preparation Kit for Illumina (Refer to **Notes 11–15** before proceeding).

3.2.1 RNA Fragmentation and Priming Starting from Purified RNA

1. Add 8 μL (<20 ng) extracted RNA to a 0.2 mL PCR tube (*see* **Note 16**).
2. Add 11 μL Elute, Prime, Fragment Mix. Total volume is 19 μL.
3. Gently pipette the entire volume up and down 6 times to mix thoroughly.
4. Using a thermocycler incubate at 94 °C for 4 min and cool to 4 °C (*see* **Note 17**).
5. Quickly spin (*see* **Note 18**).

3.2.2 First Strand cDNA Synthesis

1. In a 0.2 mL PCR tube add 1 μL First Strand Master Mix plus 8 μL SuperScript II™ to make Mix A.
2. Add 8 μL of Mix A to 19 μL fragmented and primed RNA from **step 3.2.1**. Total volume is 27 μL.
3. Gently pipette the entire volume up and down 6 times to mix thoroughly.
4. Place tube into a thermocycler with a preheated lid set to 100 °C.
5. PCR Incubation: 25 °C/10 min, 42 °C/15 min, 70 °C/15 min, hold at 4 °C.

3.2.3 Second Strand cDNA Synthesis

1. Take the first strand synthesis reaction from **step 3.2.2** and add 20 μL Second Strand Master Mix.
2. Add 5 μL Resuspension Buffer. Total volume is 52 μL.
3. Mix by gentle pipetting.

4. Using a thermocycler, incubate at 16 °C for 1 h.
5. Equilibrate the tube to room temperature (*see* **Note 19**).

3.2.4 Double-Stranded cDNA Purification Using AMPure XP Beads

1. Vortex AMPure XP beads.
2. Add 94 μL (1.8×) beads to the reaction from **step 3.2.3** and mix well.
3. Incubate for 15 min at room temperature.
4. Quickly spin and place on magnetic stand for 5 min.
5. While on the stand, remove supernatant by using a pipette.
6. Add 200 μL of fresh 80% (v/v) ethanol without mixing while still on the magnetic rack.
7. Incubate for 30 s at room temperature.
8. Remove supernatant by using a pipette.
9. Add 200 μL of fresh 80% (v/v) ethanol while still on the magnetic rack.
10. Incubate for 30 s at room temperature.
11. Remove supernatant by using a pipette.
12. Air dry for 5–15 min while tube is on the magnetic rack. When drying the beads, ensure the beads are not too dry or wet (*see* **Note 20**).
13. Add 17 μL nuclease-free molecular grade water and incubate for 4 min at room temperature.
14. Place the tube on a magnetic rack for 5 min at room temperature.
15. Transfer 16 μL of the purified double-stranded cDNA to a new PCR tube. This can be stored at −20 °C (*see* **Note 21**).

3.2.5 Adenylation of 3′ Ends

1. Take the purified cDNA from **step 3.2.4** and add 9 μL of A-Tailing Mix. Total volume is 25 μL.
2. Place tube into a thermocycler with a preheated lid set to 100 °C.
3. PCR Incubation: 37 °C/30 min, 70 °C/5 min, hold at 4 °C.
4. Place tube on ice until ready to proceed.

3.2.6 Adapter Ligation

1. Take the purified cDNA reaction from **step 3.2.5** and add 2.5 μL Resuspension Buffer, 2.5 μL Ligation Mix, and 2.5 μL Adapter Index. Total volume is 32.5 μL.
2. Incubate at 30 °C for 10 min.
3. Add 3.5 μL of Stop Ligation Buffer to inactivate the ligation. Total volume is 36 μL.
4. Gently pipette the entire volume up and down to mix thoroughly.

3.2.7 Ligation Purification Using AMPure Beads

1. Vortex AMPure XP beads.
2. Add 36 μL of beads to the 36 μL reaction from **step 3.2.6** and mix well.
3. Incubate for 10 min at room temperature.
4. Quickly spin and place on magnetic stand for 5 min.
5. Remove supernatant by using a pipette and place it into new tube.
6. Add 200 μL of fresh 80% (v/v) ethanol while still on the magnetic rack.
7. Remove supernatant by using a pipette.
8. Add 200 μL of fresh 80% (v/v) ethanol while still on the magnetic rack.
9. Remove supernatant by using a pipette.
10. Air dry for 5–15 min while tube is on the magnetic rack. When drying the beads, ensure they are not too dry or wet (*see* **Note 20**).
11. Add 21 μL nuclease-free water to allow the DNA to elute off the beads and incubate for 4 min at room temperature.
12. Quickly spin and place on a magnetic rack.
13. Transfer 20 μL to a new 0.2 mL PCR tube.

3.2.8 Enrichment of DNA Fragments

1. Take the 20 μL purified cDNA reaction from **step 3.2.7** and add 5 μL PCR Primer Cocktail and 25 μL PCR Master Mix. Total volume is 50 μL.
2. Place tube into a thermocycler and run the program shown in Table 1 with a preheated lid set to 100 °C.

Table 1
Detailed PCR protocol for amplification of cDNA fragments

	Cycles	Temperature (°C)	Time
Polymerase activation	1	98	30 s
Amplification	11		
Denaturation		98	10 s
Annealing		60	30 s
Extension		72	30 s
Final extension	1	72	5 min
Hold	1	10	∞

3.2.9 PCR Purification Using AMPure Beads

1. Vortex AMPure XP beads.
2. Add 50 μL beads to the 50 μL reaction from **step 3.2.8** and mix well.
3. Incubate for 10 min at room temperature.
4. Quickly spin and place on magnetic stand for 5 min.
5. Remove supernatant by using a pipette and place it into a new tube.
6. Add 200 μL of fresh 80% (v/v) ethanol while still on the magnetic rack.
7. Remove supernatant by using a pipette.
8. Add 200 μL of fresh 80% (v/v) ethanol while still on the magnetic rack.
9. Remove supernatant by using a pipette.
10. Air dry for 5–15 min while tube is on the magnetic rack. When drying the beads, ensure they are not too dry or wet (*see* **Note 20**).
11. Add 11 μL nuclease-free molecular grade water to elute DNA.
12. Incubate for 4 min at room temperature to allow DNA to elute off the beads.
13. Quickly spin and place on a magnetic rack.
14. Transfer 10 μL to a new 0.2 mL PCR tube.
15. This is the next-generation sequencing (NGS) library (*see* **Note 22**).

3.3 xGen® Lockdown® Probe Enrichment of DNA Library (Refer to Note 23 before proceding)

3.3.1 Probe Preparation

1. Hydrate the dried down pool of xGen® Lockdown® Probes to 1.5 pmol/μL in IDTE (10 mM Tris–HCl, 0.1 mM EDTA) buffer, pH 8.0.
2. Aliquot the probe stock into 10 μL aliquots and store at −20 °C (*see* **Note 24**).

3.3.2 Blocking of Library

1. Pooling of NGS libraries is possible but be sure to keep the quantity ≤500 ng. Reduce the total volume to 20 μL by lyophilization.
2. Add 20 μL (<500 ng) pooled, barcoded Illumina TruSeq LT Library, 5 μL of 1 mg/mL Cot-1 DNA, and 2 μL xGen Universal Blockers to a low-bind 1.7 mL tube (*see* **Note 25**). Total volume is 27 μL.
3. Dry the contents of the tube using a Freeze Dryer (Thermo Fisher Scientific, Heto PowerDry LL3000) (*see* **Note 26**).

3.3.3 Probe Hybridization

1. Thaw all SeqCap EZ hybridization buffers and equilibrate to room temperature.
2. To the dried DNA pellet, add 7.5 μL Nimblegen 2× Hybridization buffer, 3 μL Nimblegen Hybridization Component A, and 2.5 μL nuclease-free water. Total volume is 13 μL.
3. Leave the solution in the tube for 10 min to allow the pellet to go into solution.
4. Transfer the 13 μL reaction to a 0.2 mL PCR tube.
5. Incubate in a thermocycler at 95 °C for 10 min.
6. Cool on ice and add 2 μL Lockdown Probe to the tube. Pipette to mix. Total volume is 15 μL.
7. Incubate hybridization reaction at 65 °C (set heated lid at 75 °C) for 4 h.
8. Store this hybridized sample at 65 °C until needed in **step 3.3.6**.

3.3.4 Prepare Bead Wash Buffers

1. Dilute 10× Wash Buffers (I, II, III, and Stringent) and 2.5× Bead Wash Buffer to create 1× working solutions (*see* **Note 27**).
2. For 1× Wash Buffer I and 1× Stringent Wash Buffer equilibrate buffers at 65 °C for at least 2 h before starting wash steps.

3.3.5 Prepare Streptavidin Dynabeads™ (See **Note 28**)

1. Allow Dynabeads™ M-270 Streptavidin to equilibrate to room temperature for 30 min before use.
2. Mix the beads thoroughly by vortexing for 15 s.
3. Aliquot 100 μL streptavidin beads into a single 1.7 mL low-bind tube (*see* **Note 29**).
4. Place the tube in a magnetic separation rack (*see* **Note 30**). Carefully remove and discard the supernatant ensuring that all the beads remain in the tube.
5. Add 200 μL 1× Bead Wash Buffer per 100 μL beads. Vortex for 10 s.
6. Place the tube in a magnetic separation rack. Carefully remove and discard the supernatant ensuring that all the beads remain in the tube.
7. Add 200 μL 1× Bead Wash Buffer per 100 μL beads. Vortex for 10 s.
8. Place the tube in a magnetic separation rack. Carefully remove and discard the supernatant ensuring that all the beads remain in the tube.
9. After removing the buffer following the second wash, add 1× the original volume of beads of 1× Bead Wash Buffer and resuspend by vortexing.

10. Transfer 100 μL of the resuspended beads into a new low-bind tube for capture reaction (*see* **Note 31**).
11. Place the tube in a magnetic rack to bind the beads. Allow the beads to separate from the supernatant. Carefully remove and discard the clear supernatant ensuring that all the beads remain in the tube (*see* **Notes 32** and **33**).

3.3.6 Hybridization of Target to the Streptavidin Beads

1. Transfer the 15 μL hybridization sample from **step 3.3.3** to the tube containing streptavidin beads prepared in **step 3.3.5**.
2. Mix thoroughly by pipetting up and down 10 times. Ensure all beads are reconstituted (*see* **Note 34**).
3. Place the tube into a ThermoMixer (2000 rpm) set to 65 °C (with a lid temperature set at 75 °C) for 45 min to bind the DNA to the beads.
4. Vortex the tubes for 3 s, then vortex briefly every 15 min to ensure that the beads remain in suspension. Do not spin down.

3.3.7 Streptavidin Bead Wash (See **Notes 35–37**)

1. Add 100 μL preheated 1× Wash Buffer I to the tube and mix by pipetting (*see* **Note 38**).
2. Place the tube in the magnetic separation rack. Allow the beads to separate from the supernatant (*see* **Note 39**).
3. Using a pipette, remove the supernatant containing unbound DNA and discard.
4. Add 200 μL preheated 1× Stringent Wash Buffer and pipette up and down 10 times to mix. Incubate at 65 °C for 5 min.
5. Place the tube in the magnetic separation rack. Allow the beads to separate from the supernatant. Using a pipette, remove the supernatant containing unbound DNA and discard.
6. Add 200 μL preheated 1× Stringent Wash Buffer and pipette up and down 10 times to mix (*see* **Note 40**).
7. Incubate at 65 °C for 5 min.
8. Place the tube in the magnetic separation rack. Allow the beads to separate from the supernatant (*see* **Note 39**).
9. Using a pipette, remove the supernatant containing unbound DNA and discard.
10. Add 200 μL room temperature 1× Wash Buffer I and vortex for 2 min to mix.
11. Place the tube in the magnetic separation rack. Allow the beads to separate from the supernatant. Using a pipette, remove the supernatant and discard.
12. Add 200 μL room temperature 1× Wash Buffer II and vortex for 1 min to mix.

13. Place the tube in the magnetic separation rack. Allow the beads to separate from the supernatant. Using a pipette, remove the supernatant and discard.
14. Add 200 μL room temperature 1× Wash Buffer III and vortex for 30 s to mix.
15. Place the tube in the magnetic separation rack. Allow the beads to separate from the supernatant. Using a pipette, remove the supernatant and discard.
16. Remove the tube from the magnetic rack and add 20 μL nuclease-free molecular grade water to resuspend the beads.
17. Mix thoroughly by pipetting up and down 10 times (*see* **Note 34**).
18. Transfer the 20 μL reaction (with the beads) to a new 0.2 mL PCR tube.

3.3.8 Post-capture PCR

1. Prepare a PCR reaction mix by adding 5 μL PCR Primer Cocktail and 25 μL PCR master mix (TruSeq Stranded mRNA Sample Preparation Kit) to the 20 μL cDNA from **step** 7. Total volume is 50 μL.
2. Briefly vortex the mixture and quickly spin.
3. Place tube into a thermocycler and run the following program shown in Table 2 with a preheated lid set to 100 °C.

3.3.9 PCR Purification

1. Transfer the 50 μL PCR reaction to a new 1.7 mL low-binding tube.
2. Place the tube in a magnetic rack to bind the beads. Transfer the supernatant into a new 1.7 mL low-binding tube.
3. Add 40 μL (0.8× volume) Agencourt® AMPure® XP beads to the supernatant.
4. Mix well by pipetting up and down.
5. Incubate for 10 min at room temperature.

Table 2
Detailed PCR protocol for amplification of captured cDNA fragments

	Cycles	Temperature (°C)	Time
Polymerase activation	1	98	30 s
Amplification	12		
Denaturation		98	10 s
Annealing		60	30 s
Extension		72	30 s
Final extension	1	72	5 min
Hold	1	10	∞

6. Place on magnetic stand for 5 min.
7. Transfer supernatant into a new 1.7 mL low-binding tube.
8. Add 200 μL of fresh 80% (v/v) ethanol while the tube is still on the magnetic stand.
9. Remove supernatant by using a pipette.
10. Add 200 μL of fresh 80% (v/v) ethanol while the tube is still on the magnetic stand.
11. Remove supernatant by using a pipette.
12. Air dry the beads for 5–15 min while the tube is on the magnetic stand (*see* **Note 20**).
13. Elute DNA with 11 μL nuclease-free water.
14. Mix well by pipetting up and down and incubate for 4 min at room temperature.
15. Quickly spin and place the tube on the magnetic stand (*see* **Note 41**).
16. Transfer 10 μL to a new 0.2 mL PCR tube. This is the enriched NGS library.
17. This is a safe stopping point. The enriched may be stored at 4 °C for 1–2 weeks, or at −20 °C long term before sequencing.

3.4 Post-enrichment Quantification and Sequencing

1. Measure the DNA concentration for each bait capture reaction using the Qubit dsDNA HS Assay Kit or similar method to detect double-stranded DNA.
2. Send the bait capture reactions to a next-generation sequencing facility for Illumina sequencing (*see* **Note 42**).
3. To evaluate on-target, near-target, and off-target reads and the uniformity of coverage across pooled libraries (Picard Metrics), sequence each target capture pool on MiSeq using MiSeq v2 Nano Reagent Kit, with 10% PhiX spike-in.

3.5 Bioinformatic Analysis

1. Trim library adaptors off sequences using Trimmomatic.
2. Assembly: Assemble NGS reads into genomes using Galaxy platform.
3. Gap filling: Use PCR and Sanger sequencing to fill the genome gaps.
4. Annotation: Interrogate all genomes for ORFs using ORFfinder. Set the search parameters to ignore nested ORFs and filter out ORFs less than 150 bp. Select the standard genetic code and the "ATG only" rule. Identify each ORF and annotate through BLASTN and BLASX using the NCBI database.
5. Verification: Verify novel ORFs by read mapping or PCR re-sequencing.
6. Genome annotation:

a. Genome annotation: Assess read depth by mapping reads from direct or enriched NGS to their respective genomes using CLC Genomics Workbench 12 v12.0.

b. Calculate bait positions by aligning baits to each genome by BLASTN.

c. Prepare schematic diagrams of CoV genomes including bait position and read depth of NGS using Circos (v0.69.8).

d. Generate graphs displaying the data size and viral read ratio using Prism (GraphPad Prism 7).

4 Notes

1. RNA extraction can be done manually with Trizol or with any commercial extraction kit designed for viral RNA or total nucleic acid.
2. Ethanol should always be prepared fresh with nuclease-free molecular grade water.
3. This method was designed for the xGen Lockdown baits (Integrated DNA Technologies) but works for any hybridization bait set (Arbor Bioscience or other).
4. If it is difficult for the participant to expectorate, have them gently massage their cheeks to stimulate salivary flow.
5. If the patient is a small child, elderly, or a hyposalivator, they will have difficulty producing a sample. Alternative samples include an oral swab (e.g., FLOQSwab; Copan Italia SpA) placed in 1 mL viral transport media (VTM) or phosphate buffered solution (PBS). Saliva can also be pipetted from an intubated patient.
6. Use any sterile container large enough to easily spit into.
7. Salivary enzymes are still active at −80 °C. Therefore, it is essential to chill and process saliva as soon as possible, avoiding repeated freezing and thawing prior to stabilizing or purifying the nucleic acid, especially when working with RNA.
8. For viscous samples, centrifuge at 13,000 × *g*, 4 min, 4 °C and extract the supernatant.
9. Samples can be stored short term (i.e., 24–48 h) at 2–4 °C but must be stored long term at −80 °C.
10. Using the NanoDrop, the ideal 260/280 ratio for "pure" RNA is ~2.0. The 260/230 ratio provides a secondary measure of purity and should be in the range of 2.0–2.2. If the readings are significantly lower, it may be indicative of contamination with residual organic compounds from the extraction step.
11. There are many other options for adaptor sequences including Illumina and Nextera. It is important to discuss this protocol with your NGS facility of choice when deciding.

12. For the cDNA preparation, reagents used are largely from the TruSeq Stranded mRNA preparation kit.
13. All steps are performed at room temperature unless otherwise specified.
14. Optional: Include a negative/blank sample alongside library preparations to rule out any contamination.
15. If using Arbor Bioscience myBaits, always discuss library preparation and enrichment protocols with the manufacturer to ensure the proper reagents can be provided.
16. Do not try to process too many samples at once or the timing will be difficult to maintain.
17. Always remove the samples immediately after the thermocycler has cooled.
18. "Quickly spin" in this protocol denotes spinning the sample with a microcentrifuge or centrifuge just long enough to bring the sample to speed and then stop.
19. Keeping the sample at room temperature for 10 min will suffice.
20. The beads should look like they are starting to crack, but still slightly shiny. Leaving ethanol can inhibit downstream reactions, but the beads should not be overdried. Overdrying the beads will result in a dramatic loss in yield.
21. The procedure can be paused at this point.
22. This library contains viral and cellular derived DNA and is ready for enrichment using bait capture. The library can be sequenced at this point without enrichment. The library may also be split in half if enrichment success is to be tested. It is important to remember that if the library is split, the barcode is still the same, so enriched and unenriched samples cannot be sequenced on the same run.
23. This protocol also works for Arbor Biosciences myBaits as the principle is the same. Consult with Arbor Biosciences prior to experimentation.
24. Prepare the probes in small aliquots to avoid freezing and thawing. If bait pools (different viruses, for example) are regularly combined, freeze in the working solution.
25. Low-bind tubes are not essential, but when used the procedure is easier.
26. Alternatively, a centrifuge vacuum concentrator such as a Thermo Savant Speedvac can be used to dry down the sample.
27. Prepare the buffers during the 4 h incubation.
28. Streptavidin Dynabeads should be prepared immediately before use.

29. If sequencing multiple samples, streptavidin beads for up to 4 captures (400μL) can be combined into a single low-binding 1.7 mL centrifuge tube for the initial bead preparation.
30. To enhance bead separation, pipette buffer up and down once.
31. Each capture should be performed in its own low-binding tube.
32. Do not remove the supernatant until you are ready to add your hybridization sample.
33. Do not allow beads to dry out. Small amounts of remaining Bead Wash Buffer will not interfere with downstream binding of the DNA to beads.
34. While mixing the beads and the hybridization mixture, "scrape" the beads on the sides of the tube with a pipette tip.
35. All steps for the streptavidin bead wash should be performed at 65 °C to minimize nonspecific binding of the off-target DNA sequences to the capture probes. To maintain temperature, it is ideal to perform this step near the thermocycler and ThermoMixer.
36. Preheat empty 1.7 mL tubes (one for each capture reaction) to prevent drops in temperature.
37. Do not leave samples at room temperature. Keep tubes on the ThermoMixer set to 65 °C whenever possible.
38. The Wash Buffer was prepared in Sect. 3.3.4 and kept at 65 °C.
39. Bead separation should be immediate. To prevent temperature from dropping below 65 °C, quickly remove the clear supernatant with a pipette.
40. Pipette gently to prevent the formation of bubbles.
41. Due to the small elution volume, do not push the tubes all the way down into the magnetic rack. This will help keep the bead pellet intact during separation.
42. NGS methods are beyond the scope of this protocol. Please contact your chosen NGS facility and get their advice prior to attempting this protocol.

Acknowledgments

D.J.S. was supported by McMaster University's Michael G. DeGroote Initiative for Innovation in Healthcare. J.A.N. was supported by funds from the Comprehensive Antibiotic Resistance Database. D.E.A. is supported by a National Medical Research Council grant (COVID19RF2-0001). We thank Ben Tan Kiang Thong and Tanu Chawla (Duke-NUS Medical School, Singapore) for preparing Fig. 1 and Kathy Luinstra (St. Joseph's Healthcare Hamilton, Canada) for technical advice.

References

1. Liuzzi G, Nicastri E, Puro V, Zumla A, Ippolito G (2016) Zika virus in saliva-new challenges for prevention of human to human transmission. Eur J Intern Med 33:e20–e21. https://doi.org/10.1016/j.ejim.2016.04.022
2. Khurshid Z, Zohaib S, Joshi C, Moin SF, Zafar MS, Speicher DJ (2020) Saliva as a non-invasive sample for the detection of SARS-CoV-2: a systematic review. medRxiv:2020.2005.2009.20096354. https://doi.org/10.1101/2020.05.09.20096354
3. Speicher DJ, Wanzala P, D'Lima M, Johnson KE, Johnson NW (2015) Detecting DNA viruses in oral fluids: evaluation of collection and storage methods. Diagn Microbiol Infect Dis 82(2):120–127. https://doi.org/10.1016/j.diagmicrobio.2015.02.013
4. Coronaviridae Study Group of the International Committee on Taxonomy of Viruses (2020) The species Severe acute respiratory syndrome-related coronavirus: classifying 2019-nCoV and naming it SARS-CoV-2. Nat Microbiol 5(4):536–544. https://doi.org/10.1038/s41564-020-0695-z
5. Xu R, Cui B, Duan X, Zhang P, Zhou X, Yuan Q (2020) Saliva: potential diagnostic value and transmission of 2019-nCoV. Int J Oral Sci 12(1):11. https://doi.org/10.1038/s41368-020-0080-z
6. Xu H, Zhong L, Deng J, Peng J, Dan H, Zeng X et al (2020) High expression of ACE2 receptor of 2019-nCoV on the epithelial cells of oral mucosa. Int J Oral Sci 12(1):8. https://doi.org/10.1038/s41368-020-0074-x
7. Wyllie AL, Fournier J, Casanovas-Massana A, Campbell M, Tokuyama M, Vijayakumar P et al (2020) Saliva is more sensitive for SARS-CoV-2 detection in COVID-19 patients than nasopharyngeal swabs. medRxiv:2020.2004.2016.20067835. https://doi.org/10.1101/2020.04.16.20067835
8. To KK, Tsang OT, Leung WS, Tam AR, Wu TC, Lung DC et al (2020) Temporal profiles of viral load in posterior oropharyngeal saliva samples and serum antibody responses during infection by SARS-CoV-2: an observational cohort study. Lancet Infect Dis 20(5):565–574. https://doi.org/10.1016/S1473-3099(20)30196-1
9. Azzi L, Carcano G, Gianfagna F, Grossi P, Gasperina DD, Genoni A et al (2020) Saliva is a reliable tool to detect SARS-CoV-2. J Infect. https://doi.org/10.1016/j.jinf.2020.04.005
10. Su YCF, Anderson DE, Young BE, Linster M, Zhu F, Jayakumar J et al (2020) Discovery and genomic characterization of a 382-nucleotide deletion in ORF7b and ORF8 during the early evolution of SARS-CoV-2. mBio 11(4). https://doi.org/10.1128/mBio.01610-20
11. Gaudin M, Desnues C (2018) Hybrid capture-based next generation sequencing and its application to human infectious diseases. Front Microbiol 9:2924. https://doi.org/10.3389/fmicb.2018.02924
12. Hodges E, Xuan Z, Balija V, Kramer M, Molla MN, Smith SW et al (2007) Genome-wide in situ exon capture for selective resequencing. Nat Genet 39(12):1522–1527. https://doi.org/10.1038/ng.2007.42
13. Gnirke A, Melnikov A, Maguire J, Rogov P, LeProust EM, Brockman W et al (2009) Solution hybrid selection with ultra-long oligonucleotides for massively parallel targeted sequencing. Nat Biotechnol 27(2):182–189. https://doi.org/10.1038/nbt.1523
14. Nasir JA, Kozak RA, Aftanas P, Raphenya AR, Smith KM, Maguire F et al (2020) A comparison of whole genome sequencing of SARS-CoV-2 using amplicon-based sequencing, random hexamers, and bait capture. Viruses 12(8). https://doi.org/10.3390/v12080895
15. Phillippy AM, Deng X, Zhang W, Salzberg SL (2009) Efficient oligonucleotide probe selection for pan-genomic tiling arrays. BMC Bioinformatics 10:293. https://doi.org/10.1186/1471-2105-10-293
16. Campana MG (2018) BaitsTools: software for hybridization capture bait design. Mol Ecol Resour 18(2):356–361. https://doi.org/10.1111/1755-0998.12721
17. Metsky HC, Siddle KJ, Gladden-Young A, Qu J, Yang DK, Brehio P et al (2019) Capturing sequence diversity in metagenomes with comprehensive and scalable probe design. Nat Biotechnol 37(2):160–168. https://doi.org/10.1038/s41587-018-0006-x
18. Zhang Z, Schwartz S, Wagner L, Miller W (2000) A greedy algorithm for aligning DNA sequences. J Comput Biol 7(1–2):203–214. https://doi.org/10.1089/10665270050081478
19. Rouillard JM, Zuker M, Gulari E (2003) OligoArray 2.0: design of oligonucleotide probes for DNA microarrays using a thermodynamic approach. Nucleic Acids Res 31(12):3057–3062. https://doi.org/10.1093/nar/gkg426
20. Guitor AK, Raphenya AR, Klunk J, Kuch M, Alcock B, Surette MG et al (2019) Capturing the resistome: a targeted capture method to

reveal antibiotic resistance determinants in metagenomes. Antimicrob Agents Chemother 64(1). https://doi.org/10.1128/AAC.01324-19

21. Nasir JA, Speicher DJ, Kozak RA, Poinar HN, Millar MS, McArthur AG (2020) Rapid design of a bait capture platform for culture- and amplification-free next-generation sequencing of SARS-CoV-2. Preprints. https://doi.org/10.20944/preprints202002.0385.v1
22. Briese T, Kapoor A, Mishra N, Jain K, Kumar A, Jabado OJ et al (2015) Virome capture sequencing enables sensitive viral diagnosis and comprehensive virome analysis. mBio 6 (5):e01491–e01415. https://doi.org/10.1128/mBio.01491-15
23. Chalkias S, Gorham JM, Mazaika E, Parfenov M, Dang X, DePalma S et al (2018) ViroFind: a novel target-enrichment deep-sequencing platform reveals a complex JC virus population in the brain of PML patients. PLoS One 13(1):e0186945. https://doi.org/10.1371/journal.pone.0186945
24. Li B, Si HR, Zhu Y, Yang XL, Anderson DE, Shi ZL et al (2020) Discovery of bat coronaviruses through surveillance and probe capture-based next-generation sequencing. mSphere 5(1). https://doi.org/10.1128/mSphere.00807-19
25. Lim XF, Lee CB, Pascoe SM, How CB, Chan S, Tan JH et al (2019) Detection and characterization of a novel bat-borne coronavirus in Singapore using multiple molecular approaches. J Gen Virol 100(10):1363–1374. https://doi.org/10.1099/jgv.0.001307

Chapter 9

Assessing the Relationship Between Nitrate-Reducing Capacity of the Oral Microbiome and Systemic Outcomes

Charlene E. Goh, Bruno Bohn, and Ryan T. Demmer

Abstract

The significance of the oral microbiome in the generation of the nitric oxide (NO) via the enterosalivary nitrate-nitrite-nitric oxide pathway is increasingly recognized, directly linking the oral microbiome to cardiometabolic outcomes influenced by NO. The objective of this chapter is to outline a strategy of identifying pathway-specific bacterial taxa or predicted genes of interest from 16S rRNA data, specifically in the enterosalivary pathway of nitrate reduction, and analyzing their relationship with cardiometabolic outcomes using multivariable regression models.

Key words Oral microbiome, Nitrate, Nitrite, Nitric oxide pathway, Pathway activity prediction

1 Introduction

Nitric oxide (NO) is an important signaling molecule involved in many physiological processes and its deficiency has been implicated in the pathogenesis of hypertension and insulin resistance [1–3]. Thus, NO bioavailability has garnered much attention as an increase in NO production can potentially reduce high blood pressure and blood glucose levels. As NO was thought to be produced only by NO synthases in the endothelium, immune cells, and other tissues, previous research has focused on enhancing NO bioavailability produced through this synthase pathway.

However, nitrate resulting from NO oxidation metabolism or from dietary nitrate consumption has since been found to provide an important storage pool for NO. The physiological recycling of nitrate to produce NO takes place via the enterosalivary nitrate-nitrite-nitric oxide (NO_3-NO_2-NO) pathway. In this alternative pathway, oral bacteria play a crucial role by reducing salivary nitrate to nitrite, which is then swallowed and made systemically available for further reduction into NO in the blood vessels and tissues [1]. The direct role of oral bacteria in this NO_3-NO_2-NO pathway was demonstrated by several experimental studies that use

Guy R. Adami (ed.), *The Oral Microbiome: Methods and Protocols*, Methods in Molecular Biology, vol. 2327,
https://doi.org/10.1007/978-1-0716-1518-8_9,

antibacterial mouthwash to reduce oral bacteria, resulting in decreased nitrate-reducing capacity and a corresponding increase in blood pressure and plasma glucose [4–8]. Not all oral bacteria contribute to the reduction of salivary nitrate, and specific taxa with nitrate-reduction capacity have been identified [9, 10]. More recently, studies have observed a correlation between baseline nitrate-reducing bacteria abundance and differential blood pressure responses to dietary nitrate supplementation [11]. Associations of bacteria taxa and genes coding for bacterial enzymes involved along the NO_3-NO_2-NO pathway with blood pressure levels have also been observed [12, 13], further emphasizing the importance of oral microbiome composition in NO generation.

The enterosalivary pathway of NO generation has gained significant attention in recent years [14], and presents an alternative target for manipulation to improve NO bioavailability-associated cardiometabolic outcomes. While many have examined this pathway through the effects of dietary nitrate supplementation (i.e., increasing the storage pool of nitrate) on cardiometabolic outcomes [15, 16], fewer population-based studies have explored the association of the oral microbiome involved in the NO_3-NO_2-NO pathway with cardiometabolic outcomes.

Most microbiome analyses seek to identify differentially abundant taxa between disease states. The highly dimensional oral microbiome, with a large number of taxa to be compared, results in multiple comparisons and an increased false discovery rate [17]. The identification of a pathway-specific hypothesis linking the oral microbiome and cardiometabolic outcomes, such as the NO_3-NO_2-NO pathway, allows us to narrow our focus on specific bacteria or bacterial genes of interest, resulting in a priori hypothesis-driven analyses.

While whole genome shotgun metagenomic sequencing can directly yield information on the functional capacity of nitrate metabolism pathways in the oral microbiome, the relatively high cost of metagenomic sequencing and data handling may be prohibitive. Moreover, there are currently available 16S rRNA sequencing data within large cohorts that will yield valuable prospective data on the incidence of cardiometabolic outcomes and it would be a missed opportunity to not fully capitalize on those data.

The aim of this chapter is to provide explicit step-by-step examples demonstrating the identification of pathway-specific taxa or genes of interest, and their operationalization from 16S rRNA sequencing data. This exposure construct can then be leveraged in traditional statistical analysis workflows exploring the association between microbial nitrogen metabolism capacity and cardiometabolic outcomes.

2 Materials

2.1 Next-Generation High-Throughput 16S rRNA Sequencing

Sequence-based microbiome analysis consists of several steps: sample collection, storage, DNA extraction, library preparation, next-generation high-throughput microbial sequencing, quality control, sequence identification, and finally statistical analysis [18]. Briefly, bacteria from the samples collected are lysed and the DNA extracted. In library preparation, the 16S rRNA gene—the most commonly used marker gene in oral microbiome studies and the gold standard for sequence-based bacterial analyses—is amplified from the extracted DNA. These amplicons are then sequenced on the sequencing platform of choice (e.g., Illumina) to produce sequence reads. Quality control is then carried out to filter out short reads or sequences with lower quality scores before assigning the sequences to taxonomic classifiers. Several useful papers and references are available discussing the best practices and considerations at each stage to reduce the potential biases introduced [17–22].

2.2 Taxonomic Classification of Sequence Reads

In general, sequence reads obtained from sequencing are referenced against known microbial reference databases (e.g., SILVA, RDP, or Greengenes database) to assign a taxonomic identifier. A description of the different methods for sequence identification is beyond the scope of this chapter, and other papers provide an overview of the process [17]. Our previous analysis [12] utilized the Human Oral Microbiome Identification using Next-Generation Sequencing (HOMINGS) methodology specifically designed for the oral microbiome to generate species-level information, which uses a customized BLAST program (ProbeSeq for HOMINGS) [23, 24]. Other methods of taxonomic classification of 16S rRNA sequence reads are available, such as using operational taxonomic units (OTU) clustering [25] or the newer DADA2-corrected amplicon sequence variants (ASV) [26]. The HOMINGS methodology has been shown to be largely equivalent to the tree-based OTU clustering approach [27], but increasingly the use of ASVs is recommended as the standard unit of marker-gene analysis and reporting [17, 26]. Researchers have a choice of bioinformatics pipeline and reference database and are recommended to document the software versions used and all commands run [17].

2.3 Required Data for this Chapter

For this chapter, we will assume the following:

1. Standard 16S rRNA next-generation sequencing has been carried out on microbial DNA.
2. The necessary bioinformatics quality controls have been performed (e.g., filtering and trimming) [19–27].

Table 1
Truncated example of OTU/ASV output table after 16S rRNA sequencing and taxonomic alignment in the long format

Taxa	Relative_abundance	ID
Actinomyces_johnsonii	0.00001	1
Actinomyces_massiliensis	0.00389	1
Actinomyces_meyeri	0.00184	1
Actinomyces_naeslundii	0.05540	1
Actinomyces_odontolyticus	0.00001	1
...	...	...
Actinomyces_johnsonii	0.000008	2
Actinomyces_massiliensis	0.000823	2
...		2

Note the taxa is different in each row but with the same ID, and each ID will have relative abundance data for each individual taxon

Variable Key: *Taxa* refers to OTUs/ASVs relating to taxa at the species level; *ID* refers to the unique participant or sample ID. This dataset has one sample per participant; *Relative_abundance* is calculated from absolute counts of that taxa sequence divided by the total counts across all taxa in the individual sample

3. An appropriate technique for sequence inference and taxonomic alignment has been used.
4. A final table of OTUs/ASVs relating to taxa at the species level and their relative abundances in each sample has been produced. If using predicted gene abundances from 16S rRNA data, we will assume that PICRUSt2 analysis has been successfully conducted, using 16S data to infer KEGG ortholog abundances.

2.4 Required Datasets

1. Taxonomic table with relative abundances, such as shown in Table 1.
2. Metadata of participants, including the clinical outcomes of interest, e.g., systolic blood pressure, insulin resistance, such as shown in Table 2.
3. (Optional) Output from PICRUSt2 with KEGG ortholog functional abundance. (*See* Table 3 example)

2.5 Software

1. Statistical software to be used for analysis, e.g., SAS and R.
2. (Optional) PICRUSt2 or Piphillin software packages, if using predicted gene abundance from 16S rRNA sequences.

Table 2
Example of participant metadata containing cardiometabolic outcomes of interest in wide format

ID	Age	Sex	Glucose	MeanSBP	...
1	25	F	85	95.5	
2	31	F	78	116.5	
3	41	F	94	111.0	
4	30	M	78	123.5	
5	22	F	90	117.5	
...					

Study participants are not repeated across rows and each new variable (e.g., age, sex, mean systolic blood pressure) is represented as a new column variable
Summary scores (taxa or predicted gene-based) are added as a new column in the final dataset used in linear regression analyses
Variable Key: *ID* refers to the unique participant ID; *Glucose* refers to the fasting plasma glucose levels of the participant; *MeanSBP* refers to the mean systolic blood pressure of the participant

Table 3
Truncated example of PICRUSt2 output predicted absolute gene abundances using KEGG orthologs (KOs) conducted on 16S rRNA sequencing data, with KOs in rows and subject IDs in columns

KEGG_orthology	ID1	ID2	ID3	...
K00367	34	205	29	
K00370	10141	25010	31940	
K00371	10388	7487	10450	
...				

Variable Key: Column heads have *ID numbers*, referring to the unique participant ID. Each row has one KEGG Ortholog (KO), defined as functional orthologs containing groups of genes. Cells contain absolute counts of that KO in the participant's samples. Relative gene/KO abundances will be calculated before summing into a summary score

3 Methods

An important statistical issue in microbiome analysis is the high dimensionality of microbiome data which includes thousands of taxa. With the enterosalivary nitrate-nitrite-NO pathway of interest, the selection of certain bacteria a priori is possible based on existing knowledge [9, 10]. While individual taxa can be modeled one-by-one in regression models, statistical hypothesis testing may need to be adjusted for the false discovery rate, which reduces statistical power. In addition, since individual taxa analysis may fail to capture the many complex interactions between bacteria coexisting in a microbial community, a summary score can be a useful feature to give an overall picture of the microbiome community's nitrate-reducing capacity.

There are two general approaches that can be used to create a summary score from 16S sequencing data: (1) a taxa-based score, using taxa a priori identified to be associated with nitrate-reducing capacity; and (2) a predicted metagenomic (gene)-based score in which scores are based on the estimated number of genes relevant to nitrate reduction.

An advantage of the method of creating a summary score of bacteria taxa previously identified in the literature to be of importance is the specificity of the taxa selected. Numerous oral bacteria are thought to contain nitrate reductase genes, and incorporation of all bacteria containing any nitrate reductase gene into the exposure summary score may result in greater variability and noise, thus masking the effect of the important species and biasing the effect estimate toward the null.

An alternative to using taxa already associated with a pathway of interest is available. In situations where key bacterial species are not in the literature, but a functional gene(s) of interest has been identified (e.g., nitrate reductase), the use of predicted metagenomic content to operationalize the exposure may be useful. It should be noted, however, that both methods still rely on taxonomic classification from 16S rRNA marker gene sequencing and that microbial traits, such as horizontal gene transfer between bacteria or strain-level variation within species (e.g., differential nitrate-reduction capacity between strains), make misclassification of the individual's true nitrate-reducing capacity possible.

3.1 Summary Score of Bacterial Taxa Associated with Nitrate-Reducing Capacity

3.1.1 Identification of Bacteria Associated with Nitrate-Reducing Capacity

To identify the bacterial species of interest and create a summary score, a literature review was used to identify bacterial species associated with the nitrate-reducing capacity of oral microbiome samples [12]. Two reference papers were used to identify the bacterial species of interest in nitrate-reduction [9, 10]. Doel et al. used culture-based techniques to isolate and identify only nitrate reductase-positive bacteria [9], and all bacteria identified regardless of rate of nitrate-reduction was included. Additionally, Hyde et al. compared samples with high versus low nitrate-reducing capacity and used next-generation sequencing methods to identify species that were differentially abundant in the highest nitrate-reducing sample [10]. The latter sought to provide a whole community picture, including species indirectly helping in nitrate reduction; therefore, while most of the candidate species identified have a nitrate reductase gene, some like *P. melaninogenica* do not but contain a nitrite reductase gene instead. As our goal was to optimize the measurement of nitrate-reducing capacity, all species identified as candidate species, whether directly or indirectly contributing to nitrate-reducing capacity as "helper" species, were included in the summary score.

3.1.2 Operationalization Using the Arcsine-Square Root Transformation of Taxa Relative Abundance

From a taxonomic table of OTU/ASVs with absolute counts (i.e., counts of sequence hits), the relative abundance of each taxa is calculated by dividing the number of counts observed for that taxa sequence by the total counts across all taxa in the individual sample. The resulting relative abundance measure is a proportion (i.e., compositional) that is highly skewed and constrained to the range of zero to one with many zeros present (i.e., zero-inflated).

The arcsine-square root transformation has been widely used on taxa relative abundance to examine differentially abundant taxa between groups [28–33]. This transformation reduces the skewed distribution, creating a more normally distributed continuous variable that can range in the negative, stabilizing the variance, and allowing it to be effectively used in linear regression models. Unlike studies that use the microbiome as the dependent variable of interest, we use microbial relative abundance as the exposure or independent variable. Therefore, we do not employ statistical methods that model absolute counts, instead of relative abundance with zero-inflated Poisson [34] or negative binomial models [35], as others have. The arcsine-square root provides a simple transformation that can be easily performed in all software and is often used as a baseline comparison with the newly developed methods [36, 37].

3.1.3 Standardization and Creation of a Summary Score for Bacteria

Before summing the selected taxa in a summary score, standardization is first carried out. This gives equal weight to each taxon in the score, preventing very high relative abundance taxa from dominating the score. This is important especially when it is unknown whether the actual nitrate-reducing capacity of each taxon is directly correlated with its relative abundance. For example, it is plausible that nitrate-reducing capacity might vary by taxa. A sum of the relative abundances of taxa of interest without standardization would simply represent the total relative abundance of the selected bacteria in the sample. Therefore, the arcsine-square root transformed relative abundance of each taxa for an individual is standardized via division by the taxon's standard deviation across all the samples. The standardized values for the selected bacteria are then summed to create a summary score for each individual.

3.2 Summary Score of Predicted Metagenomic Genes from 16S rRNA Sequencing on the NO_3-NO_2-NO Pathways

3.2.1 Prediction of Metagenomic Content from 16S rRNA Sequencing

Bioinformatics tools can be used to predict metagenomic content from 16S rRNA marker gene sequencing. Piphillin and PICRUSt2 are the two most well-known tools for inferring metagenomic content [38–40].

These tools use the taxonomic identification, the relative abundances of the taxa, and a reference database of known bacteria genomes. The output is a functional-gene-count matrix, providing an estimate of the count of each functional gene in each sample. Comprehensive tutorials are available on how to use these tools [41].

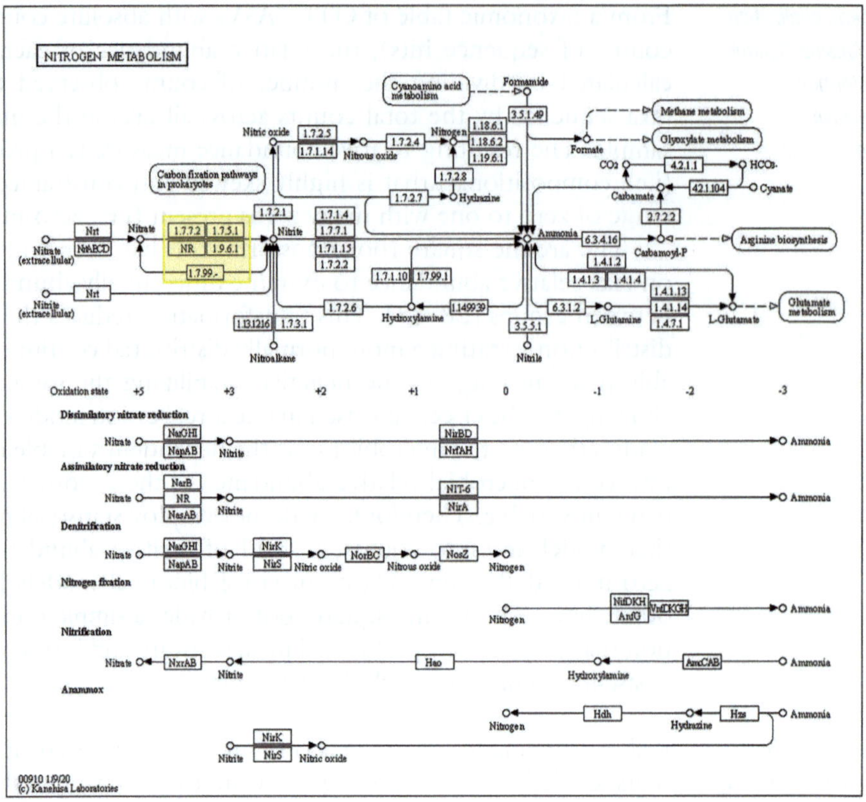

Fig. 1 KEGG pathway map00910—Nitrogen metabolism in bacteria. Publicly accessible from https:/www.genome.jp/kegg-bin/show_pathway?map00910. Nitrate-reduction enzymes are highlighted in the shaded rectangle and include EC 1.7.7.2, 1.7.5.1, 1.9.6.1, NR, and 1.7.99. Selecting nitrate reductase enzyme "EC 1.7.5.1," for example, brings up K00370, K00371, and K00374 as KO gene groups, containing *narG*, *narZ*, *nxrA*, *narH*, *narY*, *nxrB*, *narI*, and *narV* genes. KOs are defined as functional orthologs, and may consist of a single bacterial gene or genes from closely related species

3.2.2 Identification of Functional Gene Orthologs on the Enterosalivary Pathway

Genes of interest can be identified by searching the Kyoto Encyclopaedia of Genes and Genomes (KEGG) Pathways for the pathway of interest, in this case the bacterial nitrogen metabolism pathway (Fig. 1). Enzymes involved in each step of the NO_3-NO_2-NO pathway are mapped out. From Fig. 1, nitrate reduction involves nitrate reductase enzymes EC 1.7.7.2, 1.7.5.1, 1.9.6.1, NR, and 1.7.99. Selecting the respective enzyme, for example, EC 1.7.5.1, will show the associated KEGG Ortholog (KO)s—functional orthologs containing a group of bacterial genes coding for that molecular-level function. These groups of genes include nitrate reductase genes, such as *narG*, *narH*, *narI*, *napA*, *napB*, *narB*, *nasA*, *and nasB*. More details on the structure, organization, and uses of the KEGG encyclopaedia are available elsewhere [42].

3.2.3 Operationalization of Predicted Gene Abundance Summary Score

From bioinformatics tools PICRUSt2 and Piphillin, absolute counts of all possible KOs in each sample are output. As with taxa, the absolute counts of specific KOs can be converted into relative KO gene abundance by division with the total counts across all possible KOs in that individual sample. The relative gene abundances of interest are then added into a summary score and normalized using the arcsine-square root transformation as performed on the taxa relative abundance shown in Subheading 3.1.2.

Examining the NO_3-NO_2-NO pathway using predicted metagenomic content can be operationalized in several ways. As a start, the summary score containing all nitrate-reducing genes can be calculated. Our ongoing methodological work explores incorporating competition from other bacterial genes, such as nitrite reductase into the summary score [10], and creating summary scores based on bacterial metabolic pathways (e.g., respiratory denitrification) to further examine the different parts of the NO_3-NO_2-NO enterosalivary pathway in relation to cardiometabolic outcomes. It should, however, be emphasized that estimating metagenomic content from 16S rRNA sequencing does not directly measure the bacterial genes in the microbiome, and the specific strains present in the samples may not have the same functions as mapped in the bacteria reference database.

3.3 Analysis of the Association Between the Taxa and the Outcome of Interest

Multiple methods for analyzing microbiome data have been developed and the standards for microbiome analyses are rapidly evolving [17, 28, 29]. Developments include the use of alternative parameters, such as the change in ratios of taxa, to address biases introduced by comparing relative abundances between samples [43–45]. In this chapter, we analyze microbial relative abundance as the exposure or independent variable in linear regression models, adjusting for potential confounders. Patient selection criteria may be broader in population-based cross-sectional studies. Importantly, those who took antibiotics less than 30 days before recruitment were excluded, and microbiome studies often exclude those who report taking medications such as proton pump inhibitors. The patient selection criteria would likely be more restricted for smaller studies with a different study design, where control of confounding through design is necessary, or where smaller sample sizes limit the power for multivariate regression analysis.

3.4 Examples of Data Sets Used in the Analysis

See Tables 1, 2, and 3.

3.5 Code for Running Bacterial Taxa Summary Score Analyses in SAS

In this section, we provide step-by-step instruction for operationalizing a nitrate-reducing taxa summary score as discussed in Subheading 3.1. We also provide the SAS code for analyzing the created summary score with a cardiometabolic outcome of interest, as in our previous work using multivariable regression models [12, 46]. We have used CAPS for SAS keywords and mixed case for user-supplied text in keeping with typographical conventions used in SAS textbooks.

1. /* Import the OTU/ASV output from 16S rRNA sequencing and sequencing analysis with relative abundances */ (*see* **Note 1**).

```
PROC IMPORT DATAFILE= "C:\Users\CG\16SOTUrelativeabundance.
csv" OUT=microbiome DBMS=CSV REPLACE;
GETNAMES=YES;
RUN;
```

2. /* Import the participants metadata including Subject ID, socio-demographics, and clinical outcomes. Example of such a dataset is provided in Table 2 */

```
PROC IMPORT DATAFILE= "C:\Users\CG\Participantsclinicaldata.
csv" OUT=metadata DBMS=CSV REPLACE;
GETNAMES=YES;
RUN;
```

3. /* Arcsin-square root transformation on the relative abundance of each taxa */

```
DATA noarc;
SET microbiome;
transra= ARSIN (SQRT(relativeabundance));
RUN;
```

4. /* Creating a variable called ztransRA, which will be standardized in the next step */

```
DATA arcsinz;
SET noarc;
ztransRA=transRA;
RUN;
```

5. /* SAS function that carries out the *z* score standardization, creating a mean of 0 and standard deviation of 1 across all samples. The dataset has to first be sorted by taxa in order for the standardization to be performed correctly.*/

```
PROC SORT DATA= arcsinz;
BY taxa;
RUN;

PROC STANDARD DATA=arcsinz MEAN=0 STD=1 OUT=zscore;
VAR ztransRA;
BY taxa; /*see Note 2*/
RUN ;
```

6. /* Creating the nitrate-reducing taxa summary score "sumarcz" by including only the a priori identified taxa */

```
PROC SORT DATA= zscore;
BY id;
RUN;

PROC MEANS data=zscore NOPRINT ;
WHERE taxa in ("Actinomyces_naeslundii", "Actinomyces_odonto-
lyticus", "Actinomyces_viscosus", "Capnocytophaga_sputigena",
 "Corynebacterium_durum", "Corynebacterium_matruchotii", "Ei-
kenella_corrodens",
 "Haemophilus_parainfluenzae",
 "Neisseria_flavescens",
 "Neisseria_sicca",
 "Neisseria_subflava",
 "Prevotella_melaninogenica",
 "Prevotella_salivae",
 "Propionibacterium_acnes",
 "Rothia_dentocariosa",
 "Rothia_mucilaginosa",
 "Selenomonas_noxia",
 "Veillonella_dispar",
 "Veillonella_parvula",
 "Veillonella_atypica"); /*see Note 3*/
VAR ztransRA; OUTPUT OUT=sumzscore (drop= _TYPE_ _FREQ_)
sum(ztransRA)=sumarcz;
BY id;
RUN;
```

7. /* Merging the datasets to add the nitrate-reducing taxa summary score to the participant metadata */

```
PROC SORT data=sumzscore; BY id; RUN;
PROC SORT DATA= metadata; BY id; RUN;
```

```
DATA final;
MERGE sumzscore (IN=a) metadata (IN=b);
IF a AND b;
BY id;
RUN;
```

8. /* Multivariable regression models regressing systolic blood pressure outcome (MeanSBP) on the exposure summary score of nitrate-reducing capacity (sumarcz) controlling for other covariates—age, sex, ethnicity, body mass index (BMI), smoking status, etc.*/

```
PROC GENMOD DATA=final; /*see Note 4*/
CLASS sex(ref=last) ethnicity(ref="d Hispanic") smoking(re-
f="never");
MODEL MeanSBP=sumarcz age sex ethnicity bmi smoking /LINK=i-
dentity DIST=normal;
RUN;
```

3.6 Code for Running Bacterial Taxa Summary Score Analyses in R

The following sections generate results identical to those in Subheading 3.5, which creates a bacterial taxa summary score as described in Subheading 3.1; only now we are using R software, and provide the R code to perform the analyses.

1. # Import the OTU/ASV output from 16S rRNA sequencing and sequencing analysis with taxa relative abundances (*see* **Note 1**).

```
microbiome <- read.csv("C:\Users\CG\16SOTUrelativeabundance.
csv")
```

2. # Import the participants metadata including Subject ID, socio-demographics, and clinical outcomes.

```
metadata <- read.csv("C:\Users\CG\Participantsclinicaldata.
csv")
```

3. # Conduct the arcsin-square root transformation on the relative abundance of each taxa, using the function created above.

```
microbiome$arsin <- asin(sqrt(microbiome$relativeabundance))
```

4. # Conduct taxa-specific *z* score standardization, creating a mean of 0 and standard deviation of 1 across all samples (*see* **Note 5**).

```
microbiome$taxaZscore <- ave(x = microbiome$arsin, group =
microbiome$taxa, FUN = scale)
```

5. # Create vector with the selected taxa, and subset to include only these taxa (*see* **Note 3**).

```
taxaNames <- c("Actinomyces_naeslundii",
"Actinomyces_odontolyticus",
"Actinomyces_viscosus",
"Capnocytophaga_sputigena",
"Corynebacterium_durum",
"Corynebacterium_matruchotii",
"Eikenella_corrodens",
"Haemophilus_parainfluenzae",
"Neisseria_flavescens",
"Neisseria_sicca",
"Neisseria_subflava",
"Prevotella_melaninogenica",
"Prevotella_salivae",
"Propionibacterium_acnes",
"Rothia_dentocariosa",
"Rothia_mucilaginosa",
"Selenomonas_noxia",
"Veillonella_dispar",
"Veillonella_parvula",
"Veillonella_atypica")

taxaSubset <- microbiome[microbiome$taxa %in% taxaNames, ]
```

6. # Add the *z* scores for each sample, and create new data set with appropriate column names

```
sumZscore <- as.data.frame(aggregate(taxaSubset$taxaZscore,
by = taxaSubset$ID, FUN = sum)
names(sumZscore) <- c("ID", "sumarcz").
```

7. # Merge dataset with the nitrate-reducing taxa summary score to the metadata.

```
final <- merge(metadata, sumZscore, by = "ID")
```

8. # Multivariable regression model, regressing systolic blood pressure outcome (MeanSBP) on the exposure nitrate-reducing taxa summary score (sumarcz) controlling for other covariates.

```
fit <- lm(MeanSBP ~ sumarcz + age + sex + ethnicity + bmi +
smoking, data = final)

summary(fit)
```

3.7 Code for Running Predicted Gene Abundance Summary Score Analyses on SAS

In this section, we provide step-by-step instruction using SAS software to operationalize the PICRUSt2 output predicted gene abundance data, based on 16S rRNA data as discussed in Subheading 3.2 above. A nitrate-reducing gene abundance summary score is created and used as an exposure in multivariable linear regressions with a cardiometabolic outcome of interest-mean systolic blood pressure.

1. /* Import the output from PICRUSt2 analysis using KEGG Orthologs (KOs), conducted on 16S rRNA sequencing data. This data set has KOs in the rows and Subject IDs as columns. See Table 3 */

```
PROC IMPORT DATAFILE= "C:\Users\CG\PICRUSt2OutputKO.csv"
OUT=picrust2_wide DBMS=CSV REPLACE;
RUN;
```

2. /* Transpose PICRUSt2 output using the SAS procedure PROC TRANSPOSE. The NAME=ID command is to create a new variable containing all the participants IDs previously li sted as columns */ (*see* **Note 6**).

```
PROC TRANSPOSE DATA=picrust2_wide OUT=picrust2_long NAME=ID;
ID kegg_orthology;
RUN;
```

3. /* Create a variable "TotalCounts" that records the total counts of all KOs per sample. Be sure to check the names of the first and last KO variables that appear in the data set (in this example, K00360 and K15876) */

```
DATA picrust2v2;
SET picrust2_long;
TotalCounts= SUM (OF K00360--K15876); /* see Note 7*/
RUN;
```

4. /* Create a subset of data keeping only Subject ID, TotalCounts, and KOs related to nitrate reduction genes */

```
DATA picrustNO3only;
SET picrust2v2;
KEEP ID TotalCounts
K00367 K00370 K00371 K00374 K02567 K02568 K00372 K00360; /*see
Note 8*/
RUN;
```

5. /* Calculating relative abundances of the individual predicted genes by division with the "TotalCounts" of the sample*/

```
DATA picrustNO3only;
SET picrustNO3only;
K00367_rel= K00367/TotalCounts;
K00370_rel=K00370/TotalCounts;
K00371_rel=K00371/TotalCounts;
K00374_rel= K00374/TotalCounts;
K02567_rel= K02567/TotalCounts;
K02568_rel= K02568/TotalCounts;
K00372_rel=K00372/TotalCounts;
K00360_rel=K00360/TotalCounts;
RUN;
```

6. /* Create a NO_3 reduction relative gene abundance summary score, "NO3_rel", by adding the predicted relative abundances of genes involved in nitrate reduction.*/

```
DATA picrustNO3only;
SET picrustNO3only;
NO3_rel= K00367_rel + K00370_rel + K00371_rel + K00374_rel +
K02567_rel + K02568_rel + K00372_rel + K00360_rel;
RUN;
```

7. /* Arcsin-square root transformation on the NO_3 reduction gene abundance summary score */

```
DATA picrustNO3only;
SET picrustNO3only;
NO3_arsin=arsin(sqrt(NO3_rel));
RUN;
```

8. /* Import the participants metadata including Subject ID, socio-demographics, and clinical outcomes. Example of such a dataset is provided in Table 2 */

```
PROC IMPORT DATAFILE= "C:\Users\CG\Participantsclinicaldata.
csv" OUT=metadata DBMS=CSV REPLACE;
GETNAMES=YES;
RUN;
```

9. /* Merging the datasets to add the NO_3 reduction gene abundance summary score to the participant metadata */

```
PROC SORT DATA=picrustNO3only; BY id; RUN;
PROC SORT DATA= metadata; BY id; RUN;

DATA final;
MERGE picrustNO3only (IN=A) metadata (IN=B);
IF A AND B;
BY id;
RUN;
```

10. /* Multivariable regression model, regressing systolic blood pressure outcome (MeanSBP) on the exposure, NO_3 reduction gene abundance summary score (NO3_arsin), controlling for other covariates */

```
PROC GENMOD DATA=final;
 CLASS sex(ref=last) raceethn(ref="d Hispanic") cigcurr(re-
f="never");
 MODEL MeanSBP=NO3_arsin age sex raceethn bmi cigcurr /
LINK=identity DIST=normal;
RUN;
```

3.8 Code for Running Predicted Gene Abundance Summary Score Analyses on R

The following section generates results identical to those in Subheading 3.7, using predicted gene abundance data output from PICRUSt2 to create a nitrate-reducing gene abundance summary score, as described in Subheading 3.2; only now we are using R software, and provide the R code to perform the analyses.

1. # Import the output from PICRUSt2 analysis using KEGG Orthologs (KOs), conducted on 16S rRNA sequencing data. This data set has KOs in the rows and Subject IDs as columns.

```
PICRUSt2_wide <- read.csv("C:\Users\CG\PICRUSt2OutputKO.
csv")
```

2. # Transpose PICRUSt2 output.

```
PICRUSt2_long <- gather(data = PICRUSt2_wide, key = ID, value
= Count, 2:ncol(PICRUSt2_wide))
```

3. # Record total number of reads per sample as a new data set, and label columns appropriately.

```
Total <- as.data.frame(aggregate(PICRUSt2_long$Count, by =
list(PICRUSt2_long$ID), FUN = sum)
names(Total) <- c("ID", "TotalCounts")
```

4. # Create vector with the KOs of interest. Using the approach outlines above, we will select the KOs corresponding to enzymes involved in nitrate reduction to nitrite (*see* **Note 8**).

```
KOs <- c(
"K00367", #Corresponds to the gene narB
"K00370", #Corresponds to the genes narG, narZ, and nxrA
"K00371", #Corresponds to the genes narH, narY, and nxrB
"K00374", #Corresponds to the genes narI and narV
"K02567", #Corresponds to the gene napA
"K02568", #Corresponds to the gene napB
"K00372", #Corresponds to the gene nasA
"K00360", #Corresponds to the gene nasB
)
```

5. # Create a subset of data, including only the KOs selected above.

```
NO3 <- PICRUSt2_long[PICRUSt2_long$KEGG_orthology %in% KOs,]
```

6. # Convert NO_3 data set to long format.

```
NO3 <- spread(NO3, key = KEGG_orthology, value = Count)
```

7. Add the total sample counts to the NO_3 data set.

```
NO3 <- merge(NO3, Total, by = "ID")
```

8. # Transform absolute abundances into relative abundances, by dividing each value by the sample total (*see* **Note 9**).

```
NO3$K00367_rel <- NO3$K00367/ NO3$Total
NO3$K00370_rel <- NO3$K00370/ NO3$Total
NO3$K00371_rel <- NO3$K00371/ NO3$Total
NO3$K00374_rel <- NO3$K00374/ NO3$Total
NO3$K02567_rel <- NO3$K02567/ NO3$Total
NO3$K02568_rel <- NO3$K02568/ NO3$Total
NO3$K00372_rel <- NO2$K00372/ NO3$Total
NO3$K00360_rel <- NO2$K00360/ NO3$Total
```

9. # Create a NO_3 reduction relative gene abundance summary score "NO3_rel", by adding the predicted relative abundances of genes involved in nitrate reduction.

```
NO3$NO3_rel <- NO3$K00367_rel + NO3$K00370_rel + NO3
$K00371_rel + NO3$K00374_rel + NO3$K02567_rel + NO3
$K02568_rel + NO3$K00372_rel + NO3$K00360_rel
```

10. # Conduct the arcsin-square root transformation on the NO_3 reduction gene abundance summary score created.

```
NO3$NO3_arsin <- asin(sqrt (NO3$NO3_rel))
```

11. # Import the participants metadata including Subject ID, socio-demographics, and clinical outcomes.

```
metadata <- read.csv("C:\Users\CG\Participantsclinicaldata.
csv")
```

12. # Merge data sets with inferred metagenome proportions to the metadata.

```
final <- merge(metadata, NO2, by = "ID")
```

13. # Multivariable regression model, regressing systolic blood pressure outcome (MeanSBP) on the exposure, NO_3 reduction gene abundance summary score (NO3_arsin), controlling for other covariates.

```
fit <- lm(meansbp ~ NO3_arsin + age + sex + raceethn + bmi +
cigcurr, data = final)

summary(fit)
```

4 Notes

1. Often the output after 16S rRNA sequencing and sequence assignment will be in the long format, with multiple rows representing different OTUs/ASVs for the same subject. The code assumes this format as shown in Table 1.
2. It is important to perform the standardization by taxa. The command "BY taxa" tells SAS to use the standard deviation of that particular taxa across all samples for the standardization.
3. From a list of 28 putative taxa associated with nitrate-reducing species derived from prior literature [9, 10], only 20 were identified in the ORIGINS data [12]. These earlier studies looked at bacteria from tongue dorsum, supragingival plaque, and/or saliva.
4. PROC GENMOD is the SAS procedure fitting a generalized linear model to the data by maximum likelihood estimation of the parameter vector β. The "LINK=identity DIST=normal" indicates that a linear model with a normal distribution and continuous response variable is being used.
5. There are many different Z-score standardization functions in R, including the *scale* function in the base package. However, since we need to conduct a taxa-specific standardization, we use the *ave* function to apply the *scale* function to groups of data.
6. More information on PROC TRANSPOSE can be found at https://support.sas.com/resources/papers/proceedings/proceedings/forum2007/046-2007.pdf
7. The procedure SUM(OF K00360--K15876) tells SAS to sum over a range of variables in the order they are listed in the dataset. Therefore, K00360 is the first column variable and K15876 the last variable to be included in the count.
8. The list of KOs included in the nitrate reductase gene abundance summary score as derived from the KEGG pathway map are K00367, K00370, K00371, K00374, K02567, K02568,

K00372, and K00360. K10534, which corresponds to NR in the KEGG pathway, was not present in our dataset. The KOs present were selected based on their identification in KEGG as being directly involved in conversion of nitrate to nitrite (*see* Fig. 1).

9. Note that this step was done individually for each taxon, but the "for loop" approach in R can also be used to conduct the standardization.

References

1. Lundberg JO, Weitzberg E, Gladwin MT (2008) The nitrate-nitrite-nitric oxide pathway in physiology and therapeutics. Nat Rev Drug Discov 7(2):156–167
2. Sansbury BE, Hill BG (2014) Regulation of obesity and insulin resistance by nitric oxide. Free Radic Biol Med 73:383–399
3. Koch CD, Gladwin MT, Freeman BA, Lundberg JO, Weitzberg E, Morris A (2017) Enterosalivary nitrate metabolism and the microbiome: intersection of microbial metabolism, nitric oxide and diet in cardiac and pulmonary vascular health. Free Radic Biol Med 105:48–67
4. Beals JW, Binns SE, Davis JL, Giordano GR, Klochal AL, Paris HL et al (2017) Concurrent beet juice and carbohydrate ingestion: influence on glucose tolerance in obese and nonobese adults. J Nutr Metab 2017:6436783
5. Govoni M, Jansson EA, Weitzberg E, Lundberg JO (2008) The increase in plasma nitrite after a dietary nitrate load is markedly attenuated by an antibacterial mouthwash. Nitric Oxide 19:333–337
6. Woessner M, Smoliga JM, Tarzia B, Stabler T, Van Bruggen M, Allen JD (2016) A stepwise reduction in plasma and salivary nitrite with increasing strengths of mouthwash following a dietary nitrate load. Nitric Oxide 54:1–7
7. Kapil V, Haydar SM, Pearl V, Lundberg JO, Weitzberg E, Ahluwalia A (2013) Physiological role for nitrate-reducing oral bacteria in blood pressure control. Free Radic Biol Med 55:93–100
8. Bescos R, Ashworth A, Cutler C, Brookes ZL, Belfield L, Rodiles A et al (2020) Effects of chlorhexidine mouthwash on the oral microbiome. Sci Rep 10(1):5254
9. Doel JJ, Benjamin N, Hector MP, Rogers M, Allaker RP (2005) Evaluation of bacterial nitrate reduction in the human oral cavity. Eur J Oral Sci 113(1):14–19
10. Hyde ER, Andrade F, Vaksman Z, Parthasarathy K, Jiang H, Parthasarathy DK et al (2014) Metagenomic analysis of nitrate-reducing bacteria in the oral cavity: implications for nitric oxide homeostasis. PLoS One 9(3):e88645
11. Vanhatalo A, Blackwell JR, L'Heureux JE, Williams DW, Smith A, van der Giezen M et al (2018) Nitrate-responsive oral microbiome modulates nitric oxide homeostasis and blood pressure in humans. Free Radic Biol Med 124:21–30
12. Goh CE, Trinh P, Colombo PC, Gerkinger JM, Mathema B, Uhlemann A-C et al (2019) Association between nitrate-reducing oral bacteria and cardiometabolic outcomes: results from ORIGINS. J Am Heart Assoc 8(23):e013324
13. Tribble GD, Angelov N, Weltman R, Wang B-Y, Eswaran SV, Gay IC et al (2019) Frequency of tongue cleaning impacts the human tongue microbiome composition and enterosalivary circulation of nitrate. Front Cell Infect Microbiol 9:39
14. Kapil V, Khambata RS, Jones DA, Rathod K, Primus C, Massimo G et al (2020) The noncanonical pathway for in vivo nitric oxide generation: the nitrate-nitrite-nitric oxide pathway. Pharmacol Rev 72(3):692–766
15. Jackson JK, Zong G, MacDonald-Wicks LK, Patterson AJ, Willett WC, Rimm EB et al (2019) Dietary nitrate consumption and risk of CHD in women from the Nurses' Health Study. Br J Nutr 121(7):831–838
16. Jackson JK, Patterson AJ, MacDonald-Wicks LK, Oldmeadow C, McEvoy MA (2018) The role of inorganic nitrate and nitrite in cardiovascular disease risk factors: a systematic review and meta-analysis of human evidence. Nutr Rev 76(5):348–371
17. Knight R, Vrbanac A, Taylor BC, Aksenov A, Callewaert C, Debelius J et al (2018) Best

practices for analysing microbiomes. Nat Rev Microbiol 16(7):410–422
18. Tyler AD, Smith MI, Silverberg MS (2014) Analyzing the human microbiome: a "how to" guide for physicians. Am J Gastroenterol 109(7):983–993
19. Pollock J, Glendinning L, Wisedchanwet T, Watson M (2018) The madness of microbiome: attempting to find consensus "best practice" for 16S microbiome studies. Appl Environ Microbiol 84(7):e02627–e02617
20. Willis JR, Gabaldón T (2020) The human oral microbiome in health and disease: from sequences to ecosystems. Microorganisms 8 (2):308
21. Morgan XC, Huttenhower C (2012) Chapter 12: Human microbiome analysis. PLoS Comput Biol 8(12):e1002808
22. Sinha R, Abu-Ali G, Vogtmann E, Fodor AA, Ren B, Amir A et al (2017) Assessment of variation in microbial community amplicon sequencing by the microbiome quality control (MBQC) project consortium. Nat Biotechnol 35(11):1077–1086
23. Gomes BP, Berber VB, Kokaras AS, Chen T, Paster BJ (2015) Microbiomes of endodontic-periodontal lesions before and after chemomechanical preparation. J Endod 41 (12):1975–1984
24. Mougeot JL, Stevens CB, Cotton SL, Morton DS, Krishnan K, Brennan MT et al (2016) Concordance of HOMIM and HOMINGS technologies in the microbiome analysis of clinical samples. J Oral Microbiol 8:30379
25. Kopylova E, Navas-Molina JA, Mercier C, Xu ZZ, Mahe F, He Y et al (2016) Open-source sequence clustering methods improve the state of the art. mSystems 1(1):e00003
26. Callahan BJ, McMurdie PJ, Holmes SP (2017) Exact sequence variants should replace operational taxonomic units in marker-gene data analysis. ISME J 11(12):2639–2643
27. Palmer RJ, Cotton SL, Kokaras A, Gardner P, Grisius M, Pelayo E et al (2019) Analysis of oral bacterial communities: comparison of HOMINGS with a tree-based approach implemented in QIIME. J Oral Microbiol 11(1):1586413
28. Tsilimigras MC, Fodor AA (2016) Compositional data analysis of the microbiome: fundamentals, tools, and challenges. Ann Epidemiol 26(5):330–335
29. Li H (2015) Microbiome, metagenomics, and high-dimensional compositional data analysis. Annu Rev Stat Appl 2:73–94
30. Morgan XC, Kabakchiev B, Waldron L, Tyler AD, Tickle TL, Milgrom R et al (2015) Associations between host gene expression, the mucosal microbiome, and clinical outcome in the pelvic pouch of patients with inflammatory bowel disease. Genome Biol 16:67
31. Morgan XC, Tickle TL, Sokol H, Gevers D, Devaney KL, Ward DV et al (2012) Dysfunction of the intestinal microbiome in inflammatory bowel disease and treatment. Genome Biol 13(9):R79
32. Gevers D, Kugathasan S, Denson LA, Vazquez-Baeza Y, Van Treuren W, Ren B et al (2014) The treatment-naive microbiome in new-onset Crohn's disease. Cell Host Microbe 15(3):382–392
33. Zhou W, Sailani MR, Contrepois K, Zhao Y, Ahadi S, Leopold SR et al (2019) Longitudinal multi-omics of host–microbe dynamics in prediabetes. Nature 569(7758):663–671
34. Xu T, Demmer RT, Li G (2020) Zero-inflated Poisson factor model with application to microbiome read counts. Biometrics 77:91–101
35. Chen J, King E, Deek R, Wei Z, Yu Y, Grill D et al (2018) An omnibus test for differential distribution analysis of microbiome sequencing data. Bioinformatics 34(4):643–651
36. Ho NT, Li F, Wang S, Kuhn L (2019) metamicrobiomeR: an R package for analysis of microbiome relative abundance data using zero-inflated beta GAMLSS and meta-analysis across studies using random effects models. BMC Bioinformatics 20(1):188
37. Chen EZ, Li H (2016) A two-part mixed-effects model for analyzing longitudinal microbiome compositional data. Bioinformatics 32 (17):2611–2617
38. Narayan NR, Weinmaier T, Laserna-Mendieta EJ, Claesson MJ, Shanahan F, Dabbagh K et al (2020) Piphillin predicts metagenomic composition and dynamics from DADA2-corrected 16S rDNA sequences. BMC Genomics 21 (1):56
39. Langille MG, Zaneveld J, Caporaso JG, McDonald D, Knights D, Reyes JA et al (2013) Predictive functional profiling of microbial communities using 16S rRNA marker gene sequences. Nat Biotechnol 31 (9):814–821
40. Douglas GM, Maffei VJ, Zaneveld J, Yurgel SN, Brown JR, Taylor CM et al (2019) PICRUSt2: an improved and extensible approach for metagenome inference. bioRxiv. https://doi.org/10.1101/672295
41. PICRUSt2 Tutorial (v2.1.4 beta) (2019) https://github.com/picrust/picrust2/wiki/PICRUSt2-Tutorial-(v2.1.4-beta). Accessed 2 Nov 2020

42. Kanehisa M, Furumichi M, Tanabe M, Sato Y, Morishima K (2016) KEGG: new perspectives on genomes, pathways, diseases and drugs. Nucleic Acids Res 45(D1):D353–D361
43. McLaren MR, Willis AD, Callahan BJ (2019) Consistent and correctable bias in metagenomic sequencing experiments. elife 8:e46923
44. Williamson BD, Hughes JP, Willis AD (2019) A multi-view model for relative and absolute microbial abundances. bioRxiv. https://doi.org/10.1101/761486
45. Morton JT, Marotz C, Washburne A, Silverman J, Zaramela LS, Edlund A et al (2019) Establishing microbial composition measurement standards with reference frames. Nat Commun 10(1):2719
46. Demmer RT, Trinh P, Rosenbaum M, Li G, LeDuc C, Leibel R et al (2019) Subgingival microbiota and longitudinal glucose change: the oral infections, glucose intolerance and insulin resistance study (ORIGINS). J Dent Res 98:1488–1496

Chapter 10

Identification of Oral Bacterial Biosynthetic Gene Clusters Associated with Caries

Jonathon L. Baker and Anna Edlund

Abstract

Small molecules are a primary communication media of the microbial world, and play crucial, yet largely unidentified, roles in microbial ecology and disease pathogenesis. Many small molecules are produced by biosynthetic gene clusters, which can be predicted and analyzed computationally given a genome. A recent study examined the biosynthetic repertoire of the oral microbiome and cross-referenced this information against the disease status of the human host, providing leads for biosynthetic gene clusters, and their natural products, which may be key in the oral microbial ecology affecting dental caries and periodontitis. This chapter provides a step-by-step tutorial to bioinformatically to locate biosynthetic gene clusters within genomes, predict the type of natural products that are produced, and cross-reference the identified biosynthetic gene clusters to microbiomes associated with disease or health.

Key words Oral microbiome, Bioinformatics, Biosynthetic gene clusters, Mutacin, *Streptococcus mutans*, Dental caries

1 Introduction

Small molecules (SMs) are incredibly structurally and functionally diverse, and represent a crucial communication medium of the microbial world [1]. Many bacterially produced SMs are constructed by biosynthetic gene clusters (BGCs). These BGCs are frequently organized into operon modules within the genome, and work as assembly lines to produce and tailor the final SMs. SMs have been documented to serve a variety of roles in microbe-microbe and host-microbe interactions, including biofilm formation, immune modulation, stress protection, the killing of competitors, and many more [1]. Common classes of SMs produced by BGCs include lipids and glycolipids, oligosaccharides, polyketides, terpenoids, and nonribosomal peptides. The BGCs themselves are frequently grouped into classes based upon the chemical structure of their SM products; for example, many oligosaccharide-producing BGCs have a similar sequence and gene layout

Guy R. Adami (ed.), *The Oral Microbiome: Methods and Protocols*, Methods in Molecular Biology, vol. 2327,
https://doi.org/10.1007/978-1-0716-1518-8_10,

[1]. Therefore, using computational tools, an investigator can predict the presence of BGCs and the putative structural class of their products given only a genome sequence [2–4]. This development has been particularly significant for drug discovery, as a large portion of the drugs on the market are natural products of microbial origin. Several studies in recent years have investigated bacterial genomes from the human microbiome and have discovered a vast number of putative BGCs, which are predicted to produce many novel SMs, including antibiotics [4–7]. Furthermore, the abundance and/or expression of these BGCs can be examined in the context of health-associated versus disease-associated microbiomes [7, 8]. This cross-referencing provides leads as to specific BGCs that may play roles in inducing or preventing disease. While recently developed tools and techniques have greatly accelerated BGC discovery and elucidation, the vast majority of BGCs and their products remain uncharacterized. Therefore, their roles in microbial ecology and their potential to contribute to, or prevent, disease pathogenesis remain unknown.

Dental caries is the most common chronic infectious disease worldwide [9]. It is caused by a disruption of the normal, health-associated microbial ecology (i.e., dysbiosis) within the dental plaque adjacent to the tooth surface [9–11]. Acid-producing and acid-tolerant bacteria drop the local pH such that the tooth enamel is demineralized, which will lead to irreparable damage to the tooth if the process continues unchecked [10, 11]. Historically, *Streptococcus mutans* has received the most attention as a caries-causing pathogen, due to its well-characterized abilities to produce and tolerate acid, its exceptional capacity to form biofilms in the presence of sucrose (table sugar), and its ability to cause disease in animal models [12]. Knowledge regarding the caries-associated oral microbiota was significantly improved with the development of culture-independent detection methods, such as second- and third-generation sequencing techniques [13–15]. Caries is now understood to have a complex etiology and be multifactorial, with both bacterial and host factors playing important roles in development or prevention of the disease [9, 16]. Although *S. mutans* is associated with caries in many cases, caries does occur in the absence of detectable levels of *S. mutans.* The other species involved, and how the overall dental plaque ecology functions in situ and contributes to pathogenesis, remain poorly understood [17, 18]. As it has become increasingly clear that caries is the result of ecological changes, examination of the BGCs and their SM products in the oral microbiota is likely to yield useful insights into caries pathogenesis, and provide promising leads for development of novel anti-caries therapeutics.

In this chapter, analysis tools and a pipeline to mine bacterial genomes for BGCs, and subsequently cross-reference these BGCs with metadata, such as disease status, is described. The code provided in this tutorial will identify the BGCs encoded by three strains of *S. mutans* with publicly available and complete genomes, UA159 [19], UA140 [20], and NN2025 [21], and subsequently correlate these BGCs to dental caries versus health across a recently published dataset of 45 oral metagenomes [7]. *S. mutans* has long been considered a primary etiologic agent of dental caries and has a relatively diverse pangenome which encodes several hundred BGCs [22, 23]. UA159 is very well studied and represents the *S. mutans* species archetype strain [19]. UA140 is relatively well studied and is known to produce several bacteriocins (termed mutacins in *S. mutans*) including the lantibiotic Mutacin I, which is included in the MI-BiG database of experimentally characterized BGCs [20, 24]. Finally, NN2025 encodes a BGC recently described to produce the tetramic acids, mutanocyclin and reutericyclin [25, 26]. *S. mutans* is known to utilize its BGCs to produce ribosomally synthesized, posttranslationally modified peptide (RiPP) mutacins [27], as well as reutericyclin [25], to inhibit the growth of its competitors, which are largely associated with good dental health. The underpinnings of this ecological battle and its relationship to disease are topics of current research by several groups. The BGC analysis pipeline described below is a slightly modified and updated version of the one utilized by the authors in two recent publications [7, 23]. Several other tools that are not discussed or used here are available to mine and analyze BGCs (*see* **Note 1**).

2 Materials

The computational resources needed for the pipeline described in this chapter will vary greatly depending on both the step in the pipeline and how many genomes/BGCs are being queried (*see* **Note 2**). The authors tested the tutorial pipeline on both a MacBook Pro 2.8 GHz Quad Core running MacOS 10.15.6 with 16 GB of RAM and 256 GB of storage (laptop) and on a Linux server running CentOS 6.10 with 1000 GB RAM and 64 cores. In the interest of execution time, the tutorial code provided here assumes a large computing cluster (1000+ GB RAM, 64+ cores), which is why many of the commands use options with large numbers of cores and memory.

2.1 Bioinformatics Tools

1. Python (https://www.python.org).
2. R (https://www.r-project.org).
3. Conda (https://docs.conda.io/en/latest/).
4. git 2.17.1 (https://git-scm.com).
5. antiSMASH 5.1.2 (https://antismash.secondarymetabolites.org/#!/start) [3].
6. MultiGeneBlast 1.1.0 (http://multigeneblast.sourceforge.net) [28].

7. MI-BiG Database 2.0 (https://mibig.secondarymetabolites.org) [24].
8. numpy (https://numpy.org).
9. scipy (https://www.scipy.org).
10. scikit-learn (https://scikit-learn.org/stable/).
11. hmmer (http://hmmer.org).
12. biopython (https://biopython.org).
13. fasttree (http://www.microbesonline.org/fasttree/).
14. networkx (https://networkx.github.io).
15. pfam database (ftp://ftp.ebi.ac.uk/pub/databases/Pfam/current_release/Pfam-A.hmm.gz) [29].
16. BiG-SCAPE 1.0 (https://git.wur.nl/medema-group/BiG-SCAPE) [30].
17. BWA (https://github.com/lh3/bwa) [31].
18. DeSeq2 (https://bioconductor.org/packages/release/bioc/html/DESeq2.html) [32].
19. NCBI Genome Download (https://github.com/kblin/ncbi-genome-download).
20. KneadData (https://bitbucket.org/biobakery/kneaddata/wiki/Home).

3 Methods

3.1 Installation of Tools

This section describes the installation of the main modules of the BGC analysis pipeline to identify and annotate BGCs within genome(s) of interest and subsequently examine metagenomic or metatranscriptomic datasets for representation and/or expression of the BGCs. The bioinformatics tools used in this pipeline utilize the Python and/or R programming languages, which can be installed from python.org and r-project.org, respectively. Use of an Integrated Development Environment (IDE), such as Microsoft Visual Studio Code and/or RStudio, is highly recommended. In the installation process described here, the Conda package and dependency management system is used to install and manage dependency environments for NCBI Genome Download, the antibiotics and secondary metabolites analysis shell (antiSMASH), MultiGeneBlast, and Biosynthetic Genes Similarity Clustering and Prospecting Engine (BiG-SCAPE). Conda is highly recommended as it quickly installs, runs, and updates packages and their dependencies, allowing the user to switch between environments quickly. For example, antiSMASH uses Python3, while MultiGeneBlast uses Python2. Both tools also have differing Python dependencies. With just two commands, a user can switch between the working environments of each tool. The Conda package manager is included with installation of either Anaconda (full, open-source, high-performance, and optimized Python and R distribution) or Miniconda (minimal installer for Conda), both available at conda.io. This pipeline also makes use of Git, which is a free and open-source version control system, and is available at git-scm.com. Required computational resources are discussed in **Note 2**.

3.1.1 Conda Installation of NCBI Genome Download Package and Downloading of *S. mutans* Genomes

The following section of code describes installation of the NCBI Genome Download toolbox using Conda and subsequent downloading of the *S. mutans* genomes from NCBI. NCBI Genome Download is used in this tutorial to download the *S. mutans* genomes from NCBI. The NCBI Genome Download scripts are very useful for downloading genomes, whether individually or *en masse* using a list or taxonomic level, from NCBI directly from the command line (particularly useful on a server). The authors invite the reader to explore the options within the ncbi-genome-download command, as it is useful for both exploring and downloading genomes from the NCBI database.

```
# Create Conda environment for NCBI-genome-downlood tools

conda create -n ncbi-genome-download_env
conda activate ncbi-genome-download_env
conda install -c bioconda ncbi-genome-download

# Download S. mutans UA159, UA140, and NN2025 genome from NCBI

ncbi-genome-download --taxid 210007,511691 bacteria
ncbi-genome-download -A GCF_008831365.1 bacteria

# Place all 3 genome assembly files in one working directory and then unzip

gunzip *gbff.gz
conda deactivate
```

3.1.2 Conda Installation of antiSMASH

The following section of code will install the latest version of antiSMASII (currently v. 5.0) using Conda. antiSMASH is the tool that will be used to identify BGCs within genomes. antiSMASH identifies gene clusters encoding secondary metabolites of all known broad chemical classes based on rules defining the ~60 known types of BCGs. While the tutorial described here uses antiSMASH to locate the BGCs within the three *S. mutans* genomes, any genome and in fact large databases of many genomes can be examined for BGCs in one run of antiSMASH. In the experience of the authors, very large input datasets (100s of genomes) should be split into smaller sets to avoid runtime and memory errors. antiSMASH is also available as a web-based tool, which can be useful for examining small numbers of genomes and/or for users who are not comfortable using command line tools.

```
# Create antismash environment (requires bioconda conda channel)

conda create -n antismash_env antismash

# Activate the antismash environment

conda activate antismash_env
# Download antismash databases

download-antismash-databases
conda deactivate
```

3.1.3 Conda Installation of MultiGeneBlast

The following section of code uses Conda to install the MultiGeneBlast tool and the associated MI-BiG database. MultiGeneBlast is used to compare newly identified BGCs against the MI-BiG database of experimentally characterized BGCs, which is useful for predicting the type of natural product produced by the BGCs of interest.

```
# Create MultiGeneBlast conda envioronment (Requires Python2, as opposed to the
Python3 used by the antismash conda environment)

conda create -n multigeneblast_env python=2.7.15
conda activate multigeneblast_env
conda install matplotlib
conda install biopython
# Download and unzip MultiGeneBlast

wget -O multigeneblast.tar.gz \
https://sourceforge.net/projects/multigeneblast/files/1.1.13/multigeneblast_1.1.13_linux6
4.tar.gz/download
gunzip multigeneblast.tar.gz
tar -xvf multigeneblast.tar

# Download and unzip genbank_to_fasta conversion script

wget https://rocaplab.ocean.washington.edu/files/genbank_to_fasta_v1.2.zip
unzip genbank_to_fasta_v1.2.zip # move to the bin folder of the multigeneblast_env


# Download and unzip MI-BiG Database v2.0
# GenBank files

wget https://dl.secondarymetabolites.org/mibig/mibig_gbk_2.0.tar.gz

gunzip mibig_gbk_2.0.tar.gz
tar -xvf mibig_gbk_2.0.tar

# FASTA files

wget https://dl.secondarymetabolites.org/mibig/mibig_prot_seqs_2.0.fasta
```

3.1.4 Conda Installation of BiG-SCAPE

The following code describes installation of BiG-SCAPE and its required pfam database using Conda. BiG-SCAPE is a tool used to explore BGC diversity by constructing BGC sequence similarity networks.

```
# Create MI-BiG database db files (must add Multigeneblast folder to your $PATH to access the scripts)
python Multigeneblast/makedb.py mibig_db_2 mibig_gbk_2.0/*.gbk
conda deactivate
# Create environment for BiG-SCAPE and install dependencies
conda create --name bigscape_env
conda activate bigscape_env
conda install numpy scipy scikit-learn
conda install -c bioconda hmmer biopython fasttree
conda install -c anaconda networkx
# Close BiG SCAPE repository into your conda environment folder
git clone https://git.wur.nl/medema-group/BiG-SCAPE.git
# Download the pfam database
wget ftp://ftp.ebi.ac.uk/pub/databases/Pfam/current_release/Pfam-A.hmm.gz
```

```
gunzip Pfam-A.hmm.gz
# Prepare hmm database of pfam
hmmpress Pfam-A.hmm
conda deactivate bigscape_env
```

3.1.5 Installation of BWA and KneadData

The following code will install BWA and KneadData.

```
# Installation of BWA (from https://github.com/lh3/bwa)
git clone https://github.com/lh3/bwa.git
cd bwa; make
./bwa index ref.fa
./bwa mem ref.fa read-se.fq.gz | gzip -3 > aln-se.sam.gz
./bwa mem ref.fa read1.fq read2.fq | gzip -3 > aln-pe.sam.gz
# Installation of KneadData (from https://bitbucket.org/biobakery/kneaddata/wiki/Home)
pip install kneaddata
kneaddata_database –download human_gnome bowtie2 /path/to/database/
```

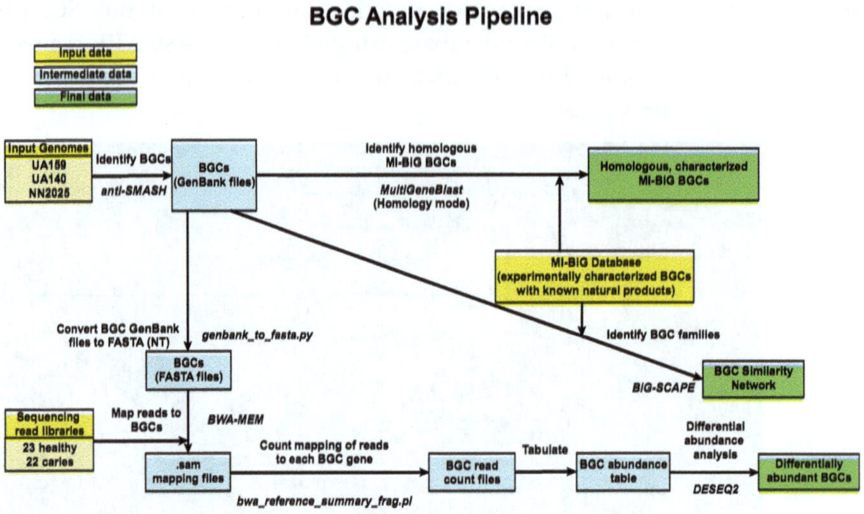

Fig. 1 Overview of BGC analysis pipeline. Flow chart illustrating the steps, files, and tools used in the pipeline described in this chapter

3.2 Running Tools

The first step of the BGC analysis pipeline will be to utilize antiSMASH to identify and annotate BGCs within the genome(s) of interest. In the tutorial, the genomes of *S. mutans* UA159, UA140, and NN2025 will be queried for BGCs using antiSMASH. Next, predictions will be made regarding the type of natural products produced from the *S. mutans* BGCs based on homology searches against the MI-BiG database of experimentally verified BGCs using MultiGeneBlast and BiG-SCAPE. Finally, we will examine the representation of the BGCs of the *S. mutans* BGCs across a panel of publicly available metagenomes from the saliva of healthy children and children with severe dental caries using DeSeq2. A flow chart of the pipeline is provided in Fig. 1.

3.2.1 Run antiSMASH to Identify BGCs

The full options of antiSMASH can be viewed by running antiSMASH with the --help flag. Although three genomes are being queried for BGCs in this tutorial, batches of many genomes can also be queried (*see* **Note 3**). However, particularly large batches (i.e., hundreds) of genomes may need to be broken into smaller subsets to be properly handled by antiSMASH (especially with modest computing resources). Note that antiSMASH can accommodate either FASTA or GenBank files as input, and any additional annotations provided in the GenBank file will be passed through to the antiSMASH output. antiSMASH does not provide exact information regarding gene boundaries of BGCs but predicts the overall gene environment, including genes involved in the biosynthesis and transport, as an example. Other programs, see below, can be applied to explore gene boundaries in detail.

```
# Activate antiSMASH conda environment (Python3)
conda activate antismash_env

# Concatenate the GenBank files of UA159, UA140, and NN2025

cat *.gbff > antismash_input.txt
mv antismash_input.txt antismash_input.gbff

# Run antismash

antismash \
  -c 36 \
  --taxon bacteria \
  --output-dir antismash_output \
  --genefinding-tool prodigal \
  antismash_input.gbff

#Deactivate antiSMASH environment

conda deactivate
```

The output folder created by antiSMASH contains a number of items. For each BGC identified, a GenBank (.gbk) file is generated (in this case 23 in total, across the 3 genomes). The results can be interactively explored by opening the index.html file in a web browser. The Overview tab provides a map of where each putative BGC occurs within each genome, as well as a list of the BGCs and their predicted BGC class (Fig. 2a). Each BGC tab contains a map of genes, which can be explored, providing further information on each gene (Fig. 2b). According to antiSMASH, the UA159 genome contains seven predicted BGCs, while the UA140 and NN2025 genomes each contain eight BGCs. In total, 14 of these BGCs are of the RiPP class (11 bacteriocins and 3 RaS-RiPPs), 3 BGCs are of the polyketide synthase (PKS) class, 2 BGCs are of the NRPS class, and 2 BGCs are of the lanthipeptide class (Fig. 2). An idea of whether some of these BGCs are redundant across the three strains and clues as to the structure of the molecule produced by each BGC will be furnished by MultiGeneBlast and BiG-SCAPE.

3.2.2 MultiGeneBlast Homology Search Against MI-BiG Database of Experimentally Characterized BGCs

Next, the BGCs of UA159, UA140, and NN2025 are compared with all the experimentally verified BGCs in the MI-BiG database. MultiGeneBlast has two run modes: homology mode and architect mode. The homology search mode is used to find homologs of a known operon or BGC, where the input file is a .gbk file of the BGC. Meanwhile, the input for the architecture search is a .fasta file of designated protein sequences, and not necessarily a known genomic region. This is useful for finding BGCs encoding specific metabolic

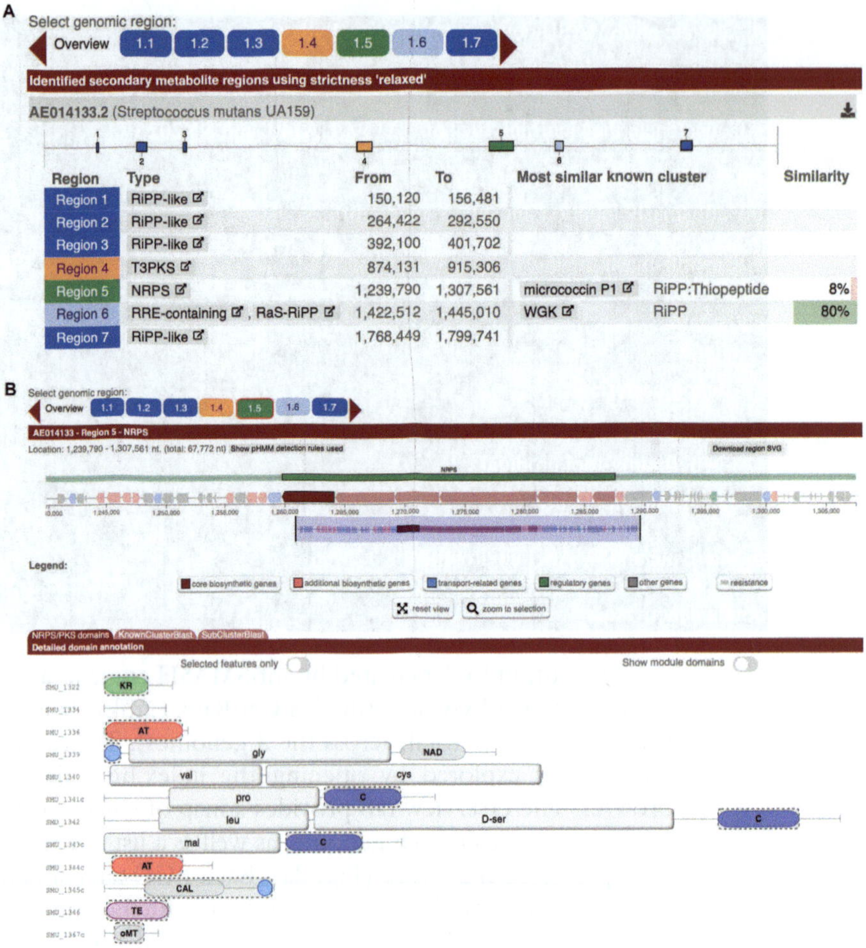

Fig. 2 Sample antiSMASH output. A screenshot of an antiSMASH output representing *S. mutans* UA159. Seven BGCs are identified and labeled with their predicted BGC class and genome location. The user can click on each BGC for more information and a detailed gene map of the BGC

pathways and combining particular metabolic steps. Only the homology search mode will be utilized in this tutorial. Users can specify the database, allowing for use of other databases (such as BLAST) or creation and use of a custom database (*see* **Note 4**). At this point, depending on the naming convention of the contigs in the original genomes, it may be useful to simplify the names of the BGCs and remove non-alphanumeric characters (other than _) which may cause downstream problems. This can be done manually or in a batch using a for loop and sed/tr commands, as performed in the code below. The full list of parameters that can be used by MultiGeneBlast can be viewed by running multigeneblast.py with the -h option. Specific cutoffs for the MultiGeneBlast search can be set with the optional parameters including minimum sequence coverage, minimum sequence identity, maximum distance between hits to be considered belonging to the same BGC locus, the number of hits per gene, and synteny weight (i.e., hits beyond the cutoff limits will not be considered in the output data). Less stringent cutoffs will provide more putative hits at the expense of computing time and more output to be examined while more stringent cutoffs will provide only highly homologous hits, but may not provide any hits if highly novel and unique BGC sequences are being examined. The commands below are designed to loop through the UA159 . gbk files of the BGCs and perform a homology search for each one.

```
# Activate MultiGeneBlast conda environment (Python2)
conda activate multigeneblast_env
# Make folders for homology files
mkdir homology_files
# Run MultiGeneBlast (homology mode)
for x in *.gbk; do
y=`echo "$x" | sed -e 's/\.gbk$//' | tr '.' '_'`
echo $y
python multigeneblast.py \
  -in "$x" \
  -from 1 \
  -to 150000 \
  -out "homology_files/$y" \
  -db path/to/mibig_db_2 \
  -cores 36
done
```

The output of MultiGeneBlast is a separate folder for each BGC. As with antiSMASH, the .xhtml file can be opened in a web browser to view the results of the homology search. Figure 3 is a screen shot example of the .xhtml output for BGC NC_004350.2.2_region002. Note that BGCs with no hits will return empty folders (and the resulting errors in creating the summary files can be safely ignored). A table of the top hits for each gene cluster from the homology search can be created using the following code. The pull-hits-1.py script is freely available at https://jonbakerlab.com/jonbakerlab/Oral_Microbiome_

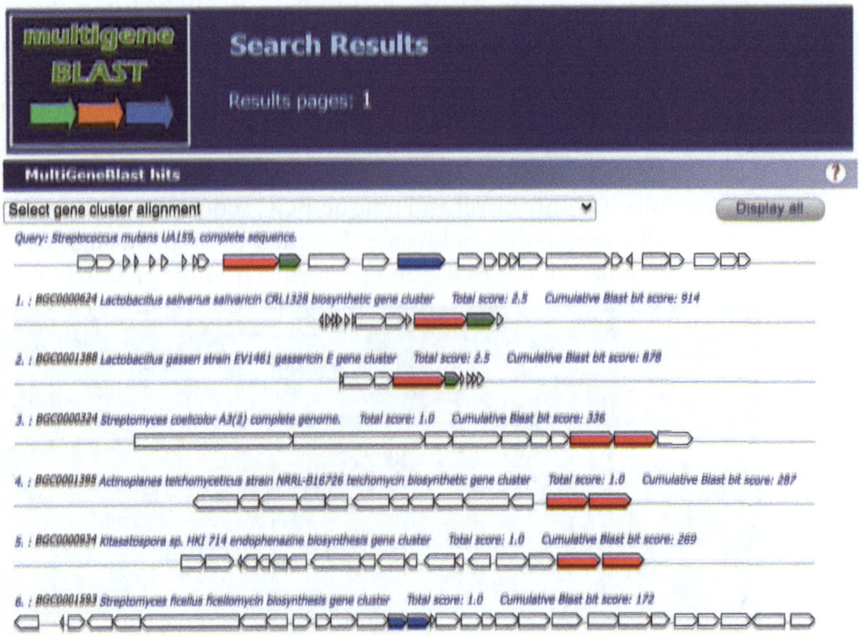

Fig. 3 Sample MultiGeneBlast output. A screenshot of MultiGeneBlast output of the *S. mutans* UA159 Region 2-BGC. The gene map of the query BGC is illustrated, as are most highly homologous BGCs found from the MI-BiG database of experimentally characterized BGCs. Homology search against already characterized BGCs is highly useful for predicting the type of natural product that an unknown BGC can produce

Chapter.

```
# Switch back to Python3
conda deactivate
conda activate antismash_env
# Pull top hits from each MultiGeneBlast homology files (some will fail since they have
no hits)
for x in N*; do
basename $x
python pull_hits-1.py \
$x/clusterblast_output.txt \
>> $x.tab
done
```

```
#To combine Tab Files into 1 excel for import into Summary File, open final file in Excel
echo -e BGC '\t' MI-BiG accession '\t' MI-BiG description '\t' Cumulative blast bit score
'\t' %identity '\t' e-values > homology_header.txt
for x in *.tab; do printf '%s\t%s\n' $x "$(cat $x)"; done > homologycombined.txt
cat homology_header.txt homologycombined.txt > homologycombined_final.txt
```

Eighteen of the 23 BGCs identified with antiSMASH were homologous to experimentally verified BGCs in the MI-BiG database, according to the MultiGeneBlast default parameters. Since the UA140 Mutacin I/III-BGC is the only BGC in these three genomes that is included in MI-BiG, the fact that it hit to itself confirms the accuracy of the analysis. Examining the MI-BiG matches can provide clues as to the structure of the products of these *S. mutans* BGCs. Note that several of the BGCs in the three genomes have actually been well characterized, and the products were previously characterized experimentally; however unfortunately they have not yet been added to the MI-BiG repository (and in several cases pre-date its existence). Specifically, in UA159, the Region 1-BGC produces Mutacin IV [33–36], the Region 2-BGC produces the NlmTE transporter, which is responsible for export of the non-lantibiotic mutacins (IV, V, and VI) [34], the Region 3-BGC produces Mutacin VI [34, 35], the Region 5-BGC produces Mutanobactin [37], and the Region 7-BGC produces Mutacin V [34, 35]. In NN2025, the Region 8-BGC is known to produce reutericyclin and mutanocyclin [25, 26].

3.2.3 BiG-SCAPE Illustrates Putative BGC Families

Next, we will use BiG-SCAPE to construct BGC similarity networks. This is particularly useful when examining large numbers of BGCs, as it can be used to group BGCs into families, explore BGC diversity linked to enzyme phylogenies, and as another method to classify novel BGCs.

BiG-SCAPE groups similar BCGs based on protein domains present, order, copy number, and DNA sequence. In this example, we will place the BGCs of UA159, UA140, and NN2025, which were identified by antiSMASH, into networks including all of the experimentally verified BGCs of the MI-BiG database. BiG-SCAPE creates networks by calculating pairwise distances between every pair of BGCs in the input dataset. In the resulting network, each BGC is represented by a node and each pairwise distance is represented by an edge (connecting line). Only pairwise distances less than the parameter given using the –cutoff flag are displayed. Therefore, with a lower (more stringent) cutoff parameter, only very closely related or homologous BGCs will cluster together (i.e., be connected) in the network. Meanwhile, if a higher (less stringent) cutoff parameter is given, more distantly related BGCs will be connected in the network (as described below for the *S. mutans* BGCs examined by this tutorial). By default, BiG-SCAPE runs a similarly networking cutoff of 0.3, in addition to further user specified cutoffs. Several cutoffs for creating the networks should be explored to get meaningful clusters that relate to predicted BGC class. BiG-SCAPE tends to crash when a large >~4 cutoffs are

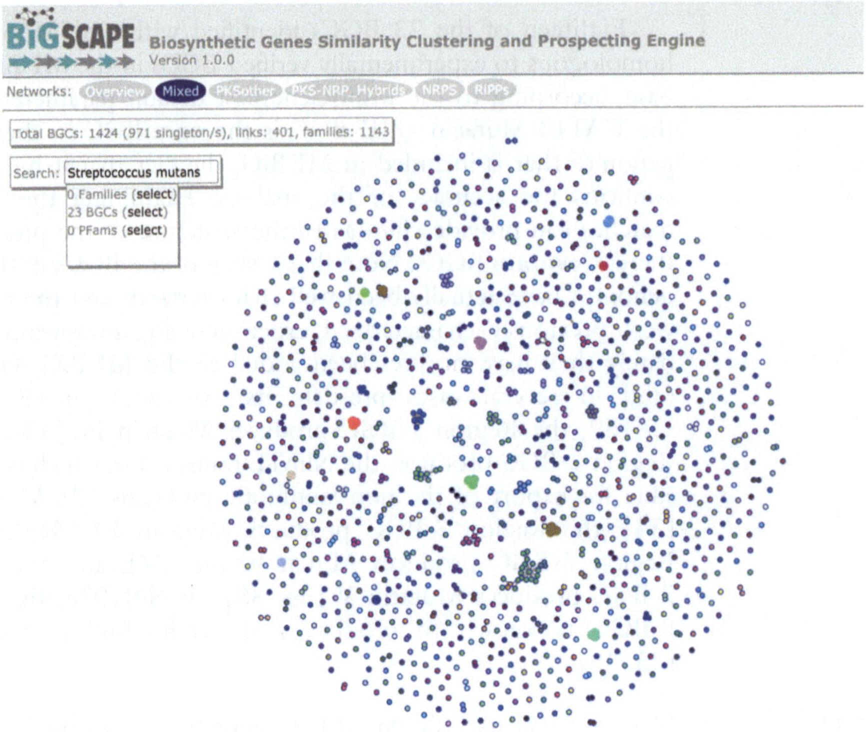

Fig. 4 Sample BiG-SCAPE output. A screenshot of the output of BiG-SCAPE analysis performed on UA159, UA140, and NN2025 against the MI-BiG database with a cutoff of 0.3. The *S. mutans* BGCs identified by antiSMASH were highlighted by large circles using the search bar. *S. mutans* BGCs that cluster together are the "same" BGC across multiple strains. Meanwhile, the mutacin K8-BGC from NN2025 and the mutacin I-BGC from UA140 form networks with closely related (and "themselves") mutacins from the MI-BiG database

specified in the same run (*see* **Note 5**). A cutoff of 0.8 was utilized in one recent inventory of oral BGCs [7], and will also be used in this tutorial, along with the default value of 0.3 (Fig. 4).

```
# Run BiG-SCAPE (input is directory containing .gbk files that were the output of antiSMASH)
conda activate bigscape_env
python bigscape.py \
  -o bigscape_results \
  -i antismash_output \
  --mibig \
  --mix \
  --pfam_dir /path/to/pfam/database \
  --cutoffs 0.2 0.8
conda deactivate bigscape_env
```

In the results folder, the index.html file can be opened in a web browser to visualize the networks. Alternatively, the network files can be opened using a program such as Cytoscape to create custom networks and have complete control over visualizations. All BGCs can be visualized together under the "Mixed" tab, while there are tabs for viewing each of the identified BGC classes individually. In the network with a cutoff of 0.3, most of the BGCs are singletons. Even with this low cutoff, several of the *S. mutans* BGCs cluster together, indicating that these are very likely the same BGC across multiple genomes. The Mutacin IV-BGC is found in the UA159 and UA140 genomes, while the *nlmTE*, the Mutacin VI, the T3PKS, the RaS RiPP, and Mutacin V-BGCs are found in all three genomes. With a cutoff of 0.3 a few of the *S. mutans* BGCs do network with MI-BiG BGCs. The NN2025 Region 3 bacteriocin-BGC networks with the Mutacin K8-BGC from *S. mutans* K8, indicating that NN2025 also produces Mutacin K8. And of course, the UA140 Mutacin I-BGC networks with itself and other closely related bacteriocins, such as Mutacin III. As expected, in the network with a cutoff of 0.8, many more BGCs network with each other. The RaS RiPP-BGC from the three *S. mutans* genomes networks with the Streptide-BGC from *Streptococcus thermophilus*, the Mutacin IV-BGCs network with Gassericin-BGCs from *Lactobacillus gasseri*, and the UA140 Region 2 NRPS-PKS BGC networks with the Nostophycin-BGC from *Nostoc* sp. 152. Examining various cutoffs can provide clues as to the products of unknown BGCs, and the results from BiG-SCAPE should largely agree with those from MultiGeneBlast.

3.2.4 Exploring BGC Representation in Microbiomes Associated with Caries Versus Good Dental Health

Lastly, representation of the *S. mutans* BGCs will be examined across a publicly available set of metagenomes, deposited in the NCBI Sequence Read Archive (SRA) with accession number SRP151559. These metagenomes were obtained from the saliva of either children with severe dental caries (23 samples), or children with healthy dentition (24 samples), for a total of 47 subjects/metagenomes [7]. The Library Name contains the sample number and group ("caries" or "healthy"). Libraries SC33_healthy and SC40_caries from SRP151559 are omitted in the tutorial analysis because they were sequenced ultra-deep to 150 million reads, and therefore have large read libraries that would create mapping files in the ~100–300 GB range (unmanageable by a user without access to a supercomputer). By mapping the 150 bp metagenomic sequencing reads to the BGCs using BWA-mem, enrichment of the BGCs in disease or health can be observed. It is important to note that when examining metagenomic data, one is only obtaining information about representation (i.e., whether a strain encoding the BGC of interest is present, and not whether that particular BGC is expressed). This is particularly useful and could be applied to any disease or metadata category. To measure actual expression and gene activity of BGCs, mRNA read mapping against individual bacterial genomes or BGC databases is also useful. At the time of writing, we decided to present an analysis from a metagenomic study because metagenomics data is much more widely available for most microbiomes as compared to metatranscriptomic datasets. When performing mRNA or DNA read mapping it is also crucial to remember that sequence library depth plays a major role in determining whether a BGC is present or not in a highly diverse microbial environment, such as the oral cavity. Due to that most human microbiome-deep sequencing libraries are relatively shallow (~5 M–10 M sequence read depth), they can only provide information of the most abundant BGCs in a diverse microbial community, while the rare BGCs may remain uncharacterized. Therefore, BGCs may still be present in a community even though they remained unidentified. It is also highly recommended to perform quality control and removal of human reads from sequencing libraries. In this tutorial, the authors recommend the use of the quality control tool KneadData, which not only filters out low-quality sequence reads but also removes reads that map to a human genome database. To circumvent tedious read mapping exercises against single bacterial genomes, it is advised to run a batch command submission including all deep sequence read libraries (here 45) in a parallel computing environment (*see* **Note 6**).

```
# Trim read libraries using KneadData (do this for each library, submitting as a batch is recommended)

kneaddata \
-i 1_S1_R1_001.fastq.gz \
-i 1_S1_R2_001.fastq.gz \
-o kneaddata_output/1_S1_R1_001 \
-db
/path/to/human/seqeuence/bowtie/database/Homo_sapiens_Bowtie2_v0.1/Homo_sapiens
\
--trimmomatic-options SLIDINGWINDOW:4:20 \
--trimmomatic-options MINLEN:90
```

Prior to read mapping with BWA-mem, the BGC nucleotide sequences must be extracted from the .gbk files produced by antiSMASH, with the file name/genome name appended to the header of each contig as follows (the sed command may need to be modified depending on the naming convention used):

```
# Extract nucleotide sequences from GenBank files (we already have the AA seqs from
above, take care not to overwrite those files)

for x in *.gbk; do python genbank_to_fasta.py -i $x -s nt; done

##loop used to append filename to fasta headers across all genomes in the database
mkdir appended
for datafile in *.fasta; do
awk '/>/{sub(">","&"FILENAME"_");sub(/\.fasta/,x)}1' "$datafile" >
appended/"$datafile"
done
```

Next, the BGCs can be concatenated to generate the database, and the database can be indexed for BWA.

```
# Generate database by concatenating BGCs
cat *.fasta > BGC_database.fa
# Index database
bwa index BGC_database.fa
```

The reads from the metagenomic dataset can then be mapped against the database. Again, it helps to do this in a parallel computing environment.

```
# Map reads to the database (do this for each of the 45 read libraries)
bwa mem \
  -t 8 \
  BGC_database.fa 1_S1_R1_001_kneaddata_paired_1.fastq
1_S1_R1_001_kneaddata_paired_2.fastq > 1.sam
```

Next, the bwa_reference_summary_frag.pl script is used to extract the mapping information from the .sam files created by BWA-mem. This script is freely available at https://github.com/jonbakerlab/Oral_Microbiome_Chapter.

```
# Get counts from all the .sam files (do this for each of the 45 read libraries)
perl bwa_reference_summary_frag.pl -f 1.sam > SC01_caries_count.txt
```

A relative abundance table can then be generated that presents each gene in the BGC database as a row, and each sample from the metagenomic dataset as a column, containing the number of reads mapped to each gene.

```
# Make sure all files have the same number of lines
wc -l *count.txt

# Make BGC abundance table
# Make row names

awk 'BEGIN {print "gene"} {print $1}' SC01_caries_count.txt > rownames.txt

# Add columns with BGC counts

for datafile in *count.txt; do
  awk 'FNR==1 {print FILENAME} {print $2}' "$datafile" > "$datafile.column"
done

# Paste together

paste rownames.txt *.column > bgc_abundance_table.txt

# Remove '.txt' from the column names

sed -e 's/_count.txt//g' bgc_abundance_table.txt > bgc_abundance_table_final.txt
```

Relative BGC gene abundances are associated with disease status (caries vs. healthy). A caveat essential to understand when analyzing sequencing data is that the data is compositional, meaning the data is provided in the form of relative, not absolute, abundances and therefore inferring absolute fold-changes or correlations using compositional data is inherently problematic [38]. Numerous microbiome studies have drawn biological conclusions based on the application of conventional statistical tools to compositional data, which has been shown to have unacceptably high false discovery rates and lead to spurious hypotheses [39]. A number of tools and approaches have been used to solve and/or circumvent these issues to varying degrees. DESeq2 is a tool used to estimate variance-mean dependence in count data from high-throughput sequencing assays and test for differential expression/abundance based on negative binomial distribution model [32]. While DESeq2 is useful in that it provides log2 fold changes and false discovery rate (FDR)-corrected *p*-values, it does not completely circumvent the issues of compositional data. No currently available tools are "perfect" for performing correlations on compositional data; therefore, readers are encouraged to experiment with multiple approaches to addressing this problem and keep the limitations of generated data in mind. DESeq2 is implemented in this pipeline using R, and the authors highly recommend the use of RStudio as an IDE for this portion of the pipeline. The R code below is a modified version of the one utilized by Aleti et al. [7], which was based upon the tutorial DESeq2 R script available at https://bioconductor.org/packages/release/bioc/vignettes/DESeq2/inst/doc/DESeq2.R.

```
## Metagenomic analysis with DESeq2
# Set working directory

setwd("/Users/jobaker/Desktop/BookChapter/7-1-20/deseq")

getwd()

# Import & pre-process ---------------------------------------------------

# Import data from BGC counts table

# Note that you will need to rename the column headers to match your library names

countdata <- read.table("bgc_abundance_table_final.txt", header=TRUE, row.names = 1)

# Save disease status (from column name) as 'condition' variable

sample_status <- grepl("caries",colnames(countdata))

sample_status[sample_status == TRUE] <- "caries"

sample_status[sample_status == FALSE] <- "healthy"

condition <- as.factor(sample_status)
```

```
# Convert to matrix and check

countdata <- as.matrix(countdata)

head(countdata)

# DESeq2 analysis---------------------------------------------------

# Import DeSeq2 Package

library(DESeq2)

# Create a coldata frame and instantiate the DESeqDataSet

coldata <- data.frame(row.names=colnames(countdata), condition)

dds <- DESeqDataSetFromMatrix(countData=countdata, colData=coldata,

design=~condition)

dds

#set healthy as reference level

dds$condition <- relevel(dds$condition, "healthy")

# Run the DESeq pipeline

dds <- DESeq(dds)
```

```
# Plot dispersions
png("qc-dispersions.png", 1000, 1000, pointsize=20)
plotDispEsts(dds, main="Dispersion plot")
dev.off()

# Regularized log transformation for clustering/heatmaps, etc
rld <- rlogTransformation(dds)
head(assay(rld))
hist(assay(rld))

# Colors for plots
## Use RColorBrewer
library(RColorBrewer)
(mycols <- brewer.pal(8, "Dark2")[1:length(unique(condition))])

# Sample distance heatmap
sampleDists <- as.matrix(dist(t(assay(rld))))
library(gplots)
png("qc-heatmap-samples.png", w=1000, h=1000, pointsize=20)
heatmap.2(as.matrix(sampleDists), key=F, trace="none",
        col=colorpanel(100, "black", "white"),
        ColSideColors=mycols[condition], RowSideColors=mycols[condition],
        margin=c(11, 11), main="Sample Distance Matrix")
```

```
dev.off()
# Principal components analysis
## Could do with built-in DESeq2 function:
## DESeq2::plotPCA(rld, intgroup="condition") + theme_bw()
rld_pca <- function (rld, intgroup = "condition", ntop = 500, colors=NULL,
legendpos="bottomleft", main="PCA Biplot", textcx=1, ...) {
  require(genefilter)
  require(calibrate)
  require(RColorBrewer)
  rv = rowVars(assay(rld))
  select = order(rv, decreasing = TRUE)[seq_len(min(ntop, length(rv)))]
  pca = prcomp(t(assay(rld)[select, ]))
  fac = factor(apply(as.data.frame(colData(rld)[, intgroup, drop = FALSE]), 1, paste,
collapse = " : "))
  if (is.null(colors)) {
    if (nlevels(fac) >= 3) {
      colors = brewer.pal(nlevels(fac), "Paired")
    }  else {
      colors = c("black", "red")
    }
  }
  pc1var <- round(summary(pca)$importance[2,1]*100, digits=1)
```

```
 pc2var <- round(summary(pca)$importance[2,2]*100, digits=1)
 pc1lab <- paste0("PC1 (",as.character(pc1var),"%)")
 pc2lab <- paste0("PC1 (",as.character(pc2var),"%)")
 plot(PC2~PC1, data=as.data.frame(pca$x), bg=colors[fac], pch=21, xlab=pc1lab,
ylab=pc2lab, main=main, ...)
 with(as.data.frame(pca$x), textxy(PC1, PC2, labs=rownames(as.data.frame(pca$x)),
cex=textcx))
 legend(legendpos, legend=levels(fac), col=colors, pch=20)
 #    rldyplot(PC2 ~ PC1, groups = fac, data = as.data.frame(pca$rld),
 #          pch = 16, cerld = 2, aspect = "iso", col = colours, main = draw.key(key =
list(rect = list(col = colours),
 #    terldt = list(levels(fac)), rep =FALSE)))
}
png("qc-pca.png", 1000, 1000, pointsize=20)
rld_pca(rld, colors=mycols, intgroup="condition", xlim=c(-50, 50))
dev.off()
# DESeq2 results
res <- results(dds)
table(res$padj<0.05)
## Order by adjusted p-value
```

```
res <- res[order(res$padj), ]
## Merge with normalized counts
resdata <- merge(as.data.frame(res), as.data.frame(counts(dds, normalized=TRUE)),
by="row.names", sort=FALSE)
names(resdata)[1] <- "Species"
head(resdata)

## Write results
write.csv(resdata, file="BGC_abundance_DeSeq2.csv")
## Examine p-values
hist(res$pvalue, breaks=50, col="grey")
## MA plot
## Could do with built-in DESeq2 function:
## DESeq2::plotMA(dds, ylim=c(-1,1), cex=1)
maplot <- function (res, thresh=0.05, labelsig=TRUE, textcx=1, ...) {
  with(res, plot(baseMean, log2FoldChange, pch=20, cex=.5, log="x", ...))
  with(subset(res, padj<thresh), points(baseMean, log2FoldChange, col="red", pch=20,
cex=1.5))
  if (labelsig) {
    require(calibrate)
```

```
    with(subset(res, padj<thresh), textxy(baseMean, log2FoldChange, labs=Species,
cex=textcex, col=2))
  }
}
png("diffexpr-maplot.png", 1500, 1000, pointsize=20)
maplot(resdata, main="MA Plot")
dev.off()

## Volcano plot with "significant" genes labeled
volcanoplot <- function (res, lfcthresh=2, sigthresh=0.05, main="Volcano Plot",
legendpos="bottomright", labelsig=TRUE, textcex=1, ...) {
  with(res, plot(log2FoldChange, -log10(pvalue), pch=20, main=main, ...))
  with(subset(res, padj<sigthresh ), points(log2FoldChange, -log10(pvalue), pch=20,
col="red", ...))
  with(subset(res, abs(log2FoldChange)>lfcthresh), points(log2FoldChange, -
log10(pvalue), pch=20, col="orange", ...))
  with(subset(res, padj<sigthresh & abs(log2FoldChange)>lfcthresh),
points(log2FoldChange, -log10(pvalue), pch=20, col="green", ...))
  if (labelsig) {
    require(calibrate)
    with(subset(res, padj<sigthresh & abs(log2FoldChange)>lfcthresh),
textxy(log2FoldChange, -log10(pvalue), labs=Species, cex=textcex, ...))
  }
```

```
legend(legendpos, xjust=1, yjust=1, legend=c(paste("FDR<",sigthresh,sep=""),
paste("|LogFC|>",lfcthresh,sep=""), "both"), pch=20, col=c("red","orange","green"))
}
png("diffexpr-volcanoplot.png", 1200, 1000, pointsize=20)
volcanoplot(resdata, lfcthresh=1, sigthresh=0.05, textcx=.8, xlim=c(-2.3, 2))
dev.off()
#Heat map for the most abundant BGCs
dds <- DESeqDataSetFromMatrix(countData=countdata, colData=coldata, design=~1)
dds
dds <- estimateSizeFactors(dds)
normalizeddata <- counts(dds, normalized=TRUE)
write.csv(normalizeddata, file="normalizeddata.csv")
library("pheatmap")
select <- order(rowMeans(counts(dds,normalized=TRUE)),
         decreasing=TRUE)[1:20]
nt <- normTransform(dds) # defaults to log2(x+1)
log2.norm.counts <- assay(nt)[select,]
df <- as.data.frame(colData(dds)[,c("condition")])
pheatmap(log2.norm.counts, cluster_rows=FALSE, show_rownames=TRUE,
cluster_cols=FALSE)
```

The output of this analysis is presented as several useful figures, including a heatmap, a PCA, and a volcano plot of the analysis, as well as a BGC-differential abundance table; BGC_abundance_DESeq2.csv. Of the 594 genes within the 23 *S. mutans* BGCs detected by antiSMASH, 20 genes, representing 14 BGCs were differentially regulated between caries and health, using an FDR adjusted *p*-value cutoff of 0.05 (Table 1). Several of the differentially represented BGC genes are the "same" gene across multiple genomes, which makes sense given that similar numbers of reads would map to such highly similar sequences. Closer examination of each gene reveals that most of the differentially represented genes are on the periphery of the predicted BGCs, with annotations suggesting that they are unlikely to be involved in the biosynthetic assembly line. However, the core-PKS gene of the UA140 Region 2-BGC, the core *lanM* lanthipeptide gene in the Mutacin K8-BGC of NN2025, and the *mucD* (core enzyme) and *mucG* (transcriptional regulator) genes of the mutanocyclin/reutericyclin-BGC of NN2025 were overrepresented in the caries-associated microbiomes compared to the health-associated microbiomes (Table 1). This suggests

Table 1
Differentially abundant BGC genes in caries-associated vs. health-associated microbiomes (genes with a yellow highlight are core biosynthetic genes)

Gene	BGC type/product	Gene name/family	Pvalue	Padj	log2FoldChange (caries/health)
NC_013928.1.region007_SMUNN2025_RS08585	NlmTE	tkt	0.00340837	0.04714908	-0.441740079
NC_013928.1.region008_SMUNN2025_RS09060	mutanocyclin	rplB	0.00049962	0.02786678	-0.370847298
NC_013928.1.region008_SMUNN2025_RS09040	mutanocyclin	rplP	0.00107275	0.02786678	-0.353919624
NZ_CP044495.1.region005_FSA28_RS03240	mutacin I	alaS	0.00381916	0.04949747	-0.278838907
NC_004350.2.region002_SMU_RS01475	NlmTE	polA	0.0014116	0.02786678	-0.273255399
NZ_CP044495.1.region003_FSA28_RS01495	NlmTE	polA	0.00143383	0.02786678	-0.271902771
NC_013928.1.region007_SMUNN2025_RS08555	NlmTE	polA	0.00167872	0.02786678	-0.271444843
NZ_CP044495.1.region002_FSA28_RS00965	NRPS	core-PKS gene	0.00253424	0.03711919	0.619897036
NC_004350.2.region007_SMU_RS08660	mutacin V	HlyD family transporter	0.00158924	0.02786678	1.000149257
NC_013928.1.region002_SMUNN2025_RS01335	mutacin V	HlyD family transporter	0.00149141	0.02786678	1.006416859
NZ_CP044495.1.region008_FSA28_RS09025	mutacin V	HlyD family transporter	0.00110764	0.02786678	1.058436714
NC_013928.1.region008_SMUNN2025_RS09135	mutanocyclin	mucD	0.0039757	0.04949747	1.061923373
NC_013928.1.region003_SMUNN2025_RS01715	mutacin K8	lanM core lanthipeptide	0.0001486	0.02786678	1.139473156
NZ_CP044495.1.region007_FSA28_RS07170	RaS RiPP	rexB	0.00092958	0.02786678	1.153595267
NC_013928.1.region008_SMUNN2025_RS09120	mutanocyclin	mucG	0.00145871	0.02786678	1.187698545
NC_004350.2.region003_SMU_RS02045	mutacin VI	ylxR family protein	0.00119057	0.02786678	1.210518266
NC_013928.1.region006_SMUNN2025_RS07900	mutacin VI	ylxR family protein	0.00119057	0.02786678	1.210518266
NZ_CP044495.1.region004_FSA28_RS02105	mutacin VI	ylxR family protein	0.00119057	0.02786678	1.210518266
NC_013928.1.region007_SMUNN2025_RS08530	NlmTE	YdcF family protein	0.00051928	0.02786678	1.22411471
NC_013928.1.region007_SMUNN2025_RS08475	NlmTE	ABC transporter permease	0.00195799	0.03047118	1.475435652

that the products of these BGCs may be associated with dysbiosis of the oral microbiome and/or caries pathogenesis and warrant further investigation. Indeed, both mutacin K8 [40] and reutericyclin [25] have been shown to be utilized by *S. mutans* to inhibit the growth of health-associated, commensal *Streptococcus* spp. It is crucial to keep in mind that this analysis represents a metagenomic perspective and not a metatranscriptomics (gene expression) perspective, although this same pipeline can be applied to metatranscriptomic read libraries as well. Starting with only a genome sequence, this pipeline can accurately predict BGCs, estimate the type of natural product(s) that they produce, and determine if these BGCs are associated with disease or a given condition.

4 Notes

1. Additional advanced tools (other than antiSMASH), which are not discussed or used here, such as BAGEL [41], ClustScan [42], NP.searcher [43], SMURF [44], ClusterFinder, PRISM [45], EvoMining [46], RODEO [47], and ARTS [48], have also been designed to perform genome mining for BGCs. These tools also implement algorithms to define BGC boundaries and to detect potential BGCs based on multiple indicators, such as signature protein domains, distant paralogs of primary metabolic enzymes, and evolutionary hallmarks [49]. Moreover, for functional characterization of biosynthetic key genes, two software programs, SBSPKS [50] and NaPDoS [51], that analyze the 3D structure and predict their natural products can be applied.
2. The computational resources needed will vary greatly depending on the step in the pipeline and how many genomes/BGCs are being queried. While this tutorial, with just three genomes, can easily be run on a laptop computer, large datasets with

many genomes will need the resources of a larger computing cluster. In the interest of saving time, the tutorial code presented here was executed by the authors on a large cluster (1000 GB RAM, 64 cores), which is why many of the commands use options with large numbers of cores and memory.

3. Although smaller numbers of genomes can easily be concatenated into one file to be fed into antiSMASH and queried for BGCs, errors may arise with very large (i.e., hundreds) sets of concatenated genomes, particularly with limited computational resources. In that case, the input genomes should be submitted to antiSMASH in smaller batches.
4. Although the MI-BiG database of experimentally characterized BGCs was used in this tutorial, users can specify the database, allowing for use of other databases (such as BLAST) or creation and use of a custom database.
5. In the experience of the authors, running more than three to four different cutoffs in the same run of BiG-SCAPE may cause the program to crash, even with a large amount of computational resources (64 cores and 1 TB of RAM).
6. The use of batch command submission in parallel computing environments is highly recommended when analyzing large dataset (here 45 samples/read libraries) to circumvent unnecessary processing time.

Acknowledgments

The research presented in this chapter was supported by NIH/-NIDCR F32-DE026947 (J.L.B.), K99-DE029228 (J.L.B.), R00-DE024534 (A.E.), and R21-DE028609 (A.E.). The bwa_reference_summary_frag.pl and pull-hits-1.py scripts were originally written by R.A. Richter at the J. Craig Venter Institute.

References

1. Donia MS, Fischbach MA (2015) Human microbiota. Small molecules from the human microbiota. Science 349:1254766
2. Sugimoto Y, Camacho FR, Wang S, Chankhamjon P, Odabas A, Biswas A et al (2019) A metagenomic strategy for harnessing the chemical repertoire of the human microbiome. Science 366(6471):eaax9176
3. Blin K, Shaw S, Steinke K, Villebro R, Ziemert N, Lee SY et al (2019) antiSMASH 5.0: updates to the secondary metabolite genome mining pipeline. Nucleic Acids Res 47:W81–W87
4. Milshteyn A, Colosimo DA, Brady SF (2018) Accessing bioactive natural products from the human microbiome. Cell Host Microbe 23:725–736
5. Cimermancic P, Medema MH, Claesen J, Kurita K, Wieland Brown LC, Mavrommatis K et al (2014) Insights into secondary metabolism from a global analysis of prokaryotic biosynthetic gene clusters. Cell 158:412–421
6. Donia MS, Cimermancic P, Schulze CJ, Wieland Brown LC, Martin J, Mitreva M et al (2014) A systematic analysis of biosynthetic gene clusters in the human microbiome reveals

a common family of antibiotics. Cell 158:1402–1414

7. Aleti G, Baker JL, Tang X, Alvarez R, Dinis M, Tran NC et al (2019) Identification of the bacterial biosynthetic gene clusters of the oral microbiome illuminates the unexplored social language of bacteria during health and disease. mBio 10(2):e00321–e00319
8. Wang S, Li N, Zou H, Wu M (2019) Gut microbiome-based secondary metabolite biosynthetic gene clusters detection in Parkinson's disease. Neurosci Lett 696:93–98
9. Pitts NB, Zero DT, Marsh PD, Ekstrand K, Weintraub JA, Ramos-Gomez F et al (2017) Dental caries. Nat Rev Dis Primers 3:17030
10. Bowen WH, Burne RA, Wu H, Koo H (2018) Oral biofilms: pathogens, matrix, and polymicrobial interactions in microenvironments. Trends Microbiol 26:229–242
11. Lamont RJ, Koo H, Hajishengallis G (2018) The oral microbiota: dynamic communities and host interactions. Nat Rev Microbiol 16:745–759
12. Lemos JA, Palmer SR, Zeng L, Wen ZT, Kajfasz JK, Freires IA et al (2019) The biology of *Streptococcus mutans*. Microbiol Spectr 7(1)
13. Methé BA, Nelson KE, Pop M, Creasy HH, Giglio MG, Huttenhower C et al (2012) A framework for human microbiome research. Nature 486(7402):215–221
14. Burne RA (2018) Getting to know "The Known Unknowns": heterogeneity in the oral microbiome. Adv Dent Res 29:66–70
15. Burne RA, Zeng L, Ahn SJ, Palmer SR, Liu Y, Lefebure T et al (2012) Progress dissecting the oral microbiome in caries and health. Adv Dent Res 24:77–80
16. Philip N, Suneja B, Walsh L (2018) Beyond *Streptococcus mutans*: clinical implications of the evolving dental caries aetiological paradigms and its associated microbiome. Br Dent J 224:219–225
17. Nascimento MM, Zaura E, Mira A, Takahashi N, Ten Cate JM (2017) Second era of OMICS in caries research: moving past the phase of disillusionment. J Dent Res 96:733–740
18. Banas JA, Drake DR (2018) Are the mutans streptococci still considered relevant to understanding the microbial etiology of dental caries? BMC Oral Health 18:129
19. Ajdic D, McShan WM, McLaughlin RE, Savic G, Chang J, Carson MB et al (2002) Genome sequence of *Streptococcus mutans* UA159, a cariogenic dental pathogen. Proc Natl Acad Sci U S A 99:14434–14439
20. Biswas I (2020) Complete genome sequences of two mutacin-producing *Streptococcus mutans* strains, T8 and UA140. Microbiol Resour Announc 9(24):e00469–e00420
21. Maruyama F, Kobata M, Kurokawa K, Nishida K, Sakurai A, Nakano K et al (2009) Comparative genomic analyses of *Streptococcus mutans* provide insights into chromosomal shuffling and species-specific content. BMC Genomics 10:358
22. Liu L, Hao T, Xie Z, Horsman GP, Chen Y (2016) Genome mining unveils widespread natural product biosynthetic capacity in human oral microbe *Streptococcus mutans*. Sci Rep 6:37479
23. Momeni SS, Beno SM, Baker JL, Edlund A, Ghazal T, Childers NK et al (2020) Caries-associated biosynthetic gene clusters in *Streptococcus mutans*. J Dent Res. https://doi.org/10.1177/0022034520914519:22034520914519
24. Medema MH, Kottmann R, Yilmaz P, Cummings M, Biggins JB, Blin K et al (2015) Minimum information about a biosynthetic gene cluster. Nat Chem Biol 11:625–631
25. Tang X, Kudo Y, Baker JL, LaBonte S, Jordan PA, McKinnie SMK et al (2020) Cariogenic *Streptococcus mutans* produces tetramic acid strain-specific antibiotics that impair commensal colonization. ACS Infect Dis 6:563–571
26. Hao T, Xie Z, Wang M, Liu L, Zhang Y, Wang W et al (2019) An anaerobic bacterium host system for heterologous expression of natural product biosynthetic gene clusters. Nat Commun 10:3665
27. Merritt J, Qi F (2012) The mutacins of *Streptococcus mutans*: regulation and ecology. Mol Oral Microbiol 27:57–69
28. Medema MH, Takano E, Breitling R (2013) Detecting sequence homology at the gene cluster level with MultiGeneBlast. Mol Biol Evol 30:1218–1223
29. Finn RD, Bateman A, Clements J, Coggill P, Eberhardt RY, Eddy SR et al (2014) Pfam: the protein families database. Nucleic Acids Res 42:D222–D230
30. Navarro-Munoz JC, Selem-Mojica N, Mullowney MW, Kautsar SA, Tryon JH, Parkinson EI et al (2020) A computational framework to explore large-scale biosynthetic diversity. Nat Chem Biol 16:60–68
31. Li H (2013) Aligning sequence reads, clone sequences and assembly contigs with BWA-MEM. arXiv https://arxiv.org/abs/1303.3997. Accessed 8 Nov 2020

32. Love MI, Huber W, Anders S (2014) Moderated estimation of fold change and dispersion for RNA-seq data with DESeq2. Genome Biol 15:550
33. Qi F, Chen P, Caufield PW (2001) The group I strain of *Streptococcus mutans*, UA140, produces both the lantibiotic mutacin I and a nonlantibiotic bacteriocin, mutacin IV. Appl Environ Microbiol 67:15–21
34. Hale JD, Heng NC, Jack RW, Tagg JR (2005) Identification of *nlmTE*, the locus encoding the ABC transport system required for export of nonlantibiotic mutacins in *Streptococcus mutans*. J Bacteriol 187:5036–5039
35. Hale JD, Ting YT, Jack RW, Tagg JR, Heng NC (2005) Bacteriocin (mutacin) production by *Streptococcus mutans* genome sequence reference strain UA159: elucidation of the antimicrobial repertoire by genetic dissection. Appl Environ Microbiol 71:7613–7617
36. Li YH, Lau PC, Lee JH, Ellen RP, Cvitkovitch DG (2001) Natural genetic transformation of *Streptococcus mutans* growing in biofilms. J Bacteriol 183:897–908
37. Joyner PM, Liu J, Zhang Z, Merritt J, Qi F, Cichewicz RH (2010) Mutanobactin A from the human oral pathogen *Streptococcus mutans* is a cross-kingdom regulator of the yeast-mycelium transition. Org Biomol Chem 8:5486–5489
38. Gloor GB, Macklaim JM, Pawlowsky-Glahn V, Egozcue JJ (2017) Microbiome datasets are compositional: and this is not optional. Front Microbiol 8:2224
39. Morton JT, Marotz C, Washburne A, Silverman J, Zaramela LS, Edlund A et al (2019) Establishing microbial composition measurement standards with reference frames. Nat Commun 10:2719
40. Robson CL, Wescombe PA, Klesse NA, Tagg JR (2007) Isolation and partial characterization of the *Streptococcus mutans* type AII lantibiotic mutacin K8. Microbiology 153:1631–1641
41. van Heel AJ, de Jong A, Montalban-Lopez M, Kok J, Kuipers OP (2013) BAGEL3: automated identification of genes encoding bacteriocins and (non-)bactericidal posttranslationally modified peptides. Nucleic Acids Res 41:W448–W453
42. Starcevic A, Zucko J, Simunkovic J, Long PF, Cullum J, Hranueli D (2008) ClustScan: an integrated program package for the semi-automatic annotation of modular biosynthetic gene clusters and *in silico* prediction of novel chemical structures. Nucleic Acids Res 36:6882–6892
43. Li MH, Ung PM, Zajkowski J, Garneau-Tsodikova S, Sherman DH (2009) Automated genome mining for natural products. BMC Bioinformatics 10:185
44. Khaldi N, Seifuddin FT, Turner G, Haft D, Nierman WC, Wolfe KH et al (2010) SMURF: genomic mapping of fungal secondary metabolite clusters. Fungal Genet Biol 47:736–741
45. Skinnider MA, Merwin NJ, Johnston CW, Magarvey NA (2017) PRISM 3: expanded prediction of natural product chemical structures from microbial genomes. Nucleic Acids Res 45: W49–W54
46. Cruz-Morales P, Kopp JF, Martinez-Guerrero-C, Yanez-Guerra LA, Selem-Mojica N, Ramos-Aboites H et al (2016) Phylogenomic analysis of natural products biosynthetic gene clusters allows discovery of arseno-organic metabolites in model streptomycetes. Genome Biol Evol 8:1906–1916
47. Tietz JI, Schwalen CJ, Patel PS, Maxson T, Blair PM, Tai HC et al (2017) A new genome-mining tool redefines the lasso peptide biosynthetic landscape. Nat Chem Biol 13:470–478
48. Alanjary M, Kronmiller B, Adamek M, Blin K, Weber T, Huson D et al (2017) The antibiotic resistant target seeker (ARTS), an exploration engine for antibiotic cluster prioritization and novel drug target discovery. Nucleic Acids Res 45:W42–W48
49. Medema MH, Fischbach MA (2015) Computational approaches to natural product discovery. Nat Chem Biol 11:639–648
50. Anand S, Prasad MV, Yadav G, Kumar N, Shehara J, Ansari MZ et al (2010) SBSPKS: structure based sequence analysis of polyketide synthases. Nucleic Acids Res 38:W487–W496
51. Ziemert N, Podell S, Penn K, Badger JH, Allen E, Jensen PR (2012) The natural product domain seeker NaPDoS: a phylogeny based bioinformatic tool to classify secondary metabolite gene diversity. PLoS One 7:e34064

Chapter 11

Usage of Metatranscriptomics to Understand Oral Disease

Takayasu Watanabe

Abstract

Metatranscriptomics is a method used to comprehensively capture bacterial activity within microbiota at the transcription level. It has become an alternative to the 16S rDNA sequencing, which uses only the 16S rRNA gene for predicting bacterial composition. By conducting metatranscriptomics, investigators can obtain substantial information about what types of genes are transcribed at the time of sampling and which bacterial taxa are responsible for their transcription. Here, I describe a protocol for metatranscriptomics for oral microbiota by using high-throughput sequencing technology. A remarkable feature of this protocol is that it uses the level of rRNA expression as the internal control for measuring transcriptional activity of each bacterial taxon. The normalized mRNA level is given by the mRNA/rRNA ratio, which indicates the extent of transcriptional activity.

Key words Metatranscriptome, Microbiota, RNA-seq, mRNA, rRNA, Transcriptional activity

1 Introduction

For several decades, the bacteriological paradigm for infectious disease has shifted from detecting single bacterial species regarded as a pathogen to considering multiple bacterial species for disease occurrence and progression. The development of high-throughput sequencing technology is one of the driving forces behind this powerful method to simultaneously capture hundreds of bacterial species within a microbiota [1]. Metagenomics for microbiota has largely been based on sequencing the 16S rDNA library, which is constructed by DNA amplification of the 16S ribosomal RNA (rRNA) gene from samples, because of the universality of the 16S rRNA gene among bacteria [2]. On the other hand, improvements in sequencing technology have enabled investigators to obtain whole genomic sequences within a microbiota—literally, as "metagenomic" data that is not limited to being derived from a particular amplicon but comprehensively includes the information of whole genomes [3]. However, these DNA-based methods potentially capture live and dead bacteria because DNA can persist in an environment without regard to

Guy R. Adami (ed.), *The Oral Microbiome: Methods and Protocols*, Methods in Molecular Biology, vol. 2327,
https://doi.org/10.1007/978-1-0716-1518-8_11,

bacterial life or death status. Instead of these methods, metatranscriptomics has recently been applied to polymicrobial diseases. Metatranscriptomics comprehensively captures bacterial RNA (i.e., a mass of transcripts) within a microbiota [4]. It has an advantage over metagenomics with respect to capturing the transcriptional activities of bacteria at the time of sampling because of the short persistence of RNA. Although the microarray has been the erstwhile mainstream of bacterial metatranscriptomics, RNA sequencing (RNA-seq) by using a high-throughput sequencer is practically the first choice today [5]. In this chapter, I describe a protocol for performing RNA-seq for oral microbiota, based on my previous investigations of the microbiota at the lesions of periodontitis and peri-implantitis, and the microbiota in dental plaque [6, 7]. A problem with DNA-based studies of the microbiome is that they do not distinguish dead from live cells. In a particular bacterial species, the presence of ribosomal RNA (rRNA) indicates potential protein synthesis and correlates well with cell proliferation in some taxa but is not necessarily a good marker of active nondormant organisms [8]. Messenger RNA (mRNA) is a short-lived molecule that is required for gene expression. We calculated the ratio of mRNA to rRNA for each taxon to identify viable taxa with in situ function. In addition, the ratio can be used for identifying taxa that are transcribing the most genes and likely to be metabolically active; they are thus more likely to be contributing to phenotype in a group of samples.

2 Materials

2.1 Sample Collection

Microbiota samples should be categorized into several groups (at least two groups) in which the characteristics of microbiota are compared. For example, the disease group is compared to the healthy group with regard to the state of a particular tissue or organ. In that situation, the healthy group is considered the control group, which is used to determine how the disease state differs from the healthy state. Another example is the comparison of two different disease groups. My previous investigation recruited individuals who had periodontitis and peri-implantitis and compared the metatranscriptome of the microbiota between the periodontitis group and the peri-implantitis group [6]. Neither of these two groups was used as the control, and they were directly compared with each other. This method was sufficient for understanding how the microbiota at the sites of periodontitis and peri-implantitis differ.

Investigators should also consider systemic health and a history of medication use when recruiting study participants. To exclude the potential effects of systemic diseases or disorders to the microbiota being investigated, study participants should be systemically healthy. Antimicrobial agents and anti-inflammatory agents may alter the ecology of microbiota and the condition of an adjacent tissue or organ; therefore, study participants should refrain from receiving these agents within at least 3 months before sample collection.

To collect samples, prepare materials or instruments that are appropriate for the sample properties. For example, swabs are convenient for collecting samples from a wide surface such as the tongue and buccal mucosa. By contrast, samples obtained from restricted locations such as plaque in the cervical margin and periodontal pocket are collected by using a sterilized curette, toothpick, or paper point.

2.2 RNA Isolation

1. RNeasy PowerMicrobiome Kit (Qiagen).
2. NucleoSpin miRNA Kit (Macherey-Nagel).
3. Ethachinmate (Nippon Gene).
4. TURBO DNase (Thermo Fisher Scientific).
5. Quant-iT RiboGreen RNA Assay Kit (Thermo Fisher Scientific).
6. QUBIT fluorometer (Thermo Fisher Scientific).
7. RNA6000 Pico Kit (Agilent Technologies).

2.3 High-Throughput RNA Sequencing

1. A-Plus Poly(A) Polymerase Tailing Kit (Cellscript).
2. SMART-Seq v4 Ultra Low Input RNA Kit for Sequencing (Takara Bio USA).
3. Nextera XT DNA Library Preparation Kit (Illumina).
4. KAPA Library Quantification Kit for Illumina Platforms (Roche).
5. High Sensitivity DNA Kit (Agilent Technologies).
6. MiSeq Reagent Kit v3 for 600 cycles (Illumina).

2.4 Data Analysis

The data obtained from a high-throughput sequencer are so large that the data analysis requires large-scale computing resources. Several tens of central processing units and at least 64 GB memory may be required for the analysis. Computing servers for personal use are available and supercomputers at various places in the world are remotely accessible.

Investigators can conduct the analysis in this protocol by using open-source software programs. Shell scripting and the R packages (https://www.R-project.org/) are useful for most operations in the protocol. Specific software programs are otherwise indicated in this protocol if they are particularly required or recommended.

3 Methods

3.1 Sample Collection

1. Collect microbiota samples from the oral cavity. In my previous investigation, subgingival plaque samples from lesions of

periodontitis and peri-implantitis were collected by inserting ten pieces of paper points into the pocket for 30 s [6] (*see* **Note 1**). Before collecting samples, the supragingival area was dried by using sterile cotton to reduce contamination by eukaryotic DNA. Alternatively, the supragingival plaque samples can be collected from the tooth surface by using a sterilized toothpick [7] (*see* **Note 2**).

2. Store the collected samples at −80 °C until their use in the following steps, described in Subheading 3.2 (*see* **Note 3**).

3.2 RNA Isolation

1. Isolate bacterial RNA from the samples by using the RNeasy PowerMicrobiome Kit, based on the manufacturer's instructions. In this kit, RNA is extracted by bead beating and purified with a membrane filter column (*see* **Note 4**). Other kits that use bead beating in a chaotropic agent followed by phase separation to reduce DNA should also be applicable.
2. Remove small RNA from the extracted RNA by using the NucleoSpin miRNA Kit. This kit is generally used for extracting RNA and for purifying it. However, the procedure of binding large RNA and not small RNA to the filter column is used in this step.
3. Remove contaminating DNA by using TURBO DNase kit. The scale of reaction mixture containing two units of DNase depends on the concentration and volume of RNA solution. After the reaction, remove DNase by ethanol precipitation or other ways; Ethachinmate can be used as a coprecipitant in ethanol precipitation [6, 7].
4. Measure RNA concentration by using the Quant-iT RiboGreen RNA Assay Kit for fluorescence-based detection using a fluorometer such as QUBIT. Perform capillary electrophoresis by using the 2100 Bioanalyzer (Agilent Technologies) and the RNA6000 Pico Kit to check RNA quality. The purified RNA may appear as a broad peak ranging from several hundred to several thousand base pairs (bp) in the electropherogram (*see* **Note 5**).

3.3 High-Throughput RNA Sequencing

3.3.1 Synthesis of Complementary DNA (cDNA)

1. Add a polyadenylate tail to the 3′ end of RNA by using the A-Plus Poly(A) Polymerase Tailing Kit. This step is required for **step 2** in which an oligo(dT) primer complementarily binds to the poly(A) tail.
2. Perform reverse transcription by using the SMART-Seq v4 Ultra Low Input RNA Kit for Sequencing, based on manufacturer's instructions. This kit allows a low amount of input RNA to be applied at the picogram level.

3.3.2 Construction of a High-Throughput Sequencing Library

1. Perform fragmentation and add the adapter sequences by using the Nextera XT DNA Library Preparation Kit, based on the manufacturer's instructions. A library solution is obtained from each sample and contains fragmented and adapter-ligated cDNA. A barcode sequence is specifically given for each library in this step to distinguish them from each other.
2. Quantify the libraries by using the KAPA Library Quantification Kit for Illumina Platforms, and perform capillary electrophoresis by using the 2100 Bioanalyzer and High Sensitivity DNA Kit to check DNA quality.

3.3.3 High-Throughput Sequencing

1. Dilute each library solution to the concentration of 2 nM.
2. Mix equal volumes of all library solutions to obtain a combined library in which the concentration of total DNA is 2 nM (*see* **Note 6**).
3. Obtain the nucleotide sequence reads from the combined library by using the high-throughput sequencing MiSeq platform and the MiSeq Reagent Kit v3 for 600 cycles by which paired end reads are obtained as a pair of forward and reverse reads from each RNA fragment (*see* **Note 7**). Read data is generated for each sample.

3.4 Data Preprocessing

1. Trim the reads in which the trailing end contains many low-quality nucleotides and remove the low-quality reads. The Trimmomatic software (Usadel Lab) has modifiable parameters for trimming and removing reads [9] (*see* **Note 8**).
2. Remove the reads that are potentially of human origin. The Deconseq software (http://deconseq.sourceforge.net/) is used for this purpose [10] (*see* **Note 9**).
3. Divide the data into paired and unpaired reads. All reads are originally in pairs of forward and reverse reads when the data are generated using MiSeq. By contrast, unpaired reads appear because of the removal of one of a pair in **steps 1** and **2**. This step is necessary for some software programs in subsequent steps (*see* **Note 10**).

3.5 Analysis for 16S rRNA Reads

3.5.1 Taxonomic Assignment for 16S rRNA Reads

1. Reconstruct nucleotide sequences of the 16S rRNA gene from putative 16S rRNA reads by using EMIRGE software (https://doi.org/github.com/csmiller/EMIRGE/) [11]. In this step, the putative 16S rRNA reads in the data are automatically identified and grouped into clusters. The clusters are similar to operational taxonomic units (OTUs) in the ordinary procedure of 16S rDNA sequencing, in which the library of only 16S rRNA gene is sequenced and clustered into OTUs.
2. Taxonomically assign the clusters by using the BLASTN search tool (National Institutes of Health, Bethesda, MD, USA) [12]. In this step, the representative read in each cluster is

used for the similarity search against a database of nucleotide sequences of the 16S rRNA gene (*see* **Note 11**). The Human Oral Microbiome Database (HOMD; http://www.homd.org) records nucleotide sequences of the 16S rRNA gene, which were identified from human oral bacteria [13], and is useful for the read data from the human oral cavity (*see* **Note 12**).

3.5.2 Estimation of Alpha and Beta Diversity

1. Estimate alpha diversity from the EMIRGE output. The EMIRGE software generates an output table of the rRNA level per taxon as a percent value of reads per kilobase of transcript per million reads (RPKMs) (*see* **Note 13**). This table is used for calculating alpha diversity indices such as Shannon's index and Simpson's index. The alpha diversity indices are used for estimating the richness and/or evenness of bacterial taxa in each sample.
2. Estimate beta diversity from the EMIRGE output. Beta diversity is a measure of the differences in the components of two or more groups of populations and is used for estimating the richness and/or evenness among samples. Drawing rarefaction curves and calculating a correlation coefficient between samples are ways to estimate beta diversity. For these purposes, software programs for 16S rDNA sequencing such as mothur (Schloss Lab, Ann Arbor, MI, USA) and QIIME (http://qiime.sourceforge.net/) are useful [14, 15]. Correlation coefficients were designed with regard to whether the population is parametric or nonparametric (*see* **Note 14**). For example, the Spearman's rank correlation coefficient is used for a nonparametric population.

3.5.3 Visualization of Similarity and Dissimilarity among Samples

1. Calculate the distance value between samples from the table of taxonomic abundance. The methods of calculation are diverse. In my previous investigations, the value of 1 − Spearman's coefficient (i.e., the mathematical difference of the correlation coefficient from 1) was used when the data of taxonomic abundance were nonparametric [6, 7]. This was used as the index of dissimilarity because Spearman's coefficient indicates the extent of similarity. The distance values for all sample pairs form a distance matrix in which the samples are listed in the row and column.
2. Draw a dendrogram by applying an algorithm of hierarchical clustering to the distance matrix. A heat map may be drawn along with the dendrogram by visualizing the taxonomic abundance as a color gradient.
3. Perform the calculation for principal coordinate analysis (PCoA) as a transformation of the table of taxonomic abundance to a new matrix. The derived matrix is formed by multidimensional coordinates in which lower dimensions hold

mathematical variance to a higher extent. This feature is used for downsizing the original multidimensional data to low-dimensional description that is easily understandable with regard to similarity and dissimilarity among samples. The coordinates of the lowest two dimensions are used for the two-dimensional PCoA plot. In the plot, the percentage values of the proportion of variance are given for two axes to indicate how each coordinate explains the data variance of population.

4. Perform the statistical test of analysis of similarity (ANOSIM) between groups. The ANOSIM is used for statistically examining whether two groups are dissimilar from each other. The test provides *R* and *P* values. A higher *R* value (with a maximum of 1) indicates a higher extent of dissimilarity between groups. The *P* value indicates the statistical significance of *R* in the same manner as in the *t* test. An *R* value close to 1 and a *P* value close to 0 indicate significant dissimilarity between groups. This test can be used for various situations of statistically examining dissimilarity in other steps in this protocol.

3.6 Analysis for mRNA Reads

3.6.1 Assignment of mRNA Reads by a Publicly Accessible Pipeline

1. Upload the preprocessed data to the Metagenomics Rapid Annotation using Subsystem Technology (MG-RAST) pipeline (https://www.mg-rast.org). The MG-RAST is an open-source tool for obtaining taxonomic and gene profiles from the high-throughput sequence data (*see* **Note 15**). Putative mRNA reads are assigned by various types of databases such as the SEED database (http://www.theseed.org/) and Kyoto Encyclopedia of Genes and Genomes (KEGG) database (Kanehisa Laboratories, Kyoto, Japan). The SEED database has four hierarchical subsystems to categorize gene functions [16]. The KEGG database is used for metabolic pathways [17]
2. Various output formats are available from the analysis using MG-RAST. The results of the assignments using the KEGG database are visualized by using iPath software (European Molecular Biology Laboratory, Heidelberg, Germany) [18].

3.6.2 Formation of mRNA Clusters

1. Generate clusters from the preprocessed reads, based on nucleotide sequence similarity. The CD-HIT software program (http://weizhongli-lab.org/cd-hit/) is used for cluster formation with modifiable parameters [19].
2. Search the representative sequence of each cluster by using the BLASTN search tool against a database of rRNA. The clusters in which the representative sequence show high similarity in the BLASTN search are considered as 16S rRNA clusters. The ARB-SILVA database (Max Planck Institute for Marine Microbiology and Jacobs University, Bremen, Germany) contains data on 16S rRNA and on other small and large subunit

rRNAs of prokaryotes and eukaryotes [20]. This database is useful for identifying putative rRNA reads.

3. Remove rRNA clusters to use the remaining clusters as the putative mRNA clusters. These steps are required for the following procedure (described in Subheading 3.6.3), which enables assigning putative mRNA reads with any database, other than those contained in the MG-RAST.

3.6.3 Assignment of mRNA Clusters

1. Search the representative sequence of each mRNA cluster by using the BLASTX search tool (National Institutes of Health) against databases of protein [12]. The Virulence Factors of Pathogenic Bacteria (VFDB; http://www.mgc.ac.cn/VFs/) is a database of protein with virulence function [21]. The National Center for Biotechnology Information GenBank nonredundant protein database (NCBI nr; Bethesda, MD, USA) covers broad protein functions with the taxonomic information [22]. This information enables the prediction of the functions and the taxonomic origins of mRNA reads. The number of reads for each function is converted to RPKM values by using the length of the transcript.
2. Draw a dendrogram, heat map, and PCoA plot to visualize similarity and dissimilarity between samples, based on the table of abundance for protein functions (*see* Subheading 3.5.3).

3.7 Uniting the Assigned Results of rRNA and mRNA Reads

3.7.1 Identification of Viable Taxa with In Situ Function (VTiFs)

1. Convert the percent value of rRNA level (*see* Subheading 3.5) to the RPKM. This step is required to compare the rRNA levels with the mRNA levels (*see* **step 1** in Subheading 3.6.3) in the following steps.
2. Visualize the rRNA-based taxonomic profiles (converted to RPKM in **step 1**) and mRNA-based taxonomic profiles as a single PCoA plot. Each sample appears to be two spots in the plot: one spot for the rRNA-based taxonomic profile and the second spot for the mRNA-based taxonomic profile. The similarity and dissimilarity between the taxonomic profiles, based on the two independent methods, are observable as the positional relation of the spots.
3. Check whether each taxon in the rRNA-based profile is present or absent in the mRNA-based taxonomic profiles. The taxa that are present in the rRNA-based and the mRNA-based taxonomic profiles are considered as viable and transcriptionally active. Hereafter, they are called "viable taxa with in situ function (VTiFs)."

3.7.2 Normalization of the mRNA Level per Taxon, Based on the rRNA Level

1. Each VTiF has two values of abundance at the rRNA level (i.e., the converted RPKM in **step 1**) and the mRNA level (*see* **step 1** in Subheading 3.6.3). Transform these two values to log2 values and calculate their mathematical difference. The obtained value of log2(mRNA level) − log2(rRNA level) is mathematically equivalent to the value of log2(mRNA/rRNA). This value indicates the ratio of mRNA level to rRNA level (i.e., mRNA/rRNA ratio). The ratio indicates transcriptional activity; a higher mRNA/rRNA ratio indicates greater activity of mRNA expression over that of rRNA expression (*see* **Note 16**).
2. Draw a bar chart for the log2 mRNA/rRNA ratio of each VTiF. The bar chart in descending order is useful for visualizing which taxa are highly active in transcription. In my previous investigations, the VTiFs with mRNA/rRNA ratios higher than a particular threshold were called "active taxa" [6, 7].
3. For each VTiF, calculate the mathematical mean of the log2-transformed rRNA and mRNA levels as follows: [log2(mRNA level) + log2(rRNA level)]/2. This step is required for **step 4**.
4. Draw a two-dimensional scatter plot by using the values calculated in **step 1** for the *y* axis and the values calculated in **step 3** for the *x* axis. This plot is called an MA plot, in which the VTiFs with high transcriptional activity (as indicated by high mRNA/rRNA ratios) are located upward (Fig. 1).

3.7.3 Visualization of the Co-occurrent Relationship Among Taxa

1. In the table of mRNA-based taxonomic abundance, the mRNA level of each sample is given for each VTiF. Calculate the correlation coefficient for the mRNA levels between two VTiFs to know how the two VTiFs co-occur among samples in terms of mRNA expression. A higher coefficient value indicates that the mRNA levels of two VTiFs are more likely in proportion among samples. The SparCC software (https://bitbucket.org/yonatanf/sparcc), which estimates correlation in a robust manner, is available for this purpose [23].
2. Visualize the correlation coefficients for VTiF pairs as the network structure. The Cytoscape software (https://cytoscape.org) is used for drawing networks from the table of correlation coefficients [24]. Each VTiF is indicated by a node. Two VTiFs are connected by an edge if they co-occur in terms of mRNA expression with a coefficient greater than a particular threshold. Excluding VTiFs with considerably low mRNA levels may help in more clearly visualizing a co-occurrent relationship. In my previous investigations, the nodes of active taxa (*see* Subheading 3.7.2, **step 2**) were indicated by bold circles, and the edges of correlation coefficients with statistical significance (provided by the SparCC software) were indicated by bold lines

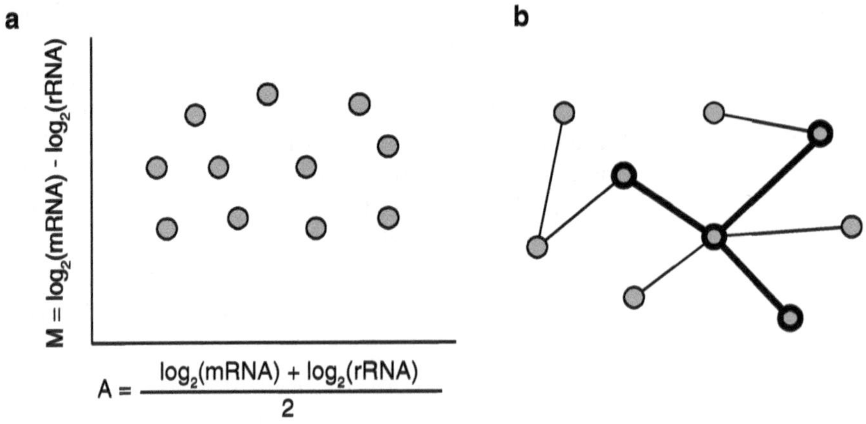

Fig. 1 Visualization of mRNA levels. (**a**) An MA plot. For each VTIF, the value of [log2(mRNA level) + log2(rRNA level)]/2 for the *x* axis and the value of log2(mRNA level) − log2(rRNA level) for *y* axis are calculated for each sample. The mean value among samples is then calculated to obtain a value of *x* coordinate and a value of *y* coordinate for each VTIF, represented as a circle in the plot. (**b**) A network structure as a visualized form of co-occurrent relationship. A correlation coefficient is calculated for each pair of VTIFs. A pair with a coefficient greater than a particular threshold is represented by circles for VTIFs and a line for a co-occurrent relationship. Interacting core taxa are represented by bold circles, and a co-occurrent relationship with statistical significance is represented by a bold line

[6, 7]. The active taxa which were detected in most samples and co-occurred with statistical significance (i.e., the nodes of bold circles that were connected by bold lines) were considered as the core in the networks and were called "interacting core taxa" (*see* **Note 17**) (Fig. 1).

4 Notes

1. Wiping or scratching the surface of a site of interest would efficiently yield samples that are sufficient in quantity for an investigation. However, these methods may disrupt the surrounding environment. For example, using a curette to collect plaque from the tooth surface in the deep part of the periodontal pocket may unintentionally disrupt the gingival epithelium. A swab is useful for this situation, but the periodontal pocket is so narrow that reaching inside is difficult with currently available swabs. Thus, a paper point was used in my previous investigation, which also took into consideration protecting the implant surface from disruption.
2. In my previous investigation, supragingival plaque was collected from all surfaces of all teeth. This method would help in understanding the characteristics of plaque microbiota as the aggregate from all teeth without considering their site-specificity with respect to the location and surface morphology

of each tooth. Collecting plaque from a particular part (e.g., buccal or mesiolingual) of a particular tooth may contribute to understanding site-specificity.

3. In my previous investigations, the samples were immersed into the buffer that appears first in the next step (*see* **step 1** in Subheading 3.2) after collected from the oral cavity, and stored as the immersed state at −80 °C [6, 7].
4. Usage of this and similar kits results in contamination with RNA that is potentially derived from the human host and is represented in the RNA-seq reads. These contaminants can be removed in the data preprocessing step (*see* **step 2** in Subheading 3.4).
5. The purified RNA should be sufficient in its amount for the input to the next step (*see* Subheading 3.3.1); however, in my experience, a lower amount of purified RNA than the required amount did not cause significant problems in the subsequent steps [6, 7].
6. Even if the concentration of particular libraries is less than 2 nM, performing the steps again for them to obtain a sufficient concentration may be unnecessary. A combined library of 2 nM may be obtained by mixing a small volume of libraries in which the concentration is much higher than 2 nM and large volume of libraries in which the concentration is less than 2 nM.
7. According to the manufacturer, the MiSeq platform can generate maximally 25 million reads, which corresponds to 15 Gb as the length of nucleotide sequence. Multiple samples can be involved in a single run of MiSeq. In my previous investigation, the libraries for 24 samples were mixed together for a single run. The number of samples involved depends on how much data are required for the analysis.
8. This software has a parameter, called "MINLEN," for controlling the minimal length of reads passing through the filtration. Setting this parameter to be near the maximal length of 300 bp obtains filtered reads in which the stretch of the nucleotide sequence is highly preserved. However, this method would be too strict to sufficiently analyze the filtered data in the subsequent steps. In my previous investigations, this parameter was set as 50 to maintain the filtered reads as much as possible, even though the reads in which the greater part was trimmed were abundant.
9. The data obtained by high-throughput sequencing should be deposited at a publicly available database and should only consist of reads of bacterial origin. After removal of eukaryotic reads, the decontaminated reads are an acceptable format to be deposited in sequence archives.

10. For example, the EMIRGE software (*see* **step 1** in Subheading 3.5.1) accepts only paired read data.
11. The criteria for considering whether a hit in the BLAST search is significant are modifiable as in the ordinary use of BLAST, such as altering setting of the thresholds of the *e*-value and sequence identity.
12. In the case of 16S rDNA sequencing, the length of the amplicon is generally too short for precisely identifying taxonomic composition, which leads to assignment at the genus level or higher ranks. Considering that this protocol is not for amplicon but is for data in which the whole length of 16S rRNA is contained even as fragmented reads, taxonomic assignment at the species level using this protocol would not seriously result in confusion in understanding bacterial composition.
13. This means that the RPKM values, the sum of which for each sample is one million reads, are converted in EMIRGE processing to be the percent values, the sum of which is 100%.
14. The term "parametric" means that the population follows a normal distribution. Some statistical tests are available for estimating whether the population is parametric; however, in my previous investigations, the data were regarded as nonparametric for the analysis, taking into consideration that there may be a certain number of outlying values (i.e., the extremely high or low values when considering the data variance) that would cause invalid statistical outcomes if parametric tests were used [6, 7].
15. The time needed for processing with the MG-RAST tool seems to depend on the server condition. I have experienced that jobs for the MG-RAST required several months to generate results. It is one of the few methods to do gene-level analysis of microbial RNA and DNA sequence that is publicly available.
16. In this protocol, the process of calculating the mRNA/rRNA ratio is called "normalization." Be aware that this term is not used in this chapter for the use in statistics such as the meaning of converting values to the range of 0 to 1 or converting the mean to 0 and the variance or standard deviation to 1 (also called "standardization").
17. When examining the co-occurrence of bacterial taxa, it is important to check the number of samples in which each taxon is detected. Taxa may not necessarily be detected in every sample; therefore, interacting core taxa would be responsible for being the core in the network because they were detected in most samples.

References

1. Loman NJ, Constantinidou C, Chan JZM, Halachev M, Sergeant M, Penn CW et al (2012) High-throughput bacterial genome sequencing: an embarrassment of choice, a world of opportunity. Nat Rev Microbiol 10 (9):599–606. https://doi.org/10.1038/nrmicro2850
2. Clarridge JE III (2004) Impact of 16S rRNA gene sequence analysis for identification of bacteria on clinical microbiology and infectious diseases. Clin Microbiol Rev 17(4):840–862. https://doi.org/10.1128/cmr.17.4.840-862.2004
3. Breitwieser FP, Lu J, Salzberg SL (2019) A review of methods and databases for metagenomic classification and assembly. Brief Bioinform 20(4):1125–1136. https://doi.org/10.1093/bib/bbx120
4. Singer E, Wagner M, Woyke T (2017) Capturing the genetic makeup of the active microbiome *in situ*. ISME J 11(9):1949–1963. https://doi.org/10.1038/ismej.2017.59
5. Lowe R, Shirley N, Bleackley M, Dolan S, Shafee T (2017) Transcriptomics technologies. PLoS Comput Biol 13(5):e1005457. https://doi.org/10.1371/journal.pcbi.1005457
6. Shiba T, Watanabe T, Kachi H, Koyangi T, Maruyama N, Murase K et al (2016) Distinct interacting core taxa in co-occurrence networks enable discrimination of polymicrobial oral diseases with similar symptoms. Sci Rep 6:30997. https://doi.org/10.1038/srep30997
7. Funahashi K, Shiba T, Watanabe T, Muramoto K, Takeuchi Y, Ogawa T et al (2019) Functional dysbiosis within dental plaque microbiota in cleft lip and palate patients. Prog Orthod 20(1):11. https://doi.org/10.1186/s40510-019-0265-1
8. Blazewicz SJ, Barnard RL, Daly RA, Firestone MK (2013) Evaluating rRNA as an indicator of microbial activity in environmental communities: limitations and uses. ISME J 7 (11):2061–2068. https://doi.org/10.1038/ismej.2013.102
9. Bolger AM, Lohse M, Usadel B (2014) Trimmomatic: a flexible trimmer for Illumina sequence data. Bioinformatics 30 (15):2114–2120. https://doi.org/10.1093/bioinformatics/btu170
10. Schmieder R, Edwards R (2011) Fast identification and removal of sequence contamination from genomic and metagenomic datasets. PLoS One 6(3):e17288. https://doi.org/10.1371/journal.pone.0017288
11. Miller CS, Baker BJ, Thomas BC, Singer SW, Banfield JF (2011) EMIRGE: reconstruction of full-length ribosomal genes from microbial community short read sequencing data. Genome Biol 12(5):R44. https://doi.org/10.1186/gb-2011-12-5-r44
12. Altschul SF, Gish W, Miller W, Myers EW, Lipman DJ (1990) Basic local alignment search tool. J Mol Biol 215(3):403–410. https://doi.org/10.1016/s0022-2836(05)80360-2
13. Chen T, Yu WH, Izard J, Baranova OV, Lakshmanan A, Dewhirst FE (2010) The Human Oral Microbiome Database: a web accessible resource for investigating oral microbe taxonomic and genomic information. Database (Oxford) 2010:baq013. https://doi.org/10.1093/database/baq013
14. Schloss PD, Westcott SL, Ryabin T, Hall JR, Hartmann M, Hollister EB et al (2009) Introducing mothur: open-source, platform-independent, community-supported software for describing and comparing microbial communities. Appl Environ Microbiol 75 (23):7537–7541. https://doi.org/10.1128/AEM.01541-09
15. Bolyen E, Rideout JR, Dillon MR, Bokulich NA, Abnet CC, Al-Ghalith GA et al (2019) Reproducible, interactive, scalable and extensible microbiome data science using QIIME 2. Nat Biotechnol 37(8):852–857. https://doi.org/10.1038/s41587-019-0209-9
16. Overbeek R, Begley T, Butler RM, Choudhuri JV, Chuang HY, Cohoon M et al (2005) The subsystems approach to genome annotation and its use in the project to annotate 1000 genomes. Nucleic Acids Res 33 (17):5691–5702. https://doi.org/10.1093/nar/gki866
17. Kanehisa M, Goto S (2000) KEGG: Kyoto encyclopedia of genes and genomes. Nucleic Acids Res 28(1):27–30
18. Darzi Y, Letunic I, Bork P, Yamada T (2018) iPath3.0: interactive pathways explorer v3. Nucleic Acids Res 46(W1):W510–W513. https://doi.org/10.1093/nar/gky299
19. Li W, Godzik A (2006) Cd-hit: a fast program for clustering and comparing large sets of protein or nucleotide sequences. Bioinformatics 22(13):1658–1659. https://doi.org/10.1093/bioinformatics/btl158
20. Quast C, Pruesse E, Yilmaz P, Gerken J, Schweer T, Yarza P et al (2013) The SILVA ribosomal RNA gene database project: improved data processing and web-based

tools. Nucleic Acids Res 41(DI):D590–D596. https://doi.org/10.1093/nar/gks1219

21. Chen L, Yang J, Yu J, Yao Z, Sun L, Shen Y et al (2005) VFDB: a reference database for bacterial virulence factors. Nucleic Acids Res 33 (DI):D325–D328. https://doi.org/10.1093/nar/gki008

22. O'Leary NA, Wright MW, Brister JR, Ciufo S, Haddad D, McVeigh R et al (2016) Reference sequence (RefSeq) database at NCBI: current status, taxonomic expansion, and functional annotation. Nucleic Acids Res 44(DI): D733–D745. https://doi.org/10.1093/nar/gkv1189

23. Friedman J, Alm EJ (2012) Inferring correlation networks from genomic survey data. PLoS Comput Biol 8(9):e1002687. https://doi.org/10.1371/journal.pcbi.1002687

24. Shannon P, Markiel A, Ozier O, Baliga NS, Wang JT, Ramage D et al (2003) Cytoscape: a software environment for integrated models of biomolecular interaction networks. Genome Res 13(11):2498–2504. https://doi.org/10.1101/gr.1239303

Chapter 12

Noninvasive Acquisition of Oral Mucosal Epithelial miRNA and Bacteria DNA/RNA from a Single Site

Guy R. Adami

Abstract

Many digestive tract microbes live adhered to tract epithelium. Work in recent years has brought the realization that these microbes and the host epithelial cells certainly must interact and that this interaction has an effect on both. One way to understand the interaction is to measure which genes are expressed in the epithelial cells and what bacteria are present. Even more informative would be to also determine what genes the bacteria express. Presented is a method to noninvasively isolate oral mucosal epithelium so to provide purified miRNA that can be used to profile miRNA expression specifically in the epithelium. miRNA is a major regulator of cell functions. Simultaneously, DNA and RNA from bacteria at the same site can be isolated to allow characterization of bacteria that coat the epithelial cells and extracellular matrix. This provides insight on the interaction between host and bacteria.

Key words Microbes, Oral mucosa, miRNA, 16s rRNA gene, Brush biopsy, RNA stabilizers

1 Introduction

Brush biopsy offers a noninvasive method of acquisition of cells and RNA from the epithelium of the oral mucosa [1]. As a result of sampling at sites within lesions that are not ulcerated, non-necrotic, and less likely to bleed [1], the samples are nearly pure epithelium: keratinocytes plus a much smaller number of immune cells. For years, this approach, while it provides information about oral squamous cell carcinoma, OSCC, [2–5], has been limited by the variability of mRNA recovery, with substantial numbers of samples having insufficient yield [6–8]. Quality is also generally lower than that of surgically obtained tissue, making RT-PCR the most reliable method of measurement [9, 10]. The variable mRNA quality in brush biopsy samples is largely specific to squamous epithelium samples [10]. In contrast, brush samples from the pseudostratified columnar epithelium of the bronchi allow reproducible measurement of mRNA, which has resulted in a clinically useful method to detect lung cancer [11, 12]. There has also been

Guy R. Adami (ed.), *The Oral Microbiome: Methods and Protocols*, Methods in Molecular Biology, vol. 2327, https://doi.org/10.1007/978-1-0716-1518-8_12, © Springer Science+Business Media, LLC, part of Springer Nature 2021 progress in the diagnosis of hard-to-detect bile duct malignancies, using brush-based sampling of mRNA from the simple cuboidal epithelium of the duct [13, 14]. The situation is different with oral microRNA (miRNA). miRNA obtained by brush biopsy from oral squamous epithelium, unlike mRNA, is of high quality similar to that of frozen tissue [6, 15, 16]. miRNA profiles in individual brush biopsy OSCC samples show approximately 50% overlap with miRNA enriched in surgically obtained tumor tissue miRNA [6]. Total overlap was not expected as surgical samples contain varying amounts of stroma, which can express miRNAs at levels different than those seen in epithelium proper. Another difference is that few miRNAs have been observed to decrease in brush biopsy OSCC samples. This is likely due to the inclusion of normal epithelium in brush biopsy sampling of smaller tumors, while normal samples are 100% normal. The method is reproducible: Comparison of one study that examined a large set of miRNAs and a separate study that examined a much smaller group found that of the 6 miRNAs in brush biopsy samples of OSCC in both studies, half showed the same statistically significant differences in the OSCC group [6]. One might conclude that by removing stroma as a contaminant, fewer samples are needed to produce a reproducible miRNA signature predictive of OSCC [17]. The simple acquisition of nearly pure cells from the epithelium is a possible large advantage when using brush biopsy, compared to tissue acquistion by scalpel biopsy. Perhaps unexpectedly, the presence of multiple and distinct cell layers in squamous epithelium and differential acquisition by the brush has not prevented its accuracy in differentiating tumor versus normal epithelium based on brush biopsy acquired miRNA profiles. Measurement of oral epithelium miRNA is not restricted to tumor samples. A study that compared brush biopsy samples from the lateral border of the tongue of tobacco smokers and never smokers revealed distinct differences in miRNA expression in the epithelium as harvested [18]. A similar study in green tea users also revealed difference in epithelium mRNA expression before and after green tea consumption, though the differences were subtle. Notably, yield of miRNA from subjects without tumor or other lesion can be lower than when obvious pathology is present but this does not affect the accuracy of the method to profile oral epithelial miRNAs.

With cytology brush collection of oral epithelium, there is simultaneous collection of bacteria on the epithelium. This results in a sample that contains cells from the epithelium, including keratinocytes and possibly innate and adaptive immune cells, cell matrix material, and the bacteria coating the exact site. Modern sample preservatives and increased sensitivity of next-generation sequencing (NGS) methods have made easy the examination of multiple biomolecules, such as RNA, and DNA from body fluid samples, such as saliva; one limitation is that it is not known where the molecules come from or where they work [19, 20]. In contrast, brush-obtained samples contain biomolecules that emanate and/or work at the site of collection—the mucosal epithelium. Included here are two approaches to study the oral epithelium and the bacteria that coat them. In one, the samples are collected serially. First, a cotton swab is used to remove bacteria. This sample will contain microbes and possibly cells from the host but few host cells that are intact. Immediately following, a cytology brush is run across the same site to collect host epithelial cells. Alternatively, both bacteria and intact epithelium cells can be harvested simultaneously with a cytology brush and placed in a single tube. How these samples are stored is determined by the needs of the experiment. Optimally, samples are collected and then immediately aliquoted and processed, but this is often not practical in a study where some samples are collected off site. However, a less demanding protocol at the time of collection can be used that will be applicable to all samples in the study. Ideally, a protocol must be chosen that provides a profile of RNA and DNA most like that of samples processed immediately on collection, and is convenient enough that it also can be carried out for all samples.

Some workers freeze oral samples, such as plaque, immediately after collection [21]. This approach is risky when applied to aqueous samples, such as saliva. For microbial DNA, this is acceptable as long as the samples are not centrifuged after thaw, when in theory bacteria of some taxa may lyse and thus be under-represented in the end [22]. For microbial RNA there is also the risk of degradation of the biomolecule during thawing and loss of RNA due to cell lysis. Solutions like DNA/RNA Shield (Zymo Research) can be used to preserve the DNA and RNA and do so in a way that inactivate pathogens. Other RNA preservatives like RNAlater (ThemoFisher) and RNAprotect Cell (Qiagen) similarly stabilize samples, but do so without disrupting cell structure and allow isolation of cells/bacteria by centrifugation post usage, thus making easy the usage of a variety of selection of isolation methods. The author has validated the usage of RNAprotect Cell Reagent for the study of oral epithelial miRNA and thus have focused on this approach in this protocol. These RNA preservatives have not been validated to maintain microbial RNA species complexity, either bacteria or fungi, though they have been used to study metatranscriptomics of oral samples as there are few good alternatives [23–27]. The possible variability between methods makes it important to follow STROBE guidelines for metagenomics studies, which recommend recording conditions of sample preservation and processing.

2 Materials

Use Milli-Q grade water (Millipore; which further purifies deionized water, to attain a sensitivity of 18 MΩ cm at 25 °C) or commercially available sterile nuclease-free distilled water. Care must be followed when discarding waste.

2.1 Equipment

1. Micropipettes (P2, P20, P200, P1000).
2. UV cabinet for DNA/RNA manipulation.
3. Fume hood cabinet when working with Trizol (Thermo Fisher Scientific Scientific)/RNAzol (Sigma Aldrich) or use an N95 mask. Wear eye protection and other safety gear.
4. Vortex mixer.
5. General purpose benchtop refrigerated microcentrifuge.
6. Fluorometer for PicoGreen and RiboGreen (both, Thermo Fisher Scientific) or other similar fluorescent dye for DNA and RNA quantification.
7. Microvolume spectrophotometer.
8. Omnidirectional shaker for bead-based homogenization.

2.2 Consumables

1. Catch-All Sample Collection Swabs (Epicentre) or sterile cotton tipped applicators.
2. Sterile barrier filter tips for pipettes.
3. Protein low binding 1.5 mL tubes with Safe-Lock system, free of human DNA, DNase, RNase, and PCR inhibitors.
4. Surgical scalpel blades.
5. Sterile cytology brush, such as Cytosoft (Medical Packing Corp) or Cytobrush Plus (Medscand).
6. RNA Clean and Concentrator-5 (Zymo Research) or similar silica-based method of RNA/nucleic acid purification. Contains Prep Buffer and Wash Buffer and spin columns.
7. miRNeasy Micro Kit (Qiagen) or similar silica-based method of RNA purification.
8. 2-mL screw-cap tubes with 0.1-mm and 0.5-mm glass and zirconia silica beads.

2.3 Reagents

1. RT-PCR grade water: non DEPC-treated, eukaryotic and bacteria DNA free, DNA/RNA nucleases free.
2. RNA stabilizer such as RNAprotect (Qiagen) cell reagent or RNAlater (Thermo Fisher Scientific), which preserve cell morphology.
3. Trizol (Thermo Fisher Scientific) or other similar guanidinium thiocyanate/phenol reagent.

4. 1-Bromo-3-Chloropropane.
5. 2000 U/mL recombinant genetically altered DNase I (grade I) Turbo DNAse (Thermo Fisher Scientific), RNAse free, in glycerol stored at −20 °C. 10× Reaction Buffer.
6. Phenol solution equilibrated in H_2O.
7. Mussel glycogen. Prepare at 20 μg/μL in RT-PCR grade water.
8. Absolute ethanol. Ethanol can be diluted to 70% and 80% in nuclease-free water.

2.4 Buffers

1. 2 M NaAc pH 4.0: Dissolve 164.0 g of sodium acetate in 500 mL of deionized Milli-Q water. Adjust the pH to 4.0 with glacial acetic acid. Allow the solution to cool overnight. Adjust the pH once again to 4.0 with glacial acetic acid and the final volume to 1 L with Milli-Q water. Can make RNAse free by adding 1 mL diethylpyrocarbonate, shake, then let sit overnight at room temperature, then sterilize by autoclaving.
2. 3 M NaAc pH 5.6: Dissolve 246.1 g of sodium acetate in 500 mL of deionized Milli-Q water. Adjust the pH to 5.6 with glacial acetic acid. Allow the solution to cool overnight. Adjust the pH once again to 5.6 with glacial acetic acid and the final volume to 1 L with Milli-Q water. Divide into aliquots and sterilize by autoclaving. This and other non-Tris based solutions can have RNAse inactivated with diethylpyrocarbonate by adding 1/1000 volume diethylpyrocarbonate and mixing well, then leaving at room temperature several hours prior to autoclaving.
3. Phosphate-buffered saline (PBS): 1×, pH 7.4. Prepare 800 mL of distilled water and add 0.2 M NaCl (11.6 g), 2.5 mM KCl (0.186 g), 8 mM Na_2HPO_4 (1.4 g), 1.5 mM KH_2PO_4 (0.2 g). Adjust the pH to 7.4 and add distilled water to prepare a 1 L solution of 1× PBS.
4. Tris–HCl Buffer: 1.0 M, pH 7.5. Prepare 100 mL of distilled water and add 60.55 g Tris to the solution. Adjust to a pH of 7.5 with HCl and add distilled water to prepare a volume 500 mL 1.0 M Tris–HCl solution.
5. EDTA 0.5 M, pH 8.0, to 400 mL distilled H_2O add 93.05 disodium ethylenediaminetetraacetate, dihydrate, then slowly add about 18 g sodium hydroxide in pellet form while mixing. Add small amounts of 10 N NaOH in liquid form till pH of 8.0 is achieved. Bring volume to 500 mL, cap, and autoclave.
6. TE Buffer: 1 mL 1 M Tris–HCl, pH 7.5, plus 200 μL 0.5 M EDTA, pH 8.0. Add distilled water to 100 mL.

3 Methods

3.1 Swab Collection of Microbes

1. Patient is screened for antibiotic usage, germicidal oral rinse usage, and time of last meal. For lateral border of tongue samples, ask the subject to protrude the tongue. The tip of the tongue is covered with a two sterile 2″ × 2″ gauze pads and grasped between index finger and thumb, while the other hand is used to run a cotton swab back and forth over 1–2 cm^2 area on the lateral border of the tongue, for about 10 s, assuring all sides of the swab are exposed.
2. The cotton swab is inserted in labeled 1.8 mL microcentrifuge tube with 800 μL TE. Shaft of swab is broken so brush can easily fit in the tube. It is closed, stored on ice for up to 2 h then frozen and stored at −20 or −80 °C as described in the top of Fig. 1.

3.2 Isolation of Microbial DNA from Frozen Sample in TE

1. Sample is removed from the freezer, thawed, and then vortexed over 30 s. Immediately 150 μL of the suspension is removed using a pipet with a barrier tip and then placed in a 1.5 mL microcentrifuge tube.
2. Sample as is without centrifugation is now ready to be subjected to full bacteria lysis and DNA extraction using any of many methods available.

3.3 Collection of Intact Epithelial Cells

1. Immediately following swab collection of biofilm, a similar procedure is done with a cytology brush. Brush is held firmly against the same site at the lateral border of the tongue site while rotated and moved back and forth over a 1 or 2 cm^2 area, preferably the same site as above, for a full 1 min. If a lesion is sampled, care is taken to avoid ulcers or areas prone to bleed (*see* **Note 1**).
2. The shaft of brush is cut above brush attachment and dropped into a 1.5 mL microcentrifuge tube with 0.8 mL Trizol or RNAzol or similar guanidium isothiocyanate plus phenol reagent. Sample is vortexed, placed in dry ice, and then stored frozen at −80 °C (*see* **Note 2**).

3.4 Host Cell miRNA Isolation from Trizol/RNAzol with Nonoptimized Microbial RNA Isolation

Prior to using this method, refer to **Note 3**.

1. Sample in 800μL Trizol/RNAzol or similar agent is thawed.
2. Vortex for 15 s, then use washed, flame-treated forceps to remove brush head, draining as best possible. Discard brush head properly. Leave closed tube on bench for 5 min.
3. Add 80μL 1-Bromo-3-Chloropropane, then shake vigorously for 15 s. Leave on bench top for 2–3 min.
4. Centrifuge for 15 min at 12,000 × *g* 4 °C.

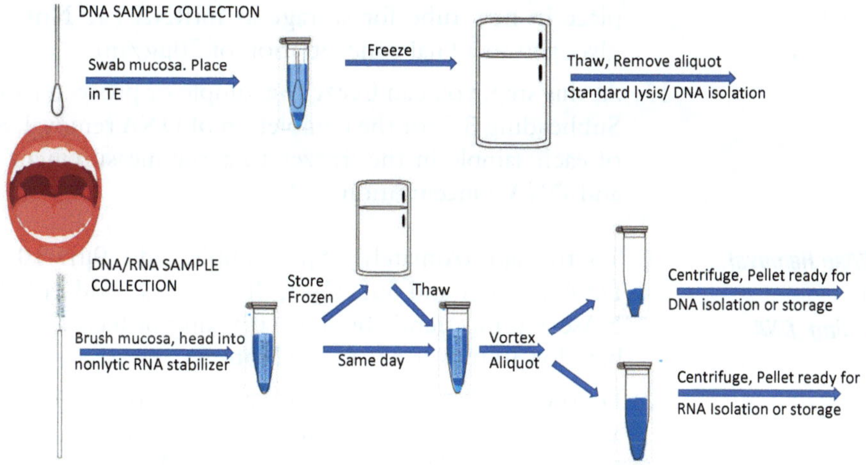

Fig. 1 Diagram top shows protocol summary for sample collection and storage after swab collection with the goal of studying the DNA. Diagram bottom shows protocol summary for sample collection and storage of brush samples collected in RNAprotect cell reagent with the goal of studying RNA and DNA from all cells. Alternative method of host miRNA isolation as described in Subheading 3.2 is not shown

5. Take upper phase with pipette, being careful not to disturb the interphase, and place in 1.8 mL microfuge tube.
6. Add 1.5 volume 100% ETOH, vortex. Add at most 700 μL at a time of the mixture to RNAeasy MinElute spin column in 2 mL collection tube.
7. Centrifuge at >8000 × *g* for 30 s. Empty 2 mL collection tube.
8. Add remaining extract + ethanol mixture to the column and repeat **step** 7.
9. Add 700μL RWT to wash the column. Centrifuge at >8000 × *g* for 15 s. Empty collection tube.
10. Add 500μL RPE solution to wash the column. Centrifuge at >8000 × *g* for 15 s. Empty collection tube.
11. Add 500μL 80% ethanol to wash the column. Centrifuge at >8000 × *g* for 15 s. Discard collection tube.
12. Place column in new tube with no lid and centrifuge at 12,000 × *g* for 5 min to remove all liquid.
13. Discard collection tube. Place column in a new microcentrifuge tube from which the lid has been removed.
14. Add 40μL H_2O to the spin column and centrifuge at 12,000 × *g* for 1 min.
15. Add 40μL H_2O to spin column, then centrifuge at 12,000 × *g* for 1 min.
16. Remove spin column and discard. Place a lid on tube. Centrifuge for 1 min at 12,000 × *g*. If there is evidence of a pellet, then remove liquid, being careful not to disturb the pellet, and

place in new tube for storage as miRNA (*see* **Note 4**). Add glycogen to a final concentration of 20μg/mL.

17. At this step, you can freeze the sample or proceed to **step 1** in Subheading 3.5 for the completion of DNA removal. Save 2μL of each sample in the freezer to allow measurement of RNA and DNA concentration.

3.5 DNAse Removal of Much of Remaining DNA

1. To the approximately 75μL sample, add 9μL 10× Turbo DNAse buffer and 5μL H_2O, then vortex. Add 1μL TurboDNAse and mix gently by slow vortexing or flicking tube. Incubate for 30 min at 37 °C (*see* **Note 5**).
2. Following this incubation, several cleanup methods are available and work well. I typically use RNA Clean and Concentrator from Zymo Research. To 90μL add 2 volume RNA binding buffer (180μL). To this add equal volume 100% ethanol (270μL), mix, and transfer to Zymo-Spin IC column. Centrifuge for 1 min at >12,000 × *g* at room temperature.
3. Perform column washes as directed by manufacturer and elute with 20μL H_2O twice. Add glycogen to final concentration 20μg/mL.

3.6 Final Concentration Determination in Preparation for cDNA Synthesis, miRNA Measurement

1. Verify loss of DNA and quantify RNA levels using fluorescence-based measurement of these molecules using QUBIT or similar fluorometer device. Samples are now ready for conversion to cDNA and analysis (*see* **Note 6**). At this point, a choice must be made to use RT-PCR or miRNAseq-based approach to quantify individual miRNAs (*see* **Note** 7).

3.7 Single Brush and Tube Method

This is for simultaneous collection of epithelial cells and bacteria with one brush used to collect the sample as diagrammed in the lower half of Fig. 1.

1. Subjects are screened as described earlier in Subheading 3.1. A single cytology brush is used to simultaneously collect microbial and epithelial samples using the same method for epithelium cell collection described above.
2. Brush head is released into tube with RNAprotect Cell Reagent (other potential options include RNAlater).
3. Tube is shaken or lightly vortexed and can then be kept at room temperature or preferably on ice for up to several hours. There are several options for preparation of the sample for storage (*see* **Notes 8** and **9**).

3.8 Isolation of Microbial DNA

1. As diagrammed in Fig. 2, thawed 100 μL sample is vortexed (*see* **Note 10**).
2. Centrifuge for 5 min at 5000 or 5500 × *g* and 4 °C.

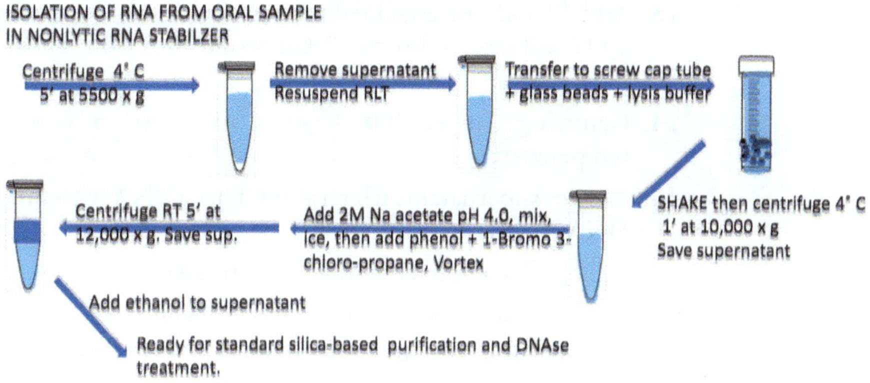

Fig. 2 Diagram shows protocol for what to do with brush biopsy samples that were originally preserved in RNAprotect cell reagent with the goal of isolating host and microbial RNA

3. Remove supernatant with a pipette, being careful not to disturb the pellet.
4. Centrifuge again for 1 min and remove remaining liquid.
5. Add 100μL PBS or TE. Sample is now ready for standard microbial DNA isolation after microbe lysis and then 16s rRNA gene analysis.

3.9 Isolation of Host Cell RNA for Maximal Yield and Microbial RNA

1. As diagrammed in Fig. 2, remove 700μL sample stored in RNAprotect Cell reagent from freezer and thaw (*see* Notes **11** and **12**).
2. Centrifuge at 5000 or 5500× g for 5 min at 4 °C.
3. Remove supernatant and discard as biohazard.
4. Centrifuge for 1–5 min and then remove remaining supernatant traces.
5. Add 1 mL RLT (Qiagen). Vortex to resuspend pellet.
6. Add to tube with silica or zirconium breaking beads (*see* **Note 13**).
7. Shake 1–1.5 min in MiniBead Beater-24 or similar homogenizer. Tube should be warm but not hot after shaking.
8. Place tubes on ice for 5 min.
9. Repeat shaking for 1–1.5 min.
10. Centrifuge at 10,000 × g for 1 min at 4 °C.
11. Collect as much of aqueous supernatant as possible without disturbing beads. The volume should be about 400 μL. If volume differs, adjust next steps accordingly.
12. To 400 μL, add 40 μL, or 1/10 volume, 2 M NaAcetate pH 4.0. Invert tube several times, then place on ice for 10 min.

13. Add 440 μL, or equal volume, phenol in H_2O, plus 88 μL, or 1/10 volume, 1-Bromo-3-Chloropropane to each tube. Vortex for 15 s.
14. Centrifuge at 12,000 × *g* or more for 5 min at room temperature.
15. Collect supernatant, which should have RNA with much of the DNA removed.
16. Add 1.25 volume ethanol (*see* **Note 14**).
17. To use Zymo RNA Cleanup and Concentrator, vortex sample and add to Zymo-Spin 1C Column in 2 mL collection tube.
18. Centrifuge at >10,000 × *g* for 30 s at room temperature. Discard flow through.
19. Wash with 400μL Prep Buffer.
20. Centrifuge at >10,000 × *g* for 30 s. Discard flow through.
21. Wash with 700μL Wash Buffer.
22. Centrifuge at >10,000 × *g* for 30 s at room temperature. Discard flow through.
23. Wash with 400μL Wash Buffer.
24. Centrifuge at >10,000 × *g* for 2 min at room temperature. Discard flow through and collection tube.
25. Place column with RNA bound in new 1.8 mL collection tube without lid.
26. Elute with 40μL H_2O. To do this add 40μL H_2O and centrifuge at >10,000 × *g* for 1 min. Repeat with another 40μL H_2O.
27. DNAase removal of remaining DNA and remainder of protocol is identical to that above for RNA purification directly from Trizol outlined in **steps 1–3** in Subheading 3.5.
28. The sample contains total RNA. For storage for future miRNA analysis, add glycogen to final concentration 20μg/mL. Store frozen in aliquots, though miRNA should show some stability to multiple thaws.
29. For storage of larger RNAs, for example in 50 μL, one can add 1/10 volume 3 M NaAcetate pH 5.2 plus 1 μL 20 mg/mL glycogen, vortex and then 2.5 volumes ethanol and vortex again (*see* **Note 15**). Store indefinitely at −20 or −80 °C.
30. There is no need to aliquot the sample (*see* **Note 16**).
31. Purified DNA is ready for PCR amplification with appropriate primer barcoding and sequencing. RNA species from host epithelium and accompanying microbes can be quantified, using PCR-based or RNAseq approaches. With careful technique quality miRNA should result (*see* **Note 17**).

4 Notes

1. The subject will feel the brush but should not feel discomfort or pain. A standard cytology brush can be used, though in some cases the brush head must be bent to accommodate collection from sites in the rear of the mouth. An alternative to the standard brush is a collector, such as the Orcellex Brush from Rovers Medical Devices, which has an advantage in collecting from sites in the posterior oral cavity, but has a large head necessitating usage of large 3–5 mL tubes for sample storage.
2. Note that any RNA preservative that leaves intact cells but inactivates nuclease can be used and the sample stored frozen.
3. This specific part of the protocol is optimized for isolation of maximal amounts of epithelial RNA but will not give a full picture of microbial RNA due to incomplete lysis of bacteria.
4. Removal of inert column fines is only necessary if a nanospectrophotometer will be used to measure RNA concentration based on UV absorbance.
5. Turbo DNAse is used as it tends to have high activity even when DNA is dilute, which is optimal for these types of samples, though other DNAse preparations or variants may in theory also work well.
6. The typical total yield is 1–2μg RNA with a fraction of that small RNA. Due to background of partially degraded RNA from dead and dying cells, the standard methods of RNA integrity analysis using a bioanalyzer that typically rely on ribosomal RNA sizing may not be useful.
7. It is advised prior to sequencing that RT-PCR-based methods be used to compare yield and quality of samples by quantifying a few miRNAs expected to be in all samples.
8. Optimally, when you return to the laboratory and prior to freezing, 100 μL can be removed and added to a separate tube for DNA isolation, while the remainder is reserved for RNA analysis. Alternatively, methodologies are available to simultaneously isolate RNA and DNA from a single sample, but it will be important to verify that they provide a sufficient yield of epithelial cell miRNA.
9. For convenience, host cells and microbes can be stored long term in RNAprotect Cell reagent frozen or other RNA preservative that preserves cell structure. Or, prior to freezing, after aliquoting, the sample can be centrifuged at 5000 or 5500 × *g*, for 5 min, the supernatant discarded, and pellets stored at −80 °C. This avoids a freeze thaw in RNAprotect Cell reagent, but has not been shown to improve recovery of all taxa present.

10. If samples are in pellet form, then resuspend in solution, TE or PBS, suitable for the DNA isolation method chosen.
11. There are several methods of preservation of mammalian biosamples for later host RNA isolation that have been well validated not to distort yield. The same is not true for sample preservation and microbial RNA. The author has used RNAprotect Cell reagent and has found that it allows concentration of brush biopsy epithelial cell samples by centrifugation even after freezing, which can remove PCR inhibitors. This also allows replacement of suspension buffer to one compatible with the chosen method of sample purification. The author has found this method to provide similar miRNA species profiles as those seen with direct and immediate extraction of mammalian RNA in Trizol/RNAzol from oral cytology samples. In contrast, the method of sample preservation of microbial RNA has not been fully validated for differential microbial RNA yields versus fresh samples. It is not known how well these preservatives work on all oral bacteria to penetrate and preserve RNA integrity. Most validation of sample complexity preservation has been on DNA, which is a more stable molecule than RNA [22, 28–30]. While these RNAlater and RNAprotect Cell reagents certainly work well on most taxa, it was not possible to find studies that verified no loss of complexity of the microbial RNA species isolated when using RNA preservative versus immediate homogenization and RNA purification of fresh samples.
12. Another option is to use a product like DNA/RNA Shield or various bead lysis buffers to immerse cytology samples immediately post collection. They have the advantage of inactivating pathogens in part by cell lysis. They are also compatible with RNA purification kits from different suppliers, as they likely contain guanidium isothiocyanate plus additional chemicals for sample denaturation [31]. The author would recommend a method that uses bead beating of the initial suspension, or some other method of homogenization, followed by the phenol extraction step (*see* Subheading 3.7). The latter is important for maximizing host epithelial RNA yields, which can be quite low from some subjects, and to minimize DNA contamination. There are a range of other RNA preservatives or stabilizers, which can be used but with which the author has no experience.
13. It is thought that 0.1 mm beads are ideal for bacteria, and 0.5 mm for fungi, though I typically use a mixture as sold preloaded in tubes by Zymo Research or other companies, or available in bulk or preloaded in tubes, from Biospec Products, Inc.

14. One can also use RNeasy kits to purify RNA, in which case the manufacturer recommends the addition of 1.5 volume ethanol. Within a project the author uses the same methods but these and other RNA cleanup methods that are silica based should work similarly as long as they are compatible with recovery of both small and large RNA.
15. Total RNA is isolated but it is aliquoted and stored according to planned future usage. Ethanol precipitation of the smallest RNAs at low concentrations, such as miRNAs, is not reliably reproducible and is therefore not recommended. Fractions for miRNA study are aliquoted and stored in water frozen. The larger RNA will include host and metagenomic RNA, though that from the host will be of variable quality. For maximal stability and ease of use the fraction to be used for large RNA analysis is stored as an ethanol/NaAcetate suspension.
16. When an aliquot of the sample in ethanol/NaAcetate is needed, bring to room temperature, vortex vigorously for 30 s, then quickly remove what you need with a pipette and place in second centrifuge tube. This volume must be centrifuged at 4 °C for 15 min. Remove supernatant, then wash invisible pellet with 100μL 75% ethanol, vortex, then centrifuge again but this time for 10 min. Remove supernatant with a pipette and if desired the 75% ethanol wash can be repeated. Let dry on bench for 5–10 min till no liquid is detectable. Dissolve pellet in desired buffer for next procedure. As long as glycogen is used as the carrier, or there is much RNA and no carrier is needed, this method works well. It does not work with acrylamide-based carriers.
17. Normal precautions should be taken for working with RNA to avoid contamination with trace ribonuclease. In addition, the PCR setup should be segregated from post-PCR work by using a UV cabinet for the former or working in separate rooms. All tips used should be barrier tips.

References

1. Adami GR, Tang JL, Markiewicz M (2017) Improving accuracy of RNA-based diagnosis and prognosis of oral cancer by using noninvasive methods. Oral Oncol 69:62–67
2. Driemel O, Kosmehl H, Rosenhahn J, Berndt A, Reichert TE, Zardi L et al (2007) Expression analysis of extracellular matrix components in brush biopsies of oral lesions. Anticancer Res 27(3B):1565–1570
3. Kolokythas A, Schwartz JL, Pytynia KB, Panda S, Yao M, Homann B et al (2011) Analysis of RNA from brush cytology detects changes in B2M, CYP1B1 and KRT17 levels with OSCC in tobacco users. Oral Oncol 47(6):532–536. https://doi.org/10.1016/j.oraloncology.2011.03.029
4. Schwartz JL, Panda S, Beam C, Bach LE, Adami GR (2008) RNA from brush oral cytology to measure squamous cell carcinoma gene expression. J Oral Pathol Med 37(2):70–77. https://doi.org/10.1111/j.1600-0714.2007.00596.x
5. Toyoshima T, Koch F, Kaemmerer P, Vairaktaris E, Al-Nawas B, Wagner W (2009) Expression of cytokeratin 17 mRNA in oral squamous cell carcinoma cells obtained by

brush biopsy: preliminary results. J Oral Pathol Med 38(6):530–534. https://doi.org/10.1111/j.1600-0714.2009.00748.x

6. Kolokythas A, Zhou Y, Schwartz JL, Adami GR (2015) Similar squamous cell carcinoma epithelium microRNA expression in never smokers and ever smokers. PLoS One 10(11): e0141695. https://doi.org/10.1371/journal.pone.0141695
7. Spira A, Beane J, Schembri F, Liu G, Ding C, Gilman S et al (2004) Noninvasive method for obtaining RNA from buccal mucosa epithelial cells for gene expression profiling. BioTechniques 36(3):484–487
8. Spivack SD, Hurteau GJ, Jain R, Kumar SV, Aldous KM, Gierthy JR et al (2004) Gene-environment interaction signatures by quantitative mRNA profiling in exfoliated buccal mucosal cells. Cancer Res 64(18):6805–6813. https://doi.org/10.1158/0008-5472.CAN-04-1771
9. Kupfer DM, White VL, Jenkins MC, Burian D (2010) Examining smoking-induced differential gene expression changes in buccal mucosa. BMC Med Genet 3:24. https://doi.org/10.1186/1755-8794-3-24
10. Sridhar S, Schembri F, Zeskind J, Shah V, Gustafson AM, Steiling K et al (2008) Smoking-induced gene expression changes in the bronchial airway are reflected in nasal and buccal epithelium. BMC Genomics 9:259. https://doi.org/10.1186/1471-2164-9-259
11. Ferguson JS, Van Wert R, Choi Y, Rosenbluth MJ, Smith KP, Huang J et al (2016) Impact of a bronchial genomic classifier on clinical decision making in patients undergoing diagnostic evaluation for lung cancer. BMC Pulm Med 16(1):66
12. Silvestri GA, Vachani A, Whitney D, Elashoff M, Smith KP, Ferguson JS et al (2015) A bronchial genomic classifier for the diagnostic evaluation of lung cancer. N Engl J Med 373(3):243–251. https://doi.org/10.1056/NEJMoa1504601
13. Kim TH, Chang JH, Lee HJ, Kim JA, Lim YS, Kim CW et al (2016) mRNA expression of CDH3, IGF2BP3, and BIRC5 in biliary brush cytology specimens is a useful adjunctive tool of cytology for the diagnosis of malignant biliary stricture. Medicine (Baltimore) 95(27): e4132. https://doi.org/10.1097/MD.0000000000004132
14. Nischalke HD, Schmitz V, Luda C, Aldenhoff K, Berger C, Feldmann G et al (2012) Detection of IGF2BP3, HOXB7, and NEK2 mRNA expression in brush cytology specimens as a new diagnostic tool in patients with biliary strictures. PLoS One 7(8):e42141. https://doi.org/10.1371/journal.pone.0042141
15. Cheng GF (2015) Circulating miRNAs: roles in cancer diagnosis, prognosis and therapy. Adv Drug Deliv Rev 81:75–93. https://doi.org/10.1016/j.addr.2014.09.001
16. John K, Wu J, Lee BW, Farah CS (2013) MicroRNAs in head and neck cancer. Int J Dent 2013:650218. https://doi.org/10.1155/2013/650218
17. Zhou Y, Kolokythas A, Schwartz JL, Epstein JB, Adami GR (2016) microRNA from brush biopsy to characterize oral squamous cell carcinoma epithelium. Cancer Med 6(1):67–78. https://doi.org/10.1002/cam4.951
18. Adami GR, Tangney CC, Tang JL, Zhou Y, Ghaffari S, Naqib A et al (2018) Effects of green tea on miRNA and microbiome of oral epithelium. Sci Rep 8(1):5873. https://doi.org/10.1038/s41598-018-22994-3
19. Cristaldi M, Mauceri R, Di Fede O, Giuliana G, Campisi G, Panzarella V (2019) Salivary biomarkers for oral squamous cell carcinoma diagnosis and follow-up: current status and perspectives. Front Physiol 10:1476. https://doi.org/10.3389/fphys.2019.01476
20. Liskova A, Samec M, Koklesȯva L, Giordano FA, Kubatka P, Golubnitschaja O (2020) Liquid biopsy is instrumental for 3PM dimensional solutions in cancer management. J Clin Med 9(9):2749. https://doi.org/10.3390/jcm9092749
21. Funahashi K, Shiba T, Watanabe T, Muramoto K, Takeuchi Y, Ogawa T et al (2019) Functional dysbiosis within dental plaque microbiota in cleft lip and palate patients. Prog Orthod 20(1):11. https://doi.org/10.1186/s40510-019-0265-1
22. Hallmaier-Wacker LK, Lueert S, Roos C, Knauf S (2018) The impact of storage buffer, DNA extraction method, and polymerase on microbial analysis. Sci Rep 8(1):6292. https://doi.org/10.1038/s41598-018-24573-y
23. Belstrom D, Constancias F, Liu Y, Yang L, Drautz-Moses DI, Schuster SC et al (2017) Metagenomic and metatranscriptomic analysis of saliva reveals disease-associated microbiota in patients with periodontitis and dental caries. NPJ Biofilms Microbiomes 3:23. https://doi.org/10.1038/s41522-017-0031-4
24. Edlund A, Yang Y, Yooseph S, He X, Shi W, McLean JS (2018) Uncovering complex microbiome activities via metatranscriptomics during 24 hours of oral biofilm assembly and maturation. Microbiome 6(1):217. https://doi.org/10.1186/s40168-018-0591-4

25. Kressirer CA, Chen T, Harriman KL, Frias-Lopez J, Dewhirst FE, Tavares MA et al (2018) Functional profiles of coronal and dentin caries in children. J Oral Microbiol 10 (1):1495976. https://doi.org/10.1080/20002297.2018.1495976
26. Mihaila D, Donegan J, Barns S, LaRocca D, Du Q, Zheng D et al (2019) The oral microbiome of early stage Parkinson's disease and its relationship with functional measures of motor and non-motor function. PLoS One 14(6): e0218252. https://doi.org/10.1371/journal.pone.0218252
27. Nowicki EM, Shroff R, Singleton JA, Renaud DE, Wallace D, Drury J et al (2018) Microbiota and metatranscriptome changes accompanying the onset of gingivitis. mBio 9(2): e00575–e00518. https://doi.org/10.1128/mBio.00575-18
28. Chen Z, Hui PC, Hui M, Yeoh YK, Wong PY, Chan MCW et al (2019) Impact of preservation method and 16S rRNA hypervariable region on gut microbiota profiling. mSystems 4(1):e00271–e00218. https://doi.org/10.1128/mSystems.00271-18
29. Vogtmann E, Chen J, Kibriya MG, Amir A, Shi J, Chen Y et al (2019) Comparison of oral collection methods for studies of microbiota. Cancer Epidemiol Biomark Prev 28 (1):137–143. https://doi.org/10.1158/1055-9965.EPI-18-0312
30. Zhou X, Nanayakkara S, Gao JL, Nguyen K-A, Adler C (2019) Storage media and not extraction method has the biggest impact on recovery of bacteria from the oral microbiome. Sci Rep 9(1):14968. https://doi.org/10.1038/s41598-019-51448-7
31. Rio DC, Ares M, Hannon GJ, Nilsen TW (2011) RNA: a laboratoy manual. Cold Spring Harbor Laboratory Press, Cold Spring Harbor, NY

Chapter 13

Bottom-Up Community Proteome Analysis of Saliva Samples and Tongue Swabs by Data-Dependent Acquisition Nano LC-MS/MS Mass Spectrometry

Alexander Rabe, Manuela Gesell Salazar, and Uwe Völker

Abstract

Analysis using mass spectrometry enables the characterization of metaproteomes in their native environments and overcomes the limitation of proteomics of pure cultures. Metaproteomics is a promising approach to link functions of currently actively expressed genes to the phylogenetic composition of the microbiome in their habitat. In this chapter, we describe the preparation of saliva samples and tongue swabs for nLC-MS/MS measurements and their bioinformatic analysis based on the Trans-Proteomic Pipeline and Prophane to study the oral microbiome.

Key words Saliva, Tongue, Metaproteomics, Human oral microbiome, nLC-MS/MS

1 Introduction

Mass spectrometry has become the method of choice in the field of proteomics when peptides, proteins, or posttranslational modifications need to be analyzed within a short time and with high accuracy [1, 2]. In recent years, improved sensitivity of mass spectrometers [3, 4] in combination with the availability of high-quality metagenomic databases [5–8] has enabled in-depth metaproteome analyses in addition to proteome analyses of pure cell or bacterial cultures [9]. The resulting field of metaproteomics offers the possibility to study bacteria and their actively expressed genes directly in their natural habitat [10]. It is therefore a promising approach not only to determine the phylogenetic composition of the microbiome but also to uncover functional aspects and their response to changing environmental influences [11, 12]. This is essential to improve our understanding of polymicrobial diseases in humans [13, 14].

Metaproteomics is an emerging scientific field, and initial studies and approaches for the investigation of the microbiome in

Guy R. Adami (ed.), *The Oral Microbiome: Methods and Protocols*, Methods in Molecular Biology, vol. 2327,
https://doi.org/10.1007/978-1-0716-1518-8_13,

different human habitats have emerged [15–19]. In a metaproteomic study in young healthy humans, we compared saliva samples and tongue swabs. Our study includes the phylogenetic composition of both microbiomes, their translated proteins as well as the human proteins [20]. In this chapter, we describe the procedure of this study from sample collection, sample preparation for mass spectrometry, through to data processing.

2 Materials

2.1 Sampling and Sample Preparation

Prepare all solutions fresh prior to usage and store them at room temperature unless otherwise specified. Follow the legal and regulatory requirements for handling biomaterials of human origin.

1. Phosphate-buffered saline (PBS): 1×, pH 7.4 with 0.2 M NaCl, 2.5 mM KCl, 8 mM Na_2HPO_4, 1.5 mM KH_2PO_4. Prepare 800 mL of distilled water and add 11.6 g NaCl, 0.186 g KCl, 1.4 g Na_2HPO_4, and 0.2 g KH_2PO_4. Adjust the pH to 7.4 and add distilled water to prepare a 1 L solution of 1× PBS.
2. Tris–HCl Buffer: 0.25 M, pH 8.0. Prepare 400 mL of distilled water and add 60.55 g Tris to the solution. Adjust a pH of 8.0 with HCl and add distilled water to prepare a volume of 0.5 L 0.25 M Tris–HCl solution.
3. Tris–HCl Buffer: 0.05 M, pH 8.0. Dilute 100 mL of the 0.25 M Tris–HCl Buffer in 400 mL distilled water.
4. Protease inhibitor cocktail: Use a protease inhibitor provided as a lyophilized powder for general use. Dissolve 1 vial of the lyophilized powder in 10 mL of 50 mM Tris–HCl Buffer (pH 8.0). Add 0.075 mL protease inhibitor cocktail solution in 1.425 mL distilled water.
5. Collection of tongue samples: Sterile timber plate (18 × 150 mm).
6. Collection of saliva samples: Paraffin gum.
7. Sterile plastic tubes with a volume of 50 and 2 mL.
8. Vortex mixer.
9. A centrifuge that can be cooled to 4 °C and that is capable of centrifuging 50 mL sample tubes at 11,500 × *g* and 2 mL sample tubes at 17,000 × *g*.
10. Ethylenediaminetetraacetic acid (EDTA): 50 mM. Prepare 50 mL distilled water, add 1.86 g EDTA and add distilled water to prepare a volume of 100 mL of 50 mM EDTA.

11. Tris-aminomethane (Tris): 100 mM. Prepare 800 mL distilled water, add 12.1 g EDTA and add distilled water to prepare a volume of 1 L of 100 mM Tris.
12. Tris-EDTA Buffer (TE-Buffer): pH 8.5. Prepare 400 mL distilled water and add 100 mM Tris and 50 mM EDTA. Adjust to pH of 8.5 and add distilled water to prepare a volume of 0.5 L TE-Buffer.
13. Cell disruption: Ultrasonic device with an ultrasonic probe.
14. Dithiothreitol (DTT): 1.3 M. Weigh 2 g DTT and add distilled water to prepare a volume of 10 mL DTT.
15. Trichloroacetic acid (TCA): 100% solution.
16. Cold acetone: 100% solution.
17. Drying process of precipitated protein pellets: Vacuum evaporator.
18. Urea-Thiourea Buffer (1× UT): 10 M (8 M urea plus 2 M thiourea). Weigh 1.92 g urea and 0.61 g thiourea, then add distilled water to prepare a volume of 4 mL 1× UT.
19. Thermomixer to cool/heat and shake sample tubes.

2.2 Protein Determination

1. Bovine serum albumin (BSA): Prepare a BSA stock solution with a concentration of 1 mg/mL.
2. Bradford reagent.
3. Vortex mixer.
4. Plastic cuvettes.
5. Spectralphotometer for optical absorption measurement at 595 nm.

2.3 Tryptic Digestion of Protein Samples

1. High-performance liquid chromatography (HPLC) water.
2. Low protein binding reaction vessels.
3. 20 mM Ammoniumbicarbonate (ABC): 0.079 g ABC in 12.5 mL HPLC water.
4. 25 mM Dithiothreitol (DTT): 0.03 g DTT in 8 mL of 20 mM ABC.
5. 100 mM Iodoacetic acid (IAA): 0.018 g IAA in 1 mL of 20 mM ABC (*see* **Note 1**).
6. Trypsin: Use 20µg lyophilized trypsin. Dissolve 1 vial of the lyophilized trypsin in 1 mL of 20 mM ABC to reach a final concentration of 20 ng/µL. For the In-Solution-Digestion, add trypsin at the ratio of 1:25, which corresponds to 8µL of a 20 ng/µL solution to a protein amount of 4µg (*see* **Note 2**).
7. Add 0.075 mL protease inhibitor cocktail solution in 1.425 mL HPLC water.

8. Urea-Thiourea Buffer (1× UT): 10 M (8 M urea plus 2 M thiourea). Weigh 1.92 g urea and 0.61 g thiourea, then add distilled water to prepare a volume of 4 mL 1× UT.
9. Thermomixer or incubator to heat sample tubes.

2.4 Purification of Peptide Samples

1. 10μL ZipTip-tip μ-C18 material with a column of a peptide binding capacity of 2μg.
2. Acetic acid: 5% solution in HPLC water, 1% solution in HPLC water, and 0.05% solution in HPLC water.
3. Acetonitrile: 100% solution; 80% solution in 1% acetic acid; 50% solution in 1% acetic acid; and 30% solution in 1% acetic acid.
4. Buffer A: 2% acetonitrile, 0.05% acetic acid in HPLC water.
5. Clear glass micro tubes for 2 mL with neutral cap.
6. Clear glass micro inserts (vial) for 0.1 mL.
7. Vacuum freeze dryer.

2.5 Buffer for HPLC

1. Buffer A: 0.1% acetic acid in HPLC water.
2. Buffer B: 0.1% acetic acid in 100% acetonitrile.

2.6 Mass Spectrometric Analysis

1. Q Exactive plus mass spectrometer (Thermo Fisher Scientific).

2.7 Software for Metaproteomic Data Analysis

1. Created binary LC-MS/MS datasets [21, 22].
2. Comet (http://comet-ms.sourceforge.net/) [23, 24].
3. Trans-Proteomic Pipeline (http://tools.proteomecenter.org/software.php) [25–27], including the following modules and tools: msconvert, PeptideProphet [28], iProphet [29, 30], Mayu [31], and ProteinProphet [32].
4. Webtool: Prophane (https://prophane.de/login).
5. Bacterial database: Human Oral Microbiome Database (http://www.homd.org/) [7, 33].
6. Human database: UniProtKB/Swissprot (https://www.uniprot.org/) [34].
7. Software for statistical computing and graphics: R (https://www.r-project.org/) [35].

3 Methods

The single steps are performed at room temperature unless otherwise described. The laboratory workflow is shown in Fig. 1.

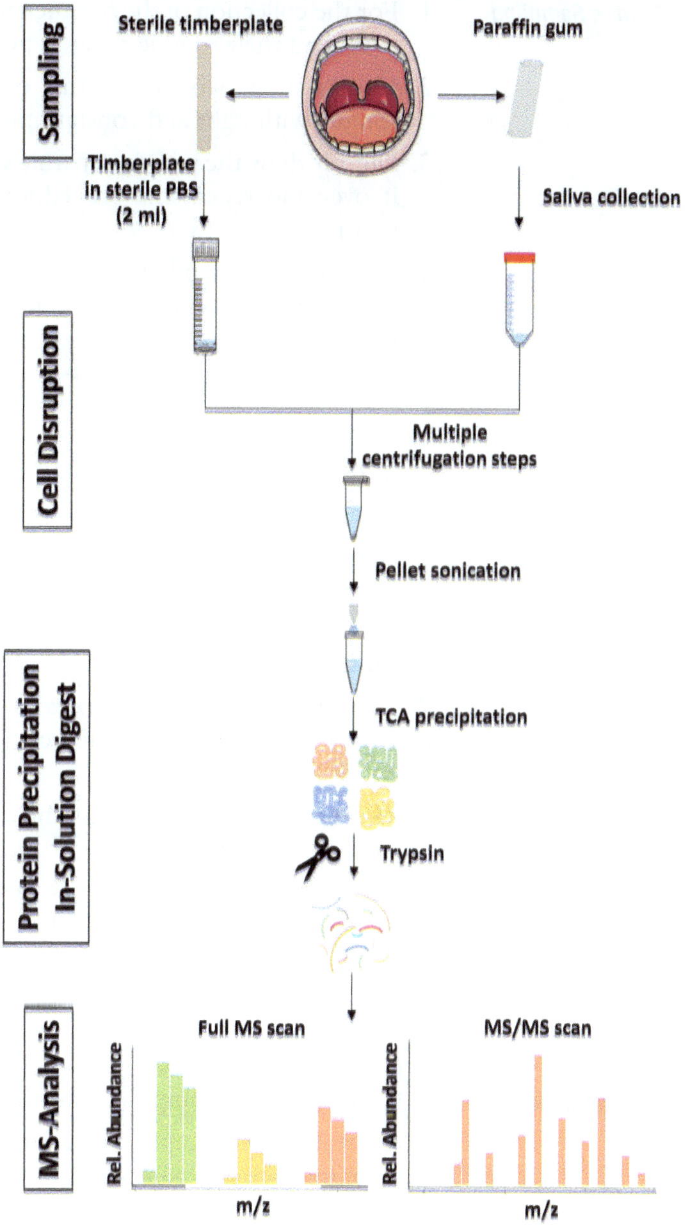

Fig. 1 Workflow for metaproteome analyses of human saliva and tongue swabs. In the first step, a tongue swab is taken with a sterile wooden spatula and transferred into 2 mL sterile PBS. The participants then chew on a paraffin gum for 1 min to stimulate the natural flow of saliva. During the chewing process, the subjects spit saliva into a Falcon Tube multiple times. The collected saliva is centrifuged, and the resulting pellet dissolved in TE buffer, followed by ultrasound treatment. The proteins precipitated by TCA are digested with trypsin. Measurement of the peptides is performed on Q Exactive Plus (LC-MS/MS). (The figure is adapted from our publication of the healthy human saliva and tongue microbiome [20])

3.1 Tongue Sampling

1. For the collection of the tongue samples kindly ask the subjects to extend their tongue as far as possible (*see* **Note 3**).
2. Place a sterile wooden spatula on the middle dorsum of the tongue with light and constant pressure for 5 s (*see* **Note 4**).
3. Slightly draw the wooden spatula ventral over the tongue, turn it over and repeat the procedure with the other side of the spatula.
4. Transfer the wooden spatula with the sample side into a prepared vessel containing 2 mL sterile 1× PBS and 40μL protease inhibitor.
5. Vortex the vessel including the spatula for 30 s.
6. Discard the wooden spatula.
7. Store the sample material on dry ice and keep at −80 °C until the next step.

3.2 Saliva Sampling

1. Provide one commercially available paraffin chewing gum for each subject. Chewing on the gum will stimulate the natural salivation and ensures a sufficiently large sample volume (*see* **Note 5**).
2. The subject chews on the paraffin chewing gum over a period of 1 min, holding a sterile vessel for collecting saliva in his hands (*see* **Note 6**).
3. During the chewing process the participants spit several times into the sample vessel.
4. Measure the collected sample volume after 1 min using the scale of the vessel and add 20μL of the protease inhibitor per 1 mL of saliva.
5. Store the sample material on dry ice and keep at −80 °C until further use.

3.3 Cell Disruption

1. Thaw the saliva samples and tongue swabs on ice.
2. Centrifuge the samples at 11,500 × *g* for 15 min. The centrifuge must be cooled down to 4 °C.
3. Discard the resulting supernatant and resuspend the pellet with at least 500μL TE buffer (*see* **Note** 7).
4. Transfer the dissolved pellet into a smaller reaction vessel.
5. The suspension is treated with an ultrasound probe for 3 × 30 s. The samples remain on ice during and after the ultrasonic treatment (*see* **Note 8**).
6. Centrifuge the samples at 4 °C and 16,200 × *g* for 30 min.
7. Pipette the supernatant into a new vessel for the next treatment steps. The remaining pellet can be discarded.

3.4 Precipitation of Proteins and Protein Assay

1. Add 0.6μL 1.3 M DTT per 100μL sample volume and vortex the sample for 10 s.
2. Incubate the sample for 30 min at 37 °C.
3. Add TCA until a final concentration of 15% and invert the tube several times.
4. Incubate the samples on ice for 60 min.
5. Centrifuge the samples for 45 min at 4 °C and 17,000 × *g*.
6. Remove the supernatant with a pipette without touching the pellet.
7. Wash the pellet with 500μL cold acetone by inverting the vessel several times.
8. Centrifuge the samples for 15 min at 4 °C and 17,000 × *g* and then remove the acetone.
9. Wash the pellet again with 500μL cold acetone by inverting the vessel several times.
10. Centrifuge the sample for 15 min at 4 °C and 17,000 × *g* and remove the excess acetone.
11. Dry the pellet in a vacuum evaporator for 1 min to completely remove the acetone.
12. Dissolve the precipitated and dried proteins in 1× UT. For the saliva pellets you need at least 50μL and for the tongue pellets at least 35μL 1× UT (*see* **Note 9**).
13. Perform protein determination according to Bradford [36]. Follow the instructions of your local supplier for Bradford reagents. The saliva protein concentration averages 6.4μg/μL (± 2.3μg/μL), which is three times as high as the tongue samples where the average concentration is 1.7μg/μL (± 1.6μg/μL) based on our study of 24 healthy subjects aged between 20 and 30 years [20].

3.5 Reduction, Alkylation, and Protein Digest

For the following steps, a protein amount of 4μg is required. The volume for the 4μg in our study including 24 healthy subjects aged 20–30 years was typically 3.4μL (± 1.3μL) for saliva and 10.9μL (± 4.9μL) for the tongue samples [20]. The total sample volume differs between the individual samples depending on the determined protein concentration. For this reason, the following steps specify the final concentrations to be achieved with the substances for reduction, alkylation, and protein digestion in relation to the total volume of the sample. The incubation of the samples in the following single steps was performed without shaking or any other movement.

1. Add DTT to a final concentration of 2.5 mM to the protein mixture and incubate the protein solution for 60 min at 60 °C,

which will reduce disulfide bonds of cysteines to sulfhydryl groups.

2. Prevent re-oxidation of the thiol groups by alkylation of the protein mixture with a final concentration of 10 mM IAA at 37 °C and an incubation time of 30 min (*see* **Note 10**).
3. Dilute the samples 1:10 with 20 mM ammonium bicarbonate, resulting in a urea/thiourea concentration of less than 2 M.
4. Add trypsin in the ratio 1:25 (trypsin/sample) and incubate the sample at 37 °C for 17 h in the dark.
5. Terminate the activity of the enzyme trypsin adding 5% acetic acid to a final concentration of 1% acetic acid to the peptide mixture.

3.6 Purification of Peptide Sample

Increase the purity of the peptide sample by desalting and decreasing the amount of hydrophilic substances with a 10μL ZipTip packed with μ-C18 material and a total binding capacity of 2μg.

1. Set the volume of the pipette to 10μL.
2. Activate the μ-C18 material by pressing the plunger button down and aspirate the 100% ACN solution into the ZipTip-tip. Discard the activation solution (*see* **Note 11**).
3. Repeat the procedure three times in total.
4. Equilibrate the μ-C18 material using a three-step decreasing concentration of 80%, 50%, and 30% ACN.
5. Start with 80% ACN, by aspirating the solution and discarding it into the waste.
6. Repeat the procedure five times in total.
7. Perform the same steps described in **steps 5** and **6** for the 50% and 30% ACN.
8. The equilibration of the μ-C18 material in the column is completed with two cycles of aspirating of 1% acetic acid and its discarding.
9. Load the peptides onto the equilibrated column, by performing 15–20 aspiration-dispense cycles of the entire sample material (*see* **Note 12**).
10. Remove salts and detergents, by washing the column with five cycles of aspirating with 1% acetic acid and discarding.
11. Elute the column-bound peptides by aspirating and dispensing 8μL of 50% ACN three times.
12. Aspirate 50% ACN a fourth time and transfer the ACN-peptide mixture into a glass micro vial.
13. Elute the column-bound peptides a second time by aspirating and dispensing 8μL of 80% ACN three times.

14. Aspirate 80% ACN a fourth time and transfer the ACN-peptide mixture into the same glass micro vial as before.
15. Lyophilize the samples in a vacuum freeze dryer.
16. Fill up the micro vials with 20µL of buffer A to reach a peptide concentration of 0.1µg/µL (*see* **Note 13**).

3.7 LC-MS/MS Measurement Performed on a Nano LC-MS/MS System

1. Reverse phase nano LC/MS-MS: Load the complex peptide mixtures onto a precolumn.
2. The subsequent 120-min separation of the tryptic peptides is performed on analytical column using a linear gradient of 2–25% with the binary buffer B.
3. The mass spectrometric analysis is performed in data-dependent acquisition mode using a high-resolution accurate-mass MS-instrument of the Q Exactive Orbitrap MS series. Detailed information for parameter of a LC-MS/MS method using an Ultimate 3000 and a Q Exactive plus mass spectrometer (Thermo Fisher Scientific) are shown in Table 1.

3.8 Bioinformatic Analysis of LC-MS/MS Raw Data for Peptide and Protein Identification Using the Trans-Proteomic Pipeline

Initially, the spectra data generated by mass spectrometry must be analyzed and interpreted. The mass spectra are searched against a decoy database. In several steps, the peptides and proteins are identified, and their probability is calculated. The data are processed with the Trans-Proteomic Pipeline (TPP) [25–27]. The TPP is Linux-based and used via command line. Figure 2 highlights the key steps of the data analysis workflow.

1. Combine the oral microbiome database (HOMD) [7, 33] and the human database (UniProtKB/Swissprot) [34] to create a database containing both bacterial and human protein sequences.
2. Add a decoy protein sequence to each human and bacterial protein sequence to create a reverse decoy database from the combined database (*see* **Note 14**).
3. Convert the result files of the mass spectrometric analysis from .raw data format to .mzML data format using the *msconvert* module of the TPP [21, 22].
4. Start the *Comet* search [23, 24] using the combined sequence decoy database to interpret the mass spectra. The settings of the search parameters are listed in Table 2.
5. Use the wrapper tool *xinteract* [25] of the TPP to run the modules *PeptideProphet* [28] and *iProphet* [29, 30] at once. *PeptideProphet* converts the individual result files of the database search into the pep.xml-format and additionally merges them into a single interact-pep.xml result file. Furthermore, it performs a spectrum-level validation followed by peptide-level

Table 1
Required materials for reversed phase liquid chromatography (RPLC) and the parameters to be set for mass spectrometric measurements

Reversed phase liquid chromatography (RPLC)	
Instrument	Ultimate 3000 RSLC (Thermo Fisher Scientific)
Trap column	75μm inner diameter, packed with 3μm C18 particles (Acclaim PepMap100, Thermo Fisher Scientific)
Analytical column	Accucore 150-C18 (Thermo Fisher Scientific) 25 cm × 75μm, 2.6μm C18 particles, 150 Å pore size
Buffer system	Binary buffer system consisting of 0.1% acetic acid water (buffer A) and 100% ACN in 0.1% acetic acid (buffer B)
Flow rate	300 nL/min
Gradient	Linear gradient of buffer B from 2% up to 25%
Gradient duration	120 min
Column oven temperature	40 °C
Mass spectrometry (MS)	
Instrument	Q Exactive plus mass spectrometer (Thermo Fisher Scientific)
Operation mode	Data-dependent
Full MS	
MS scan resolution	70,000
AGC target	3e6
Maximum ion injection time for the MS scan	120 ms
Scan range	300–1650 *m/z*
Spectra data type	Profile
dd-MS2	
Resolution	17,500
MS/MS AGC target	2e5
Maximum ion injection time for the MS/MS scans	120 ms
Spectra data type	Centroid
Selection for MS/MS	10 most abundant isotope patterns with charge ≥2 from the survey scan
Isolation window	3 *m/z*
Fixed first mass	100 *m/z*
Dissociation mode	Higher energy collisional dissociation (HCD)
Normalized collision energy	27.5%
Dynamic exclusion	30 s
Charge exclusion	1, >6

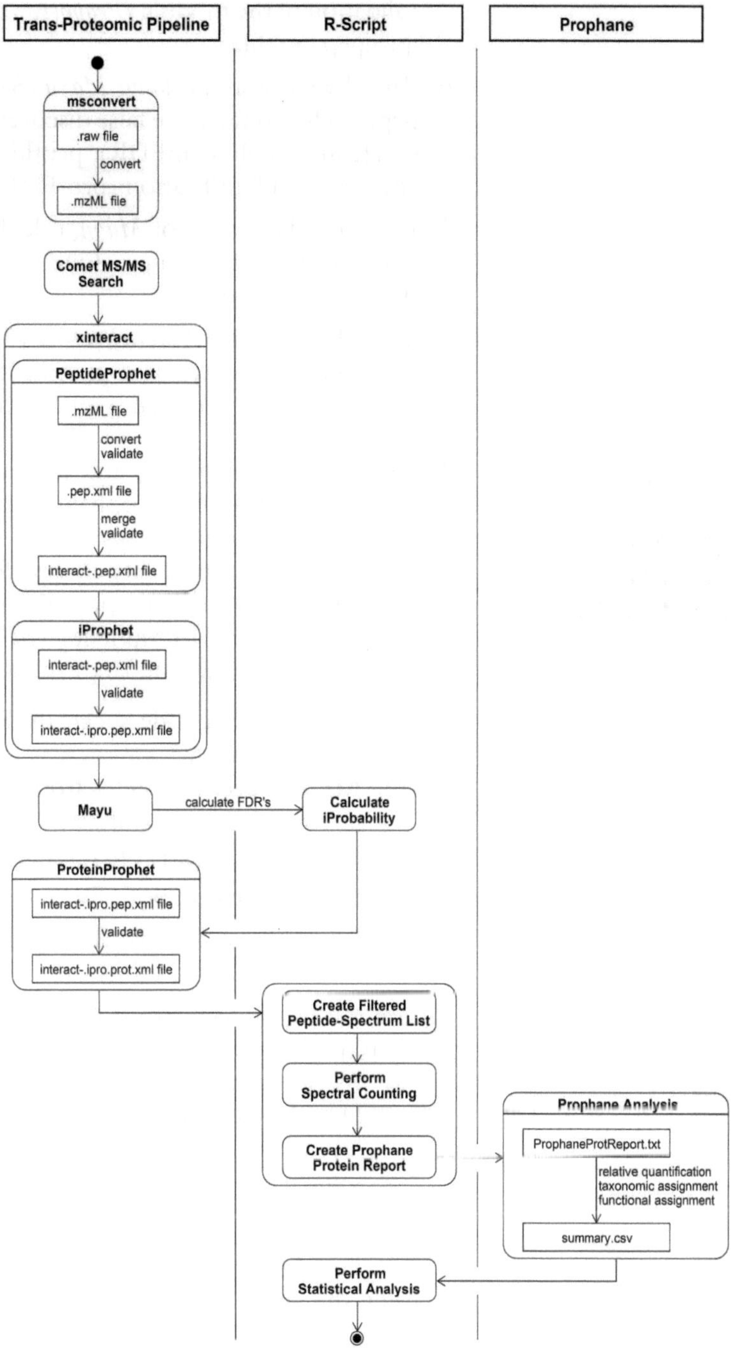

Fig. 2 The UML activity diagram summarizes the different process steps for the evaluation of the metaproteomic data. The raw data is converted into the mzML data format using msconvert. Peptide identification is performed by Comet based on a reverse decoy database containing human and bacterial protein sequences. The validation of the identified peptides is performed by the modules PeptideProphet and iProphet. With a complimentary evaluation by Mayu and the

validation of the module *iProphet*, which results in the interact-.ipro.pep.xml file.

6. Run the software package *Mayu* [31] using the interact-.ipro.pep.xml file to calculate false discovery rates (FDR) for peptide-spectrum matches (mFDR), peptide identification (pepFDR), and protein identification (protFDR) (*see* **Note 15**).
7. Based on the results of *Mayu*, calculate the iProbability (value between 0 and 1) for a protFDR = 0.05 to refine the results of the *iProphet* module.
8. Start the module *ProteinProphet* [32] and use the calculated iProbability to determine protein identification probabilities. *ProteinProphet* creates an interact-.ipro.prot.xml result file.

3.9 Prophane: Taxonomic and Functional Assignment of Identified Proteins

Identified spectra, peptides, and proteins must be appropriately prepared for the web tool Prophane to perform a relative quantification as well as taxonomic and functional assignments. The data are filtered according to quality criteria, followed by spectral counting. The proteins with their spectral counts are summarized in a report and uploaded to Prophane. The preparation of the data is done with the programming language R [35].

1. Create a filtered peptide-spectrum list based on the calculated iProbability of Mayu:
   (a) Remove decoy proteins from the *Mayu* result file and the interact-.ipro.pep.xml file.
   (b) Identify the overlap between the two files using the spectra that occur in both files.
   (c) Select all data, whose iProbability is greater than or equal to the calculated value.
2. Use the peptide-spectrum list and the interact-.ipro.prot.xml file to perform spectral counting.
   (a) Assign to each peptide and spectrum of the filtered peptide-spectrum list the corresponding protein of the interact-.ipro.prot.xml file based on the peptide sequences.
   (b) Count the number of spectra per protein.
3. Based on the requirements of Prophane, create a protein report using the result file of the spectral counting.

Fig. 2 (continued) setting of the ProtFDR to 5.0%, stricter filter criteria are set in the context of protein assignment by the module ProteinProphet. Prophane is used for the taxonomic and functional assignment of the identified proteins. (The activity diagram was created with the program UMLet in version 14.2)

Table 2
Comet was used with release 2016.01 rev. 2. Parameters, different from the default settings, are listed in the table

General	
Decoy search	0 (= no)
Num threads	8
Masses	
Peptide mass tolerance	10
Peptide mass units	2 (= ppm)
Mass type parent	1 (= monoisotopic masses)
Mass type fragment	1 (= monoisotopic masses)
Precursor tolerance type	0 (= MH+)
Isotope error	1 (= on −1/0/1/2/3 (standard C13 error))
Variable modifications	
Variable mod01	15.9949 M 0 3–1 0 0 (= methionine)
max variable mods in peptide	5
Require variable mod	0
Fragment ions	
Fragment bin tol	0.01
Fragment bin offset	0.0
Theoretical fragment ions	1 (= M peak only)
Use B ions	1 (= yes)
Use Y ions	1 (= yes)
Use NL ions	1 (= yes)
Misc parameters	
Digest mass range	600.0–5000.0
Num results	50
Skip researching	1
Max fragment charge	3
Max precursor charge	6
Nucleotide reading frame	0
Clip nterm methionine	0
Spectrum batch size	10,000
Spectral processing	
Minimum peaks	5

(continued)

Table 2
(continued)

Minimum intensity	0
Remove precursor peak	0 (= nor)
Remove precursor tolerance	1.5
Clear mz range	0.0 0.0
Additional modifications	
Add C cysteine	57.021464

4. Start the webtool Prophane and import the protein report. Prophane calculates normalized spectral abundance factor values (NSAF-values) and performs taxonomic and functional assignment of proteins.
5. Use the Prophane report for further data analysis.

4 Notes

1. During the preparation process, extended exposure to light should be avoided. For this, the vessel should be wrapped in aluminum foil. The solution should then be stored on ice.
2. For proteome analysis, we have established in our laboratory the sequencing grade modified porcine trypsin (# V5111) from Promega [37]. A high purity of the trypsin is guaranteed by the manufacturer using affinity chromatography. The trypsin is provided by the company in 5 × 20μg ampules in lyophilized [37] or liquid frozen form in 50 mM acetic acid [38], whereby we use the lyophilized form. High stability and activity, as well as the prevention of autolytic digestion of the native trypsin, is ensured by modified lysins through reductive methylation [39]. The specificity of trypsin is further increased by treatment with tosyl phenylalanyl chloromethyl ketone (TPCK) [40]. Another advantage is its improved resistance to denaturation by chemicals, such as SDS, urea, acetonitrile, or guanidine HCl, which are commonly used in proteomics [39]. For further details regarding the handling of the trypsin, storage conditions, and other applications, please refer to the manufacturer's protocols [37, 38].
3. We have always started by collecting tongue swabs first and then saliva samples to keep the contamination of the tongue samples with saliva as minimal as possible.
4. To avoid triggering the gag reflex of the subjects, care should be taken to not insert the spatula too far into the oral cavity.

5. We used commercially available paraffin chewing gum (1.5 g) from the company Ivoclar Vivadent GmbH (Germany), which were delivered individually packed in blister packages. We also recommend using commercially available chewing gums, as these are available in standardized packages. The taste of the paraffin gum and the sensation during the chewing process is described by some subjects as unpleasant. It is possible that a little paraffin gets stuck to the teeth, but it is completely harmless for the subjects.
6. We recommend using a vessel with an opening large enough for the subjects to spit into.
7. Pellets can vary greatly regarding their stability and size. Pellets can be of low density and will loosen even with small movements. On the other hand, it can happen that the pellet is exceptionally large, and more than 500µL are necessary to bring it completely in solution.
8. We recommend testing beforehand at which strength the ultrasound treatment must be performed, as there are differences between the manufacturers' devices. We suggest determining the optimal settings of the ultrasound device directly for the sample material. The material of different test persons should be pooled to eliminate individual differences of the samples. Several combinations of ultrasonic intensities and durations should be compared by protein determination to determine the optimal combination.
9. The vacuum dried protein pellets can be dissolved very easily in 1× UT by pipetting up and down several times. After this step, the sample may be stored at −80 °C and processed later.
10. The alkylation step must be performed in the dark
11. During the entire purification process, ensure that no air is drawn into the ZipTip-tip, as this will reduce the quality of the purification. This is best accomplished by pipetting at a steady and gentle speed.
12. According to the manufacturer's instructions, the equilibrated column should be loaded with peptides using 15–20 aspiration-dispense cycles. Usually, we transfer the sample volume from the original reaction vessel to a new vessel to ensure that the entire sample volume has passed the column.
13. Subsequently, peptides can be measured by mass spectrometry directly or stored at −80 °C. Depending on the used measuring method, mass spectrometer, precolumn, and other conditions, the sample volume could be sufficient for several measurements, but here we recommend preparing the sample again by protein digestion and purification to ensure high quality of the sample measurement.

14. The Decoy database was created using an R-script. All target proteins were inverted and read from right to left. Furthermore, each inverted protein was tagged with DECOY and incremented by one (DECOY1 <protein sequence>, DECOY2 <protein sequence>, ...). The application of a decoy database of nonsense proteins of reversed sequences is necessary to estimate the number of incorrect and correct peptide and protein identifications, which enables us to conclude on the quality of the data set [41].
15. In general, in proteomics experiments the quality of peptide-spectrum matches (PSMs) is determined based on a false discovery rate. A cutoff is defined, which is usually PSMs FDR ≤ 0.05. With Mayu [31], we aim to raise the qualitative assignment to the level of protein identification (protFDR). The reason for this is that the protFDR is a more informative quality dimension than the PSMs FDR since further analyses are performed at the protein level and not at the spectra level. Another positive side effect is that the use of protFDR as a cutoff leads to a reduction in the PSMs FDR as multiple PSMs contribute to a single protein identification and reinforce or do not reinforce each other. This therefore contributes to an increase in the quality of the filtered data set, which is of great relevance in metaproteomics, since exceptionally large protein databases with a wide variety of species and different domains are used [42, 43].

Acknowledgments

There are no conflicts of interest to declare.

This study was supported by an unrestricted educational grant of the "Deutsche Gesellschaft für Parodontologie" to Alexander Rabe.

The authors gratefully acknowledge Thomas Kocher for providing the laboratory capacities and making the funding of the project possible. We also express gratitude to Alexander Welk and all the volunteers from his student course, who allowed us to collect saliva samples and tongue swabs. The authors address special thanks to Stephan Fuchs, the developer of Prophane, who was always available with advice and support, as well as Stephan Michalik for his support in data analysis and his helpful feedback. We also show appreciation to Ulrike Lissner for her technical assistance in the laboratory.

Thanks are also due to Les Laboratoires Servier and their service Servier Medical Art based on the Creative Commons Attribution 3.0 Unported License. For Fig. 1, we used images from their series "Digestive System," "People," and "Intracellular Components."

References

1. Larance M, Lamond AI (2015) Multidimensional proteomics for cell biology. Nat Rev Mol Cell Biol 16:269–280
2. Altelaar AFM, Munoz J, Heck AJR (2013) Next-generation proteomics: towards an integrative view of proteome dynamics. Nat Rev Genet 14:35–48
3. Gillet LC, Leitner A, Aebersold R (2016) Mass spectrometry applied to bottom-up proteomics: entering the high-throughput era for hypothesis testing. Annu Rev Anal Chem (Palo Alto, Calif) 9:449–472
4. Snyder DT, Pulliam CJ, Ouyang Z, Cooks RG (2016) Miniature and fieldable mass spectrometers: recent advances. Anal Chem 88:2–29
5. Wu S, Sun C, Li Y, Wang T, Jia L, Lai S et al (2020) GMrepo: a database of curated and consistently annotated human gut metagenomes. Nucleic Acids Res 48:D545–D553
6. Pruesse E, Quast C, Knittel K, Fuchs BM, Ludwig W, Peplies J et al (2007) SILVA: a comprehensive online resource for quality checked and aligned ribosomal RNA sequence data compatible with ARB. Nucleic Acids Res 35:7188–7196
7. Chen T, Yu W-H, Izard J, Baranova OV, Lakshmanan A, Dewhirst FE (2010) The Human Oral Microbiome Database: a web accessible resource for investigating oral microbe taxonomic and genomic information. Database (Oxford) 2010:baq013
8. Muth T, Renard BY, Martens L (2016) Metaproteomic data analysis at a glance: advances in computational microbial community proteomics. Expert Rev Proteomics 13:757–769
9. Wilmes P, Wexler M, Bond PL (2008) Metaproteomics provides functional insight into activated sludge wastewater treatment. PLoS One 3:e1778
10. Hanson BT, Hewson I, Madsen EL (2014) Metaproteomic survey of six aquatic habitats: discovering the identities of microbial populations active in biogeochemical cycling. Microb Ecol 67:520–539
11. Jagtap PD, Blakely A, Murray K, Stewart S, Kooren J, Johnson JE et al (2015) Metaproteomic analysis using the Galaxy framework. Proteomics 15:3553–3565
12. Abram F (2015) Systems-based approaches to unravel multi-species microbial community functioning. Comput Struct Biotechnol J 13:24–32
13. Stacy A, McNally L, Darch SE, Brown SP, Whiteley M (2016) The biogeography of polymicrobial infection. Nat Rev Microbiol 14:93–105
14. Brogden KA, Guthmiller JM, Taylor CE (2005) Human polymicrobial infections. Lancet 365:253–255
15. Jagtap P, McGowan T, Bandhakavi S, Tu ZJ, Seymour S, Griffin TJ et al (2012) Deep metaproteomic analysis of human salivary supernatant. Proteomics 12:992–1001
16. Velsko IM, Fellows Yates JA, Aron F, Hagan RW, Frantz LAF, Loe L et al (2019) Microbial differences between dental plaque and historic dental calculus are related to oral biofilm maturation stage. Microbiome 7:102
17. Wang Y, Zhou Y, Xiao X, Zheng J, Zhou H (2020) Metaproteomics: a strategy to study the taxonomy and functionality of the gut microbiota. J Proteomics 219:103737
18. Gurdeep Singh R, Tanca A, Palomba A, Van der Jeugt F, Verschaffelt P, Uzzau S et al (2019) Unipept 4.0: functional analysis of metaproteome data. J Proteome Res 18:606–615
19. Muth T, Kohrs F, Heyer R, Benndorf D, Rapp E, Reichl U et al (2018) MPA portable: a stand-alone software package for analyzing metaproteome samples on the go. Anal Chem 90:685–689
20. Rabe A, Gesell Salazar M, Michalik S, Fuchs S, Welk A, Kocher T et al (2019) Metaproteomics analysis of microbial diversity of human saliva and tongue dorsum in young healthy individuals. J Oral Microbiol 11:1654786
21. Pedrioli PGA, Eng JK, Hubley R, Vogelzang M, Deutsch EW, Raught B et al (2004) A common open representation of mass spectrometry data and its application to proteomics research. Nat Biotechnol 22:1459–1466
22. Deutsch EW (2012) File formats commonly used in mass spectrometry proteomics. Mol Cell Proteomics 11:1612–1621
23. Eng JK, Hoopmann MR, Jahan TA, Egertson JD, Noble WS, MacCoss MJ et al (2015) A deeper look into comet—implementation and features. J Am Soc Mass Spectrom 26:1865–1874
24. Eng JK, Jahan TA, Hoopmann MR (2013) Comet: an open-source MS/MS sequence database search tool. Proteomics 13:22–24
25. Deutsch EW, Mendoza L, Shteynberg D, Farrah T, Lam H, Tasman N et al (2010) A guided tour of the trans-proteomic pipeline. Proteomics 10:1150–1159

26. Deutsch EW, Mendoza L, Shteynberg D, Slagel J, Sun Z, Moritz RL (2015) Trans-proteomic pipeline, a standardized data processing pipeline for large-scale reproducible proteomics informatics. Proteomics Clin Appl 9:745–754
27. Keller A, Eng J, Zhang N, Li X, Aebersold R (2005) A uniform proteomics MS/MS analysis platform utilizing open XML file formats. Mol Syst Biol 1:2005.0017
28. Keller A, Nesvizhskii AI, Kolker E, Aebersold R (2002) Empirical statistical model to estimate the accuracy of peptide identifications made by MS/MS and database search. Anal Chem 74:5383–5392
29. Shteynberg D, Deutsch EW, Lam H, Eng JK, Sun Z, Tasman N et al (2011) iProphet: multi-level integrative analysis of shotgun proteomic data improves peptide and protein identification rates and error estimates. Mol Cell Proteomics 10:M111.007690
30. Shteynberg D, Nesvizhskii AI, Moritz RL, Deutsch EW (2013) Combining results of multiple search engines in proteomics. Mol Cell Proteomics 12:2383–2393
31. Reiter L, Claassen M, Schrimpf SP, Jovanovic M, Schmidt A, Buhmann JM et al (2009) Protein identification false discovery rates for very large proteomics data sets generated by tandem mass spectrometry. Mol Cell Proteomics 8:2405–2417
32. Nesvizhskii AI, Keller A, Kolker E, Aebersold R (2003) A statistical model for identifying proteins by tandem mass spectrometry. Anal Chem 75:4646–4658
33. Dewhirst FE, Chen T, Izard J, Paster BJ, Tanner ACR, Yu W et al (2010) The human oral microbiome. J Bacteriol 192:5002–5017
34. The UniProt Consortium (2019) UniProt: a worldwide hub of protein knowledge. Nucleic Acids Res 47:D506–D515
35. R Development Core Team (2010) R: a language and environment for statistical computing. R Foundation for Statistical Computing, Vienna, Austria
36. Bradford MM (1976) A rapid and sensitive method for the quantitation of microgram quantities of protein utilizing the principle of protein-dye binding. Anal Biochem 72:248–254
37. Promega Corporation. Sequencing Grade Modified Trypsin Certificate of Analysis 9PIV511
38. Promega Corporation. Sequencing Grade Modified Trypsin, Frozen, Product Information 9PIV5113
39. Rice RH, Means GE, Brown WD (1977) Stabilization of bovine trypsin by reductive methylation. Biochim Biophys Acta 492:316–321
40. Keil-Dlouha V, Zylber N, Imhoff J-M, Tong N-T, Keil B (1971) Proteolytic activity of pseudotrypsin. FEBS Lett 16:291–295
41. Elias JE, Gygi SP (2010) Target-decoy search strategy for mass spectrometry-based proteomics. Methods Mol Biol 604:55–71
42. Muth T, Benndorf D, Reichl U, Rapp E, Martens L (2013) Searching for a needle in a stack of needles: challenges in metaproteomics data analysis. Mol BioSyst 9:578–585
43. Tanca A, Palomba A, Fraumene C, Pagnozzi D, Manghina V, Deligios M et al (2016) The impact of sequence database choice on metaproteomic results in gut microbiota studies. Microbiome 4:51

Chapter 14

Strain-Level Profiling of Oral Microbiota with Targeted Sequencing

Chiranjit Mukherjee and Eugene J. Leys

Abstract

Targeted sequencing of one or more regions of the bacterial 16S rRNA gene fragment has emerged as a gold standard for investigating taxonomic diversity in complex microbial communities, such as those found in the oral cavity. While this approach is useful for identifying bacteria up to genus level, its ability to distinguish between many closely related oral species, or explore strain-level variations within each species, is very limited. Here we present an approach based on targeted sequencing the 16S–23S Intergenic Spacer Region (ISR) in the bacterial ribosomal operon for taxonomic characterization of microbial communities at a subspecies or strain level. This approach retains the advantages of 16S-based methods, such as easy library preparation, high throughput, short amplicon sizes, and low cost of sequencing, while providing subspecies-level resolution as a result of naturally higher genetic diversity present in the ISR compared to the 16S hypervariable regions. These advantages make it an excellent tool for high-resolution oral microbiota characterization.

Key words Microbiome, Bacteria, Strains, Amplicon sequencing, Taxonomic assignment, Illumina MiSeq, DADA2, High resolution

1 Introduction

The oral cavity is a heterogenous environment, and comparison of nine distinct habitats within the oral cavity has showed that each site has a somewhat distinct community, with different predominant species [1]. This diverse range of commensal bacteria which make up the normal microbiota in the oral cavity play a major role not only in maintaining oral health, but also systemic health [2]. Shifts in community composition at the level of species have been implicated in two of the most common diseases, dental caries and periodontitis [2]. Thus, a deeper understanding of the composition of these microbial communities is required to better elucidate the complex relationship between the oral microbiome and human health. A step toward this goal is the development of improved

Guy R. Adami (ed.), *The Oral Microbiome: Methods and Protocols*, Methods in Molecular Biology, vol. 2327, https://doi.org/10.1007/978-1-0716-1518-8_14,

methods for characterizing microbial communities at highest possible resolution, to draw sufficient clinically relevant insights.

Culture-independent molecular techniques for characterizing genetic material from a source sample have provided extensive information about the enormous taxonomic diversity of microbial communities as a whole, and in particular host-associated communities such as the oral microbiota [3]. A popular strategy used is the selection of a specific "marker gene," commonly a conserved housekeeping gene, that can be used to survey members of a particular domain. Sequencing of the 16S rRNA gene from the bacterial ribosomal operon is one such marker-gene-based approach, employed by a large number of recent microbiome studies. The lowered cost of short read sequencing, along with higher accuracy rates, made short read sequencing a method of choice for most marker gene-based surveys of microbial communities. While the use of multiple hypervariable regions, state-of-the-art bioinformatic pipelines, and well-curated databases can help maximize the resolution achievable with 16S rRNA gene sequencing, the inherent variability of this region only allows up to species-level resolution at best [4]. However, an increasing realization in the field is that species-level resolution is insufficient for many applications that require a finer understanding of the structure and function of these host-associated communities [5, 6]. As most individuals share common oral species, strain-level resolution is required for studies of microbial transmission or stability, and for exploring strain variations in disease association. Thus, there is a need for high-resolution, high-throughput methods for characterizing microbial communities that can be applied for large-scale clinical studies.

One approach that has been recently explored for strain-level characterization of microbiota is whole metagenome sequencing. While metagenomics provides an excellent tool for exploring gene diversity within the community, as shown by the human microbiome project [7], efforts have been made to utilize the deep sequencing data to profile taxonomic diversity at higher resolutions [6, 8, 9]. A major problem with using metagenomic sequencing approaches for bacterial strain profiling is the requirement of very deep sequencing, especially for oral samples such as saliva where host DNA makes up over 80% of the genetic content. These methods often identify a single dominant strain for each species [10], possibly due to lack of sufficient sequencing depth to identify the less abundant rare strains. For these reasons, such metagenomic strain profiling tools have found limited use in comprehensively profiling strain diversities in exceptionally diverse communities, such as the oral microbiome. Strain-level analyses have previously been conducted with targeted methods such as RFLP and MLST, a good example being epidemiological studies of *Mycobacterium tuberculosis* [11]. However, these approaches are limited by their focus on a select set of organisms and are not suitable for community-level analysis.

An alternate approach to strain profiling is the use of a universal marker gene that is able to provide subspecies-level taxonomic resolution. Such a strategy utilizing a different region of the bacterial ribosomal operon, the 16S–23S Intergenic Spacer Region (ISR), was proposed as early as 1991 by Barry et al. [12]. This approach is based on the understanding that the spacer region being largely non coding, except for the presence of tRNA coding genes, is under comparatively less selective pressure than the 16S gene, and as a result shows greater variability that can be exploited to distinguish among closely related species and even explore subspecies variations [12]. Consequently, many groups have utilized sequence analysis of the ISR to differentiate among species which were not distinguishable using 16S-based methods [13–15]. A study by Chen et al. concluded that ISR-based sequencing approach was an improvement over 16S-based approaches for identification of species among the clinically relevant viridans group *Streptococci* [15]. Our laboratory was one of the first to explore the use of bacterial ISRs for strain identification for the oral species *A. actinomycetemcomitans* [16]. Consequently, it also showed that heteroduplex analysis using ISRs could be a tool for identifying strains of the periodontal pathogen *Porphyromonas gingivalis*, establishing a link between ISR phylogeny and disease-associated phenotypes of the strains [17, 18]. While these approaches clearly demonstrated the resolving power of the ISR, low-throughput sequencing methods available at the time, and the use of species-specific primers did not allow application of ISR-based sequencing to community-level strain analysis. More recently two groups utilized modern sequencing technologies to profile the bacterial ISRs. Ruegger et al. [19] developed an Illumina HiSeq-based amplicon sequencing approach targeting the ISRs and showed a considerable increase in resolving power compared to 16S-based sequencing. However, the absence of complimentary high-resolution bioinformatic pipelines and comprehensive well-curated ISR databases limited the application of the ISR sequencing in these studies.

Recently, a number of new bioinformatic methods have been developed that do not require assigning sequence reads into fixed threshold Operational Taxonomic Units (OTU) bins as is done for de novo OTU clustering methods [20–24]. These methods attempt to infer true biological variants among amplicon reads and resolve sequences that are as little as one nucleotide apart. These unique sequence variants are referred to as amplicon sequence variants or ASVs. One such ASV-based method that has become widely used is DADA2 [22]. DADA2 aims to "denoise" the amplicon reads by incorporating an error modeling approach that estimates the error rate within the dataset, and uses that information, in conjunction with abundance information of individual sequence reads, to determine the probability of a sequence variant having originated due to sequencing error. This is based on per base quality scores and the assumption that true biological variants are likely to be observed at a greater rate than variants arising due to random sequencing errors [25]. DADA2 has been shown to have greater sensitivity and specificity compared to most OTU-based methods [22]. These characteristics make DADA2 a perfect tool for high-resolution processing of marker genes, especially when preserving single nucleotide variations is crucial to explore strain-level differences among amplicons from the same species.

To explore strain-level communities in the oral microbiome, we developed an amplicon-based microbial characterization strategy that achieves ultra-high resolution by combining targeted sequencing of the highly variable 16S–23S ISR marker gene and processing of those sequences with the high-resolution denoising platform DADA2. An ISR sequence database was also developed by extracting the 16S–23S intergenic spacer region from publicly available genomic sequences of common oral bacteria, so as to assign taxonomic identity to the ISR amplicons. We validated this approach in a clinical study comparing microbial communities from dental plaque of five adult subjects over a 1-year period [26]. The ISR-DADA2 approach detected 5.2-fold more amplicon sequence variants than the standard 16S species-level reference database approach, and multiple genotypic variants of the ISR were identified for most oral species, demonstrating a high level of subspecies variation in the oral microbiota [26].

A generalized workflow for strain-level characterization of oral microbiota samples is described here. Overall the library preparation steps for ISR amplicon sequencing have been developed analogous to the 16S short read sequencing library preparation protocol developed by Illumina (*see* **Note 1**), with the main difference being the step of generating ISR amplicons in place of 16S. This allows our protocol to be easily adopted by laboratories already familiar with preparing 16S/ITS libraries for sequencing on Illumina MiSeq platform. A detailed workflow for generating ISR amplicons is described here, along with the specific steps of a bioinformatic pipeline for utilizing those ISR amplicon sequences to explore strain-level diversity in the samples. We generate ISR amplicons using locus-specific primers to amplify the target region (between the 3′-end of the 16S gene and the 5′-end of the 23S gene). Thereafter, these ISR amplicons can be processed as described in the Illumina 16S protocol (*see* **Note 1**), either in-house or at sequencing centers where the remaining steps can be completed to finalize library generation.

The bioinformatic pipeline consists of initial processing steps that are a variation of the DADA2 pipeline for 16S sequences (https://benjjneb.github.io/dada2/tutorial.html), adapted for

ISR reads. Post sequencing, the demultiplexed FASTQ files can be directly processed using the DADA2-based pipeline provided here, to generate a table of sample versus ISR-amplicon sequence variants (ISR-ASVs), analogous to sample versus 16S-OTUs or sample versus 16S-ASVs table. These ISR ASVs can then be directly used for diversity analysis and multidimensional ordinations. Mapping the DADA2-derived ISR-ASVs to a database of ISR reads allows us to bin the ISR-ASVs into species-level bins. Thus, the ISR-ASVs within each species bin constitute the ISR-strains of that species. For this purpose, we currently maintain the Human Oral ISR database [26]. The present version of the ISR database (ISR-db) consists of over 3000 unique ISR sequences, representing close to 300 of the most abundant oral bacteria species and is publicly available (https://github.com/cm0109/ISR_database). All software tools included in this pipeline are open source and can be implemented without the need for heavy computational power. An overview of the molecular and bioinformatic approach for ISR amplicon sequencing and related bioinformatics is presented in Fig. 1.

The ISR approach described here provides a high-throughput, high-resolution yet cost-effective method that allows subspecies-level community fingerprinting at a cost comparable to 16S rRNA gene amplicon sequencing. This new approach will be useful for a range of applications that require high-resolution identification of organisms, including microbial tracking, community fingerprinting, and identification of virulence-associated strains.

2 Materials

2.1 Sequencing Library Preparation

1. QIAamp DNA Mini Kit (Qiagen) or other (*see* **Note 2**).
2. 0.1-mm Glass beads.
3. Bead-based homogenizer for 1–2 mL tubes (*see* **Note 3**).
4. Benchtop centrifuge.
5. PCR Template: DNA extracted from samples, normalized to ~5 ng/μL (*see* **Note 4**).
6. High-fidelity DNA polymerase, such as Invitrogen AccuPrime Taq DNA Polymerase, High Fidelity.
7. Amplicon PCR primers, which are ISR locus-specific primers with included Illumina adapter (*see* **Note 5**). The sequences of the locus-specific part of the primers are as below:

   rD1f: 5′-GGCTGGATCACCTCCTT [27].

   EricM: 5′-GCCWAGGCATCCDCC [28].
8. Thermocycler.

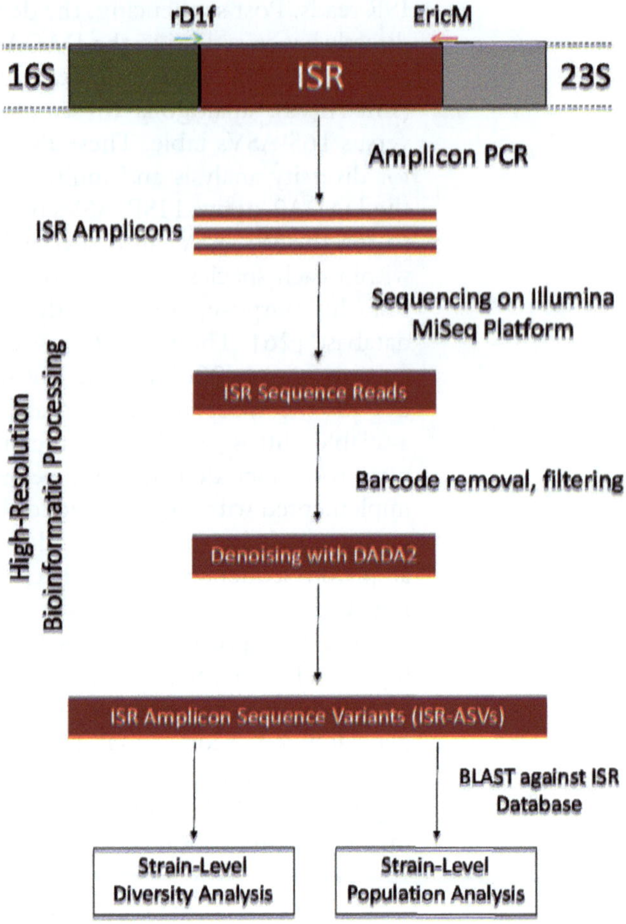

Fig. 1 Overview of the ISR amplicon sequencing approach for strain-level analysis of microbial communities

9. Agencourt AMPure XP PCR Purification system (Beckman Coulter).
10. Quant-iT High-Sensitivity dsDNA Assay Kit (Invitrogen) or similar fluorescent dye-based DNA assay kit.
11. Spectramax Microplate reader (Molecular Devices) or equivalent microplate reader.

2.2 Bioinformatic Processing

1. R [29].
2. R packages:DADA2 [22] and Phyloseq [30].

   We recommend RStudio, the integrated development environment, for working in R [31] (*see* **Note 6**).
3. Linux computing environment with access to command line interface.
4. Command line BLAST+ suite [32] (https://ftp.ncbi.nlm.nih.gov/blast/executables/blast+/LATEST/).

Human Oral ISR Database [26] (https://github.com/cm0109/ISR_database).

3 Methods

3.1 Library Preparation

The following method describes a generalized strategy to develop targeted ISR amplicons from DNA extracted from oral (saliva/plaque) samples.

3.1.1 Preparing DNA Template for Library Preparation

DNA extraction methods that have been successfully utilized for generating amplicon sequencing libraries can be used for ISR sequencing, and no special considerations applies. A standard bacterial DNA extraction protocol that has been optimized for subgingival plaque samples [26] is described here.

1. Subgingival plaque samples collected using sterile paper points should be placed in 200μL buffer ATL (QIAamp DNA Mini Kit) (*see* **Note 2**) for storage at −20 °C until DNA extraction.
2. At the time of extraction, incubate thawed samples with 300μL ATL and 40μL Proteinase K (QIAamp DNA Mini Kit) (*see* **Note 2**) at 56 °C for 2 h.
3. Separate the solution from the paper points by centrifugation, using a perforated microcentrifuge tube.
4. Homogenize the solution using 0.25 g of 0.1-mm glass beads in a Mini-Beadbeater-16, for 60 s at 3450 oscillations/min.
5. Purify the genomic DNA using QIAamp DNA Mini Kit (*see* **Note 2**) according to the manufacturer's directions and elute in 30μL of the buffer AE (*see* **Note 7**).

3.1.2 Performing Amplicon PCR

See Table 1 for reference.

1. Prepare 25μL PCR reactions using components described in Subheading 2.1. Each sequencing run should include no template reactions as negative control and model microbial community template as positive control reactions.
2. Run PCR program with initial denaturation step at 94 °C for 2 min.
3. Repeat **steps 4–6** 25×.
4. Denaturation at 94 °C for 30 s.
5. Annealing at 55 °C for 30 s.
6. Extension at 68 °C for 1 min.
7. Final extension at 72 °C for 5 min.
8. Incubation at 4 °C.

Table 1
Details of each reaction component in amplicon PCR

Reagents	Volume per reaction (μL)
Template DNA (~5 ng/μL)	2.5
PCR grade Water	17.3
10× Buffer II	2.5
AccuPrime Taq	0.2
Forward Primer	1.25
Reverse Primer	1.25
Total volume in each well	25

9. Purify PCR products with the Agencourt AMPure XP PCR Purification system using manufacturer's guidelines.
10. Proceed to the next steps of Illumina 16S library preparation, i.e., the Index PCR step, which adds the sample-specific barcodes for multiplexing (*see* **Note 1**). Alternately, at this stage purified amplicon PCR products may be sent over to the sequencing facility for the subsequent library preparation steps, which are common to amplicon sequencing for Illumina platforms. Using unique sample-specific barcodes allows inclusion of a large number of samples in the same sequencing run, thereby reducing overall sequencing cost per sample (*see* **Note 8**).
11. The finalized libraries are pooled and sequenced on the Illumina MiSeq Platform, using 300 base pair paired-end chemistry, with addition of appropriate PhiX control (*see* **Note 9**).

3.2 Bioinformatic Processing

Base call (BCL) files generated from the sequencer can be converted to "demultiplexed" FASTQ files with Illumina's bcl2fastq conversion software v2.20 (Illumina, USA), using barcode index information. This process removes multiplexing barcodes from the sequences, and generates sample-specific FASTQ files which is the input for this bioinformatic processing pipeline, and is generally performed at the sequencing center. For processing the resulting FASTQ files, we utilize the Bioconductor package *dada2* [33] for denoising and inference of exact amplicon sequence variants.

Currently, only the R1 FASTQ files are used for analysis, since the ISR vary in length among the different species of oral bacteria, and merging R1 and R2 reads are not successful for all species.

The code below describes the steps involved in processing the ISR sequences, from demultiplexed FASTQ files to generating ISR amplicon sequence variants (ISR-ASVs).

1. Load required libraries:

```
library(dada2)
library(phyloseq)
```

2. Assign path to unzipped forward Fastq files:

```
path <- "<path to your fastqs>" # directory of zipped
R1 FASTQ files
list.files(path) # Inspect to make sure all fastqs
are listed
```

3. Assign full path names:

```
fnFs <- sort(list.files(path, pattern="_R1.fq",
full.names = TRUE))
```

4. Assign sample names by extracting required characters from Fastq file names:

```
sample.names <- sapply(strsplit(basename(fnFs),
"_"), `[`, 1)
# Modify this as per your sample name
```

5. Inspect quality profiles for filtering decision:

```
# Plot quality profiles
plotQualityProfile(fnFs[1:4])
```

6. Create filtered read path:

```
filt_path <- file.path(path, "filtered") # Place
filtered files in filtered/ subdirectory
filtFs <- file.path(filt_path, paste0(sample.names,
"_F_filt.fastq.gz"))
```

7. Filter Fastq files with DADA2:

```
# Forward primer used is "GGCTGGATCACCTCCTT" which is
17 bases
# Thus, trimLeft set to 17, others are default
parameters
```

```
# Change parameters according to quality of your
fastq reads as seen in previous step
# Create filte object
fastq_filt <- filterAndTrim(fnFs, filtFs, trim-
Left=17,
 maxN=0, maxEE=2, truncQ=2, rm.phix=TRUE,
 compress=TRUE, multithread=TRUE)
```

8. Learn error rates with DADA2:

```
# Learn Error rates
errF <- learnErrors(filtFs, multithread=TRUE)
# Plot Errors
plotErrors(errF, nominalQ=TRUE)
```

9. Dereplicate the filtered sequences with DADA2:

```
derepFs <- derepFastq(filtFs, verbose=TRUE)
# Name the derep-class objects by the sample names
names(derepFs) <- sample.names
```

10. Denoise sequences with DADA2:

```
dadaFs <- dada(derepFs, err=errF, multithread=TRUE)
#Make sequence table without merging
seqtabF <- makeSequenceTable(dadaFs)
```

11. PCR amplification steps involved in the library preparation can result in chimeric sequences, which need to be removed before downstream processing. DADA2 has a built-in function to do this:

```
seqtabF.nochim <- removeBimeraDenovo(seqtabF, meth-
od="consensus",
 multithread=TRUE, verbose=TRUE)
# Compute proportion of chimeric reads
sum(seqtabF.nochim)/sum(seqtabF)*100
```

12. Generate statistics for DADA2 processing:

```
getN <- function(x) sum(getUniques(x))
track <- cbind(fastq_filt, sapply(dadaFs, getN),
rowSums(seqtabF), rowSums(seqtabF.nochim))
colnames(track) <- c("input", "filtered", "de-
noised", "tabled","nochim")
rownames(track) <- sample.names
DADA2_stats <- as.data.frame(track)
# Save stats as text file
write.table(DADA2_stats, file="DADA2_stats.txt",
sep="\t", quote=F, col.names = NA)
```

13. Make ISR ASV table with Phyloseq:

```
atab <- otu_table(seqtabF.nochim, taxa_are_rows=-
FALSE)
colnames(atab) <- paste0("seq", seq(ncol(atab)))
atab.df <- as.data.frame(atab)
# Save DADA2 assigned ISR ASVs as text file for
downstream analysis
write.table(atab.df, file="DADA2_ISR_ASVs.txt",
sep="\t", quote=F, col.names = NA)
```

14. Extract sequences for ISR ASVs:

```
atab_seqs <- colnames(seqtabF.nochim) # save se-
quences as object
# Write seqs file as output fasta file:
for (i in 1:length(atab_seqs)){
 sink("ISR_asvs.seqs.fa", append = T) # Append is set
to true, run only once!
  cat(paste(">",colnames(atab.df)[i],sep=""), atab_-
seqs[i],sep="\n")
 sink()
}
```

15. In the end an ITR ASV table is generated which is suitable for alpha and beta diversity analysis. To assign taxonomy, the approach available at this writing is to use command line interface within LINUX to implement BLASTN using the Human Oral ISR Database from the authors as the reference library.

4 Notes

1. The Illumina 16S library preparation guide can be found at: https://support.illumina.com/content/dam/illumina-support/documents/documentation/chemistry_documentation/16s/16s-metagenomic-library-prep-guide-15044223-b.pdf
2. Any method of bacteria lysis and DNA isolation that works on the species of interest can be selected, but must be used consistently for all experiments.
3. Instructions as written are for the Mini-Beadbeater-16 (BioSpec Products, USA) bead-based homogenizer.
4. Since the concentration of DNA extraction product from each sample may vary, we recommend adjusting the concentration to ~5 ng/μL in order to normalize the amount of DNA in the amplicon PCR template.
5. For sequencing on the Illumina platform, these locus-specific ISR primers have to be combined with the Illumina-specific forward and reverse adapters, TCGTCGGCAGCGTCAGATGTGTATAAGAGACAG and GTCTCGTGGGCTCGGAGATGTGTATAAGAGACAG , respectively. For detailed considerations for primer design, please refer to **Note 1**.
6. R is a scripting language optimized for statistical analysis and graphical presentation of data. It is an open-source platform, and there are multiple educational sources to learn how to use it.
7. For best results, extracted DNA concentration should be greater than 5 ng/μL. To facilitate equal weight of all samples in the library, we recommend normalizing the extracted DNA to a concentration of 5 ng/μL for all samples, and use the normalized product as template for the Amplicon PCR. For this purpose, we recommend quantification of DNA using Quant-iT High-Sensitivity dsDNA Assay Kit or equivalent, on a microplate reader, using manufacturer's guidelines.
8. In our experience, MiSeq Reagent Kit v3 average over 12 million high-quality read ($\geq$Q28) for paired-end 2 $\times$ 300 bp runs. Therefore, to achieve an average sequencing depth of 100,000 sequences, 120 samples at a time can be included in a single MiSeq run with degenerate barcodes for multiplexing. Please consult with your sequencing facility for additional considerations.
9. We recommend a conservative approach of adding 15–20% PhiX control for better quality reads, even though it results in slightly lower throughput.

References

1. Aas JA, Paster BJ, Stokes LN, Olsen I, Dewhirst FE (2005) Defining the normal bacterial flora of the oral cavity. J Clin Microbiol 43:5721–5732
2. Wade WG (2013) The oral microbiome in health and disease. Pharmacol Res 69:97–114. https://doi.org/10.1016/j.phrs.2012.11.006
3. Siqueira JF, Rôças IN (2017) The oral microbiota in health and disease: an overview of molecular findings. Methods Mol Biol 1537:127–138. https://doi.org/10.1007/978-1-4939-6685-1_7
4. Ellegaard KM, Engel P (2016) Beyond 16S rRNA community profiling: intra-species diversity in the gut microbiota. Front Microbiol 21(7):1475. https://doi.org/10.3389/fmicb.2016.0147
5. Luo C, Knight R, Siljander H, Knip M, Xavier RJ, Gevers D (2015) ConStrains identifies microbial strains in metagenomic datasets. Nat Biotechnol 33:1045–1052. https://doi.org/10.1038/nbt.3319
6. Segata N (2018) On the road to strain-resolved comparative metagenomics. mSystems. https://doi.org/10.1128/mSystems.00190-17
7. Huttenhower C, Gevers D, Knight R, Abubucker S, Badger JH, Chinwalla AT et al (2012) Structure, function and diversity of the healthy human microbiome. Nature 486:207–214. https://doi.org/10.1038/nature11234
8. Schloissnig S, Arumugam M, Sunagawa S, Mitreva M, Tap J, Zhu A et al (2013) Genomic variation landscape of the human gut microbiome. Nature 493:45–50. https://doi.org/10.1038/nature11711
9. Lloyd-Price J, Mahurkar A, Rahnavard G, Crabtree J, Orvis J, Hall AB et al (2017) Strains, functions and dynamics in the expanded human microbiome project. Nature 550:61–66. https://doi.org/10.1038/nature23889
10. Truong DT, Tett A, Pasolli E, Huttenhowe C, Segata N (2017) Microbial strain-level population structure and genetic diversity from metagenomes. Genome Res 27:626–638
11. Kato-Maeda M, Metcalfe JZ, Flores L (2011) Genotyping of Mycobacterium tuberculosis: application in epidemiologic studies. Future Microbiol 6:203–216. https://doi.org/10.2217/fmb.10.165
12. Barry T, Colleran G, Glennon M, Dunican LK, Gannon F (1991) The 16s/23s ribosomal spacer region as a target for DNA probes to identify eubacteria. Genome Res 1:51–56. https://doi.org/10.1101/gr.1.1.51
13. Graham TA, Golsteyn-Thomas EJ, Thomas JE, Gannon VP (1997) Inter- and intraspecies comparison of the 16S–23S rRNA operon intergenic spacer regions of six Listeria spp. Int J Syst Bacteriol 47:863–869
14. Dec M, Urban-Chmiel R, Gnat S, Puchalski A, Wernicki A (2014) Identification of Lactobacillus strains of goose origin using MALDI-TOF mass spectrometry and 16S–23S rDNA intergenic spacer PCR analysis. Res Microbiol 165:190–201. https://doi.org/10.1016/j.resmic.2014.02.003
15. Chen C, Teng L, Chang T (2004) Identification of clinically relevant viridans group streptococci by sequence analysis of the 16S–23S ribosomal DNA spacer region. J Clin Microbiol 42:2651–2657
16. Leys EJ, Griffen AL, Strong SJ, Fuerst PA (1994) Detection and strain identification of Actinobacillus actinomycetemcomitans by nested PCR. J Clin Microbiol 32:1288–1294
17. Leys EJ, Smith JH, Lyons SR, Griffen AL (1999) Identification of Porphyromonas gingivalis strains by heteroduplex analysis and detection of multiple strains. J Clin Microbiol 37:3906–3911
18. Griffen AL, Lyons SR, Becker MR, Moeschberger ML, Leys EJ (1999) Porphyromonas gingivalis strain variability and periodontitis. J Clin Microbiol 37:4028–4033
19. Ruegger PM, Clark RT, Weger JR, Braun J, Borneman J (2014) Improved resolution of bacteria by high throughput sequence analysis of the rRNA internal transcribed spacer. J Microbiol Methods 105:82–87
20. Eren AM, Maignien L, Sul WJ, Murphy LG, Grim SL, Morrison HG et al (2013) Oligotyping: differentiating between closely related microbial taxa using 16S rRNA gene data. Methods Ecol Evol 4:1111–1119
21. Tikhonov M, Leach RW, Wingreen NS (2015) Interpreting 16S metagenomic data without clustering to achieve sub-OTU resolution. ISME J 9:68–80. https://doi.org/10.1038/ismej.2014.117
22. Callahan BJ, McMurdie PJ, Rosen MJ, Rosen MJ, Han AW, Johnson AJA et al (2016) DADA2: high-resolution sample inference from Illumina amplicon data. Nat Methods 13:581–583. https://doi.org/10.1038/nmeth.3869

23. Edgar RC (2016) UNOISE2: improved error-correction for Illumina 16S and ITS amplicon sequencing. bioRXiv. https://doi.org/10.1101/081257
24. Amir A, McDonald D, Navas-Molina JA, Kopylova E, Morton JT, Xu ZZ et al (2017) Deblur rapidly resolves single-nucleotide community sequence patterns. mSystems. https://doi.org/10.1128/mSystems.00191-16
25. Callahan BJ, McMurdie PJ, Holmes SP (2017) Exact sequence variants should replace operational taxonomic units in marker-gene data analysis. ISME J 11:2639–2643. https://doi.org/10.1038/ismej.2017.119
26. Mukherjee C, Beall CJ, Griffen AL, Leys EJ (2018) High-resolution ISR amplicon sequencing reveals personalized oral microbiome. Microbiome 6:153. https://doi.org/10.1186/s40168-018-0535-z
27. Weisburg WG, Barns SM, Pelletier DA, Lane DJ (2019) 16S ribosomal DNA amplification for phylogenetic study. J Bacteriol 173 (2):697–703
28. Rumpf RW, Griffen AL, Wen BG, Leys EJ (1999) Sequencing of the ribosomal intergenic spacer region for strain identification of Porphyromonas gingivalis. J Clin Microbiol 37:2723–2725
29. R Core Team (2017) R: a language and environment for statistical computing. R Foundation for Statistical Computing, Vienna, Austria. http://www.R-project.org/
30. McMurdie PJ, Holmes S (2013) Phyloseq: an R package for reproducible interactive analysis and graphics of microbiome census data. PLoS One 8:e61217. https://doi.org/10.1371/journal.pone.0061217
31. Rstudio Team (2016) RStudio: integrated development for R. RStudio, PBC, Boston, MA
32. Camacho C, Coulouris G, Avagyan V, Ma N, Papadopoulos J, Bealer K et al (2009) BLAST+: architecture and applications. BMC Bioinformatics 10:421. https://doi.org/10.1186/1471-2015-10-421
33. Callahan BJ, Wong J, Heiner C, Oh S, Theriot CM, Gulati AS et al (2019) High-throughput amplicon sequencing of the full-length 16S rRNA gene with single-nucleotide resolution. Nucleic Acids Res. https://doi.org/10.1093/nar/gkz569

Chapter 15

Profiling the Human Oral Mycobiome in Tissue and Saliva Using ITS2 DNA Metabarcoding Compared to a Fungal-Specific Database

David J. Speicher and Ramy K. Aziz

Abstract

The advent of high-throughput sequencing has caused a paradigm shift from the one-pathogen one-disease model to the significance of dysbiosis of the oral microbiome, including the oral mycobiome. The oral mycobiome can be profiled by a method modified from that used to profile the bacteriome with 16S rRNA gene primers. The first modification is to include an initial fungus lysis step that ensures representative yields of fungal DNA. The second step is to use a reliable target, the ITS1 and/or ITS2 regions of the 23S rRNA, to define the oral fungal population, and modifications of library preparation required to deal with the variable sized amplicons generated. In this chapter, a proven microbiomic approach to identify fungal populations in oral tissue samples associated with cancer is described. This approach is also applicable to the study of the salivary mycobiome in both healthy and diseased individuals.

Key words Mycobiome, Fungi, Microbiome, Oral cancer, Saliva

1 Introduction

While many oral diseases are thought to be associated with a single or a few pathogens, the advent of high-throughput sequencing, also known as next-generation sequencing (NGS) highlights the significance of a dysbiotic microbiome, whether bacterial, fungal, or viral. While pathogenic fungi, such as *Aspergillus* and *Blastomyces dermatitidis/gilchristii*, commonly cause severe infections of the lower respiratory tract [1], nonpathogenic fungi also play an important role in disease. The oral microbiome is the most diverse found in the body [2, 3] and consists of a basal oral mycobiome of 74 culturable and 11 non-culturable genera, and 101 species, with *Candida*, *Malassezia*, *Aspergillus*, and *Cladosporium* being the most common [4]. *Candida* is the most common fungal pathogen, especially among the HIV-positive population [5], and has been identified with oral squamous cell carcinoma [6] and dental caries [7]. Levels of *Candida* spp. also change with levels of oral

Guy R. Adami (ed.), *The Oral Microbiome: Methods and Protocols*, Methods in Molecular Biology, vol. 2327,
https://doi.org/10.1007/978-1-0716-1518-8_15,

and systemic disease [8, 9]. Malassezia, a common skin commensal [10], was found in high levels in the oral cavity [6, 11]. Research into the role of the oral mycobiome is in its infancy, with studies just beginning to correlate fungus and disease [12]. Current models for causation are highly speculative, especially with the discovery of new fungal genera that are difficult to culture but may contribute to specific diseases [13, 14].

Historically, the detection of fungi has been primarily by culture, with molecular methods used in recent years. These methods work to detect a few species but do not give a complete picture, especially as fungi are present in low numbers and are more difficult to culture than bacteria [8]. Like bacterial 16S rRNA profiling, NGS technologies using universal fungal primers can profile the oral mycobiome, provided a few modifications are followed: (1) doing additional lysis steps to harvest the genomic DNA without significant loss or degradation; (2) choosing primers that target the internal transcribed spacer (ITS) 1 and/or ITS2 rRNA region; and (3) using alignment tools and reference libraries specific for fungi. To compare results between studies and to improve accuracy in profiling the oral mycobiome, there is a need for uniform protocols from sample collection to the identification of quantitation of fungal taxa [6, 9, 15].

Fungi have a rigid cell wall composed of glycans, chitin, mannans, and glycoprotein that makes DNA extraction much more difficult than for bacteria. Extraction of fungal DNA requires enzymatic lysis and/or vigorous bead-beating prior to DNA extraction [3]. While the various methods affect DNA yield and quality to a varying degree, their impact on diversity appears minor [16]. Lysis with enzymes, like MetaPolyzyme Multilytic Enzyme Mix (Sigma-Aldrich), achieve higher DNA yields but mechanical disruption using bead-beating tend to produce more consistent amplification of the ITS region [17]. However, it is critical that methods are used consistently between studies so data can be compared reliably. It is also important to extract negative controls to rule out contaminants, especially when dealing with low biomass samples [1, 18]. DNA extraction kits and other laboratory reagents are sources of potential contamination of fungal DNA as they may be sterile but not DNA-free, thus introducing foreign DNA at any point of sample processing [19].

While eukaryotic 18S rRNA gene closely corresponds to the prokaryotic 16S rRNA gene, the 18S rRNA gene has insufficient variability to allow differentiation, further challenged by an expected overabundance of 18S rRNA from human tissues. Therefore, fungal metabarcoding, i.e., the molecular identification of fungi using taxon-specific primers, uses the ITS region between the 18S and 28S rRNA genes [20]. As the entire ITS region (500–600 bp) is too large to sequence in a single run, the highly conserved 5.8S region is excluded from sequencing, and fungal metabarcoding uses the ITS1 and/or ITS2 rRNA operon. There is still great debate as to which ITS subunit is best as each region discriminates against certain taxa. The ITS1 is the most rapidly evolving operon and discriminates against Basidiomycota [8, 21]. The ITS2 is moderately to rapidly evolving, discriminates against Ascomycota, and showed the greatest resolution of low-abundance taxa [8, 21, 22]. The ITS2 was also found to be superior for counting the maximal number of species in a mock community of 21 fungal species from the gut, mouth, nose, skin, and vagina [23]. While some studies have shown that ITS1 and ITS2 can have significant impact on fungal profiling [24], other studies have shown that ITS1 and ITS2 yield similar results and suggest sequencing both regions to overcome bias [25]. Deciding which metabarcode to use ultimately depends on the samples used and the research question. If the research is to characterize fungal diversity, sequencing both the ITS1 and ITS2 might cover more fungal taxa. However, if the goal is to compare fungal taxa in different samples or populations, choosing ITS1 or ITS2 and using a rigid sample handling protocol will suffice.

The variable length of the ITS1 and ITS2 regions, ranging from 200 to 600 bp depending on the fungal taxa, is a challenge to fungal taxa identification. Bidirectional sequencing, followed by merging of sequence pairs, might get problematic for some taxa with larger templates because the opposing reads might not overlap. Therefore, care must be taken in the choice of amplicon length [26]. A second complication is that the ITS copy number can vary significantly from just a few to hundreds [27]. This makes quantitative comparisons between different taxa suspect. Additionally, within some species there is variability on how many copies of ITS sequence are present, further decreasing the ability to compare different studies. Another barrier is the poorer annotation of the fungal rRNA cistron across the species residing in the human mouth [8]. As a result, one must use libraries of total fungal sequences, many of which also include a range of other eukaryotic taxa and some ambiguity in nomenclature [12]. A final problem is with assigning taxa to specific reads. There is a desperate need to resolve the fungal nomenclature based on phylogeny and provide a single name for a single fungus. Under the current nomenclature a single fungus can have multiple names reflecting their asexual/sexual morphs or diverse historical or geographical discoveries [8].

In this chapter, we provide a detailed protocol from sample collection to bioinformatic analysis for profiling the oral mycobiome based on methods used by Perera et al. [6] and McTaggart et al. [1] with modifications made due to recent advances in the field (Fig. 1). This method is specifically designed for determining the oral mycobiome profile in saliva and excised tissue but can be modified for any sample type, including soil. If using this method on a non-oral sample, one must decide whether to amplify ITS1 or

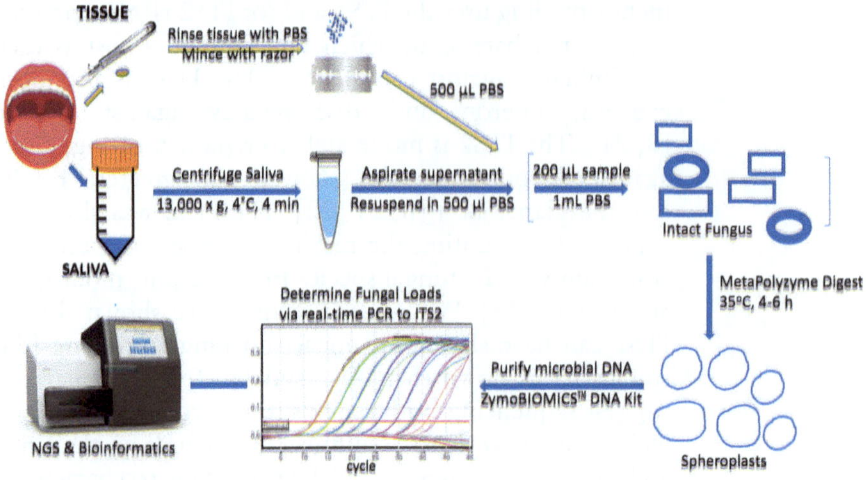

Fig. 1 Schematic of the steps from sample collection DNA isolation and sequencing required to profile the oral mycobiome

ITS2 and decide the database used to determine OTU taxonomy of the various reads. Detailed bioinformatics is beyond the scope of this chapter, but a skeletal outline of the bioinformatics pipeline is provided to give an understanding of the steps needed. We encourage anyone attempting this protocol to collaborate with a next-generation facility and a bioinformatics team to aid in proper design and analysis.

2 Materials

2.1 Sample Collection and Extraction

These and other steps are outlined in Fig. 1.

2.1.1 Collection of Saliva

1. Sterile 50 mL conical tube or urine container.
2. Ice packs, crushed ice or dry ice.
3. Refrigerated centrifuge able to accommodate 50 mL conical tubes.
4. −80 °C Freezer.

2.1.2 Collection of Oral Tissues

1. Sterile surgical blades and gauze.
2. Sterile screw-cap vials.
3. Dry ice.
4. −80 °C Freezer.

2.2 Sample Processing and Extraction

1. Phosphate-buffered saline (PBS), pH 7.5 (without EDTA). PBS: 1×, pH 7.5. Prepare 800 mL of distilled water and add 0.2 M NaCl (11.6 g), 2.5 mM KCl (0.186 g), 8 mM Na_2HPO_4 (1.4 g), 1.5 mM KH_2PO_4 (0.2 g). Adjust the pH to 7.5 with a pH meter and addition of HCl add distilled water to prepare a 1 L solution of 1× PBS.
2. Low-binding 1.5 and 2.0 mL microcentrifuge tubes (sterile and RNase/DNase-free).
3. MetaPolyzyme Multilytic Enzyme Mix (Sigma-Aldrich).
4. ZymoBIOMICS DNA Microprep Kit (Zymo Research).
5. FastPrep-24 Classic bead beating grinder and lysis system (MP Biomedicals).
6. Vortex.
7. Refrigerated centrifuge able to accommodate 50 mL conical tubes.
8. Refrigerated centrifuge able to accommodate microcentrifuge tubes.
9. NanoDrop™ ND-1000 Spectrophotometer, Qubit Fluorometer (Thermo Fisher Scientific) or Denovix Fluorometer (DeNovix).

2.3 Determining Fungal Load

1. Primers: Fungal ITS2 (*see* **Note 1**).

   ITS3-F: 5′-GCATCGATGAAGAACGCAGC-3′.
   ITS4-R: 5′-TCCTCCGCTTATTRATATGC-3′.
2. Primers: Human β-actin gene

   β-actin-gDNA-F: 5′-TCCGCAAAGACCTGTACGC-3′.
   β-actin-gDNA-R: 5′-CAGTGAGGACCCTGGATGTG-3′.
3. LightCycler 480 Instrument II (Roche Diagnostics).
4. Nuclease-free molecular grade water.
5. LightCycler 480 SYBR Green I Master (Roche Diagnostics).
6. LightCycler Multiwell Plates with optical seals (Roche Diagnostics).

2.4 Library Preparation and Sequencing

1. Adapter-linked ITS2 primers [28] (*see* **Note 2**).
   5′-tcgtcggcagcgtcagatgtgtataagagacagGCATCGATGAAG AACGCAGC-3′.
   5′-gtctcgtgggctcggagatgtgtataagagacag TCCTCCGCTTATT GATATGC-3′.
2. IDTE (10 mM Tris, 0.1 mM EDTA) buffer, pH 8.0.
3. xGen Universal Blocker-TS Mix (Integrated DNA Technologies).
4. Human Cot-1 DNA (Thermo Fisher Scientific).

5. NimbleGen 2× hybridization buffer and NimbleGen 2× hybridization solution (Roche Diagnostics).
6. Dynabeads M-270 Streptavidin magnetic beads (Thermo Fisher Scientific).
7. Multiwell plates or possibly Midi-Magnet plates (Alpaqua).
8. Microseal B plate covers (Bio-Rad).
9. ThermoMixer C shaker (Eppendorf).
10. SeqCap EZ Hybridization and Wash kit (Roche Diagnostics).
11. Agencourt AMPure XP beads (Beckman Coulter Genomics) or other similar DNA cleanup kit.
12. Nuclease-free molecular grade water.
13. A 96-well magnetic rack, e.g., EpiMag HT (EpiGentek) or Magna GrIP HT96 Rack (Sigma-Aldrich).
14. MiSeq v2 Nano Reagent Kit or iSeq 100 i1 300-cycle Reagent Kit (Illumina).
15. PhiX Control v3 Library (Illumina).
16. Optional: Agilent 2100 Bioanalyzer system (Agilent Technologies).
17. Optional: Bioanalyzer High-Sensitivity DNA Kit (Agilent Technologies).

2.5 Sequence Data Processing, Taxonomy Assignment, and Analysis

1. FastQC software at https://www.bioinformatics.babraham.ac.uk/projects/fastqc/
2. Seqtk for trimming and processing sequence files: https://github.com/lh3/seqtk
3. Trimmomatic at http://www.usadellab.org/cms/?page=trimmomatic [29].
4. The UNITE database at https://unite.ut.ee/repository.php
5. BLASTN (online on NCBI: https://blast.ncbi.nlm.nih.gov/) or part of the stand-alone BLAST+ package [30] at https://blast.ncbi.nlm.nih.gov/Blast.cgi?PAGE_TYPE=BlastDocs&DOC_TYPE=Download
6. CLC Genomics Workbench version 12.0 (Qiagen) or UniPro UGENE (free download at http://ugene.net/).
7. QIIME [31–33] at http://qiime.org/ (*see* **Note 3**).
8. Optional: MicrobiomeAnalyst at https://www.microbiomeanalyst.ca
9. Prism (GraphPad Prism 7) or R statistical environment (https://www.r-project.org/) and R-Studio for R-visualization.

10. Optional: Phyloseq R-package (can be directly installed in R or from https://www.bioconductor.org/packages/release/bioc/html/phyloseq.html).

3 Methods

3.1 Sample Collection and Extraction

3.1.1 Collection of Saliva

1. Ensure that participants have not been treated with antibiotics or antifungals for at least the past 2 months (*see* **Note 4**).
2. The oral cavity should be assessed for missing teeth, oral hygiene, and periodontal status.
3. Whenever possible collect saliva between 8:00 and 11:00 AM (*see* **Note 5**).
4. Request that the participant refrains from eating, drinking, smoking, or oral hygiene procedures for at least 1 h prior to collection.
5. Request that participants quickly rinse their mouth with distilled drinking water, which can then be swallowed or expectorated (*see* **Note 6**).
6. Collect ~5 mL unstimulated whole saliva by having the patient expectorate (*see* **Notes 7** and **8**) into a sterile container (*see* **Note 9**).
7. Immediately following collection, cool the sample on ice packs or dry ice, or put in the fridge or aliquot and freeze (*see* **Note 10**).
8. Within 2 h after collection, store saliva sample at −80 °C or process.
9. Samples should remain at −80 °C until extracted with freeze/thaw cycles avoided.

3.1.2 Collection of Oral Tissue

1. Ensure that participants have not been treated with antibiotics or antifungals for at least the past 2 months (*see* **Note 4**).
2. The oral cavity should be assessed for missing teeth, oral hygiene, and periodontal status.
3. Incisional biopsies or punch biopsies should be surgically removed aseptically and laid on a pile of sterile gauze.
4. Using a new sterile surgical blade (*see* **Note 11**) excise a small piece of tissue (~3 mm^3) and aseptically transfer the tissue into a screw-cap vial (*see* **Note 12**).
5. Place the vial on dry ice and store at −80 °C as soon as possible.

3.2 Sample Processing and Extraction

3.2.1 Processing of Salivary Pellets

1. Centrifuge large samples (>2 mL saliva) at 2,800 × *g* for 10 min at 4 °C in a 50 mL conical tube or small samples (1–2 mL saliva) at 13,000 × *g*, 4 min, 4 °C in microcentrifuge tubes (*see* **Note 13**).
2. Using a pipette, carefully aspirate the supernatant without disturbing the pellet.
3. Thoroughly resuspend the pellet in 500 μL PBS.
4. The resuspended pellet can be extracted in Subheading 3.2.3.

3.2.2 Processing of Tissue Samples

1. Rinse tissue section with sterile molecular grade water to wash off any possible external contaminant.
2. Finely chop ~100 mg of tissue with a sterile blade.
3. Place the finely chopped tissue in 500 μL PBS to be extracted in Subheading 3.2.3.

3.2.3 Extraction of Total DNA

Before proceeding, refer to **Note 14**.

1. Digest the sample with MetaPolyzyme (a mixture of six enzymes; two of these enzymes (lyticase and chitinase) target the fungal cell wall) (*see* **Note 15**) by performing the following steps:
   (a) Thoroughly mix the sample by vortexing.
   (b) Add 200 μL resuspended sample into a 2.0 mL microcentrifuge tubes.
   (c) Add 1 mL PBS pH 7.5, vortex, centrifuge at 13,000 × *g* for 1 min, and aspirate the supernatant.
   (d) Repeat **step 1c** two more times.
   (e) Resuspend the pellet in 150 μL PBS pH 7.5 and vortex thoroughly.
   (f) Add 0.02% sodium azide to prevent bacterial growth.
   (g) Add 25 μL MetaPolyzyme and incubate at 35 °C for 4–6 h.
2. Purify microbial DNA using the ZymoBIOMICS DNA Microprep Kit using a mixture of 0.1- and 0.5-mm beads (*see* **Note 16**), using the following steps:
   (a) Add 175 μL sample from **step 1** to ZR BashingBead Lysis Tubes.
   (b) Add 750 μL ZymoBIOMICS Lysis Solution to the tube and cap tightly.
   (c) Place 2.0 mL tubes in a bead beater and process at maximum speed for 5 min.
   (d) Centrifuge at 13,000 × *g* for 1 min.

(e) Transfer 400 μL supernatant to the Zymo-Spin III-F Filter in a collection tube and centrifuge at 8000 × *g* for 1 min. Discard the Zymo-Spin III-F Filter.

(f) Add 1200 μL of ZymoBIOMICS DNA Binding Buffer to the filtrate in the collection tube.

(g) Transfer 800 μL of the mixture to a Zymo-Spin IC-Z Column in a collection tube and centrifuge at 10,000 × *g* for 1 min.

(h) Discard the flow through and repeat **step 2g**.

(i) With the filter in a new collection tube add 400 μL ZymoBIOMICS DNA Wash Buffer 1 and centrifuge at 10,000 × *g* for 1 min.

(j) Repeat **step 2i** with 700 μL ZymoBIOMICS DNA Wash Buffer 2 and discard flow through.

(k) Repeat **step 2i** with 200 μL ZymoBIOMICS DNA Wash Buffer 2.

(l) Transfer column to a low bind 1.5 microcentrifuge tube and elute with 20 μL ZymoBIOMICS DNase/RNase Free Water, incubate at room temperature for 1 min and centrifuge at 10,000 × *g* for 1 min (*see* **Note 17**).

3. Using a Zymo-Spin II-μHRC Filter in a collection tube add 600 μL ZymoBIOMICS HRC Prep Solution to the filter and centrifuge at 8000 × *g* for 3 min.

   (a) Transfer the eluted DNA from **step 3l** to the Zymo-Spin II-μHRC Filter in a low bind 1.5 microcentrifuge tube and centrifuge at max speed for 3 min.

   (b) The filtered DNA can now be used for PCR analysis.

4. Determine the quantity of DNA via NanoDrop or Qubit Fluorometer (*see* **Notes 18** and **19**).

5. Store samples at −80 °C when not in use.

3.3 Determining Fungal Load

1. Assess the fungal load by quantification of the ITS2 normalized to the human β-actin gene by real-time PCR using SybrGreen and the $2^{-\Delta\Delta Ct}$ method. Assuming a similar amplification rate of the two DNA targets, and correcting for copy number, this will provide a way to compare relative amounts of host and fungal genomes in different samples.

2. Perform real-time PCR in 10 μL reactions consisting of the following: 5 μL SYBR Green master mix, 0.2 μL primer mix (10 μM), 2.8 μL water, and 2 μL template DNA.

3. Place multi-well plates into a real-time thermocycler and run the program shown in Table 1.

Table 1
Detailed PCR protocol for determining fungal load

	Cycles	Temperature (°C)	Time
Polymerase activation	1	95	10 min
Amplification	40		
Denaturation		95	15 s
Annealing		55	30 s
Extension		60	60 s
Hold	1	10	∞

Table 2
Detailed PCR protocol for amplifying community DNA using the specific ITS-2 primers

	Cycles	Temperature (°C)	Time
Polymerase activation	1	95	3 min
Amplification	25		
Denaturation		95	30 s
Annealing		55	30 s
Extension		72	30 s
Hold	1	10	∞

3.4 Library Preparation and Sequencing

1. Amplify community DNA using the specific ITS-2 primers, linked to Illumina's adapter sequences, to generate the amplicon library (*see* **Notes 1** and **2**).
2. The reaction follows Illumina's recommendation, with the possible choice of any high-fidelity Taq polymerase enzyme and the corresponding master-mix (*see* **Note 20**).
3. The amount of DNA per reaction should ideally be ~12.5 ng in a 25 μL reaction volume. The amount of DNA should be determined by Qubit or the DeNovix dsDNA High Sensitivity Assay (*see* **Note 21**).
4. Place multi-well plates into a thermocycler and run the sequence shown in Table 2.
5. Perform agarose gel electrophoresis using a 2–4% agarose gel to ensure successful amplification (*see* **Note 22**).
6. Verify the amplification success by visualization of amplicon products which typically range from 250 to 590 bp.
7. Purify the amplicon using Agencourt AMPure XP beads or similar DNA cleanup kit:

(a) Mix PCR amplicon with AMPure XP beads in 96-well microplates (*see* **Note 23**).

(b) Wash twice with 80% v/v ethanol.

(c) Elute the DNA in the Illumina buffer normally supplied with the MiSeq reagents. If this buffer is not available, IDTE buffer will suffice.

8. Perform a second high-fidelity PCR to add unique Nextera XT indexes (*see* **Note 24**). Caution: When mixing indexes ensure that each i5 and i7 index combination is unique. Perform PCR (8 cycles of amplification) to link the adapters to the purified amplicon. Include negative amplification controls to rule out any contamination (*see* **Note 25**).
9. Examine the amplicon by gel electrophoresis or with the Agilent BioAnalyzer to ensure Nextera XT indexes were added to the amplicon (*see* **Notes 26** and **27**).
10. If there is a delay, prepared libraries can be stored at this time at −20 °C for up to 5 days.
11. Determine the library concentration and pool the adapter-linked amplicons, in equimolar amounts as follows: Using a Qubit or DeNovix fluorometer determine the concentration of the library. For each library, calculate the volume containing 5 ng, and pool equal amounts. Adjust the volume with Illumina sample buffer or IDTE buffer.
12. Load the pooled library into your Illumina sequencer of choice (either a MiSeq v3 2 × 300 bp reagent cartridge or an iSeq V1 or V2 2 × 300 bp cartridge).

3.5 Sequence Data Processing, Taxonomy Assignment, and Analysis

Before proceeding, refer to **Note 28**.

1. Using FastQC check the quality of FASTQ files and discard samples with low sequence quality or PHRED scores <30.
2. Using Trimmomatic and different commands of Seqtk demultiplex and trim the Nextera XT adaptors (*see* **Note 29**).
3. Sequence read pairs can be merged based on an overlap of at least 20 bp and filtered to discard reads with >0.5 expected errors in USEARCH v9.2.64.
4. UPARSE pipeline in USEARCH v9.2.64 is applied to dereplicated, filtered reads to remove chimeras (UCHIME) and perform de novo OTUs (operational taxonomic units) picking at 97% sequence identity [34].
5. Deposit the filtered and dereplicated sequences to the NCBI Sequence Read Archive.
6. Using the BLASTn database and in the International Nucleotide Sequencing Database and modified version of FHiTINGS determine OTU taxonomy of the various reads (*see* **Note 30**).

7. Filter OTUs with <0.1% relative abundance from each sample (*see* **Notes 31** and **32**)
8. Further analysis such as alpha-diversity, beta-diversity, PCoA, and Cluster Analysis can be performed in QIIME or MicrobiomeAnalyst.
9. Detection of differentially abundant taxa between cases and controls is performed in QIIME (*see* **Note 33**) and can further be analyzed online on MicrobiomeAnalyst [35, 36].
10. Additional statistical analysis and visualization can be performed by the Phyloseq R-package [37, 38].
11. Taxon-normalized abundances can be calculated using the following formula with values expressed in arbitrary units (a.u.) [39].

$$\text{Normalized abundance} = \text{Relative abundance} \times \left(\frac{\text{Total fungal reads}}{\text{Total reads}}\right) \times \left(\text{PCR product concentration}\left(\frac{\text{ng}}{\mu\text{L}}\right)\right) \times \left(\frac{1}{2^{\text{No.PCR cycles}}}\right) \times \left(\frac{25\,\mu\text{L}}{\text{volume }(\mu\text{L})\text{DNA input}}\right) \times \left(10^{9}\right)$$

4 Notes

1. These primers amplify the Internal Transcribed Spacer 2 (ITS2) region of the fungal between 5.8S and 28S rRNA genes. In *Saccharomyces cerevisiae*, these primers amplify the entire ITS, including the end of 5.8S and the start of 28S.
2. Adaptor-linked primers are the same primers used to amplify the ITS2 region (upper case bases) attached to an adaptor sequence (lower case bases). Any adapter-link sequence will work, but it is advisable to discuss the method with the sequencing facility prior to ordering.
3. QIIME is also available through different analysis platforms, e.g., the Galaxy environment (https://usegalaxy.org/) as well as CLC Genomics Workbench and UGENE.
4. It is crucial to take a history of antibiotic/antifungal and steroidal ant-inflammatory usage from the subject, as this is critical for data interpretation.
5. Owing to diurnal variation in oral microbes, collecting samples at a consistent time is recommended, although it is not essential if not practically possible.
6. Rinsing with sterile water ensures the mouth is free of debris that can complicate the processing of samples.

7. If it is difficult for the participants to expectorate, have them gently massage their cheeks to stimulate salivary flow.
8. If the participant is a small child, elderly, or a hyposalivator, they may have difficulty producing a sample. Alternative samples include an oral swab (e.g., FLOQSwab®; Copan Italia SpA) placed in 1 mL viral transport media or phosphate-buffered solution (PBS). Saliva can also be pipetted from the mouth into a sterile container.
9. Any sterile container, which is large enough to easily spit into, will work.
10. Salivary enzymes remain active at −80 °C. Therefore, saliva samples should be chilled and processed as soon as possible, avoiding freezing prior to stabilizing or purifying the nucleic acid.
11. Oral fungal concentrations can vary greatly between subjects, making avoidance of cross contamination of samples a concern.
12. If sampling oral cancers, excise a piece of the deep tissue at the macroscopically visible advancing front of the neoplasm, avoiding contamination from the tumor surface. The rest of the biopsy can be sent in 10% buffered formalin to histopathologically confirm the oral lesion. Control tissue can either be an adjacent normal tissue or another type of oral lesion, such as intra-oral fibro-epithelial polyps (FEP).
13. When pelleting fungus from saliva, a compromise must be made when choosing between high speed which produces a tight pellet, but possible lysis of some species, and low speed which produces a looser pellet from which it is more difficult to remove the liquid component [7, 32, 33].
14. Include a non-template DNA isolation control by passing molecular grade water through the entire extraction process.
15. The manufacturer's protocol is available at https://www.sigmaaldrich.com/content/dam/sigma-aldrich/docs/Sigma/Product_Information_Sheet/1/mac4lpis.pdf
16. Further description of the purification process is available at https://files.zymoresearch.com/protocols/_d4301_d4305_zymobiomics_dna_microprep_kit.pdf
17. A brown-colored pellet may form at the bottom of the tube after centrifugation. Avoid this pellet when collecting the eluted DNA.
18. A NanoDrop™ or other micro-spectrophotometer provides the yeild of nucleic acid, but does not differentiate RNA from DNA, while a Qubit, or a similar fluorometer, does differentiate and provides yield.

19. Using the NanoDrop™ the ideal 260/280 ratio for "pure" RNA is ~2.0. The 260/230 ratio provides a secondary measure of purity and should be in the range of 2.0–2.2. If the readings are significantly lower, it may be indicative of contamination with residual organic compounds from the extraction step.
20. Some high-fidelity polymerases are provided already in a master mix, while others must be mixed with buffer, magnesium chloride, and premixed deoxynucleotide triphosphate (dNTPs). Recommended enzymes include Kapa Hi-Fi Hot-start ReadyMix (Roche), Q5 Hot Start High-Fidelity 2× MasterMix (New England Biolabs), or GeneAmp High-Fidelity PCR system (Invitrogen/Thermo Fisher Scientific).
21. DNA should be quantified with a fluorometer, rather than a spectrophotometer (e.g., NanoDrop™) to guarantee that only DNA—not RNA—is quantified by the instrument. The typical fluorometers for such use include the Qubit or DeNovix.
22. Product verification can also be performed using an Agilent BioAnalyzer, which provides higher resolution and sensitivity than gel electrophoresis and uses less sample.
23. If a plate shaker is available, this step can be performed in midi-plates covered with a micro-seal adhesive.
24. There are ≤96 possible combinations depending on the Nextera XT Index Kit used. Index kits can support 24 sequencing reactions (typical for iSeq and MiSeq reactions) or 96 reactions (optimal for MiSeq). Both Nextera XT kits (24 indices for 96 reactions, and 96 indices for 384 reactions) are commercially available from Illumina.
25. Negative amplification controls should consist of (a) master mix alone and (b) master mix with other reaction components but no DNA.
26. The amplicon will slightly increase in size (~70 bp) as it has been linked to adapters.
27. A pre-tagged amplicon control should be used for size comparison, as the size difference may not be noticeable using DNA ladders.
28. This is a brief summary of the steps that can be undertaken to go from the FASTq amplicon sequence to measure relative levels of fungal taxa. The exact approach will depend on the preferences of the analyst. Currently, several different pipelines are being validated to analyze fungal sequences using various methods to filter reads and align them to the UNITE and other curated libraries containing fungal sequences.
29. This can be done on a Linux computer, MacOSC terminal, a Windows command-line terminal, a virtual machine-enabled Windows system, or an online server.

30. The BLASTn function is used to align the reads to the UNITE database of fungal ITS high-quality DNA reference sequences in order to classify the sequences to the species level [5].
31. This is done to avoid spurious identification of low-level sequence reads that have high numbers of sequence errors.
32. OTUs are suggested here to allow comparison with prior analyses; however, the current state of the art is to use exact sequence variants (ESVs) or amplicon sequence variants (ASVs) as they offer higher resolution power and less ambiguity with using thresholds to assign taxons [40]. QIIME2, as well as other similar tools, such as DADA2 or MOTHUR, implement ASV-based classification.
33. A full, continually updated protocol/tutorial is provided at the QIIME user forum, URL: https://forum.qiime2.org/t/fungal-its-analysis-tutorial/7351

Acknowledgments

D.J.S. was supported by McMaster University's Michael G. DeGroote Initiative for Innovation in Healthcare. R.K.A. is supported by the Egyptian Academy for Scientific Research and Technology JESOR program (grant #3046 "Center for Genome and Microbiome Research"). We are grateful to Matthew Gutierrez (Oral Medicine and Diagnostic Sciences, University of Illinois at Chicago) for his assistance in preparing the figure.

References

1. McTaggart LR, Copeland JK, Surendra A, Wang PW, Husain S, Coburn B et al (2019) Mycobiome sequencing and analysis applied to fungal community profiling of the lower respiratory tract during fungal pathogenesis. Front Microbiol 10:512. https://doi.org/10.3389/fmicb.2019.00512
2. Human Microbiome Project Consortium (2012) Structure, function and diversity of the healthy human microbiome. Nature 486 (7402):207–214. https://doi.org/10.1038/nature11234
3. Tiew PY, Mac Aogain M, Ali N, Thng KX, Goh K, Lau KJX et al (2020) The mycobiome in health and disease: emerging concepts, methodologies and challenges. Mycopathologia 185(2):207–231. https://doi.org/10.1007/s11046-019-00413-z
4. Ghannoum MA, Jurevic RJ, Mukherjee PK, Cui F, Sikaroodi M, Naqvi A et al (2010) Characterization of the oral fungal microbiome (mycobiome) in healthy individuals. PLoS Pathog 6(1):e1000713. https://doi.org/10.1371/journal.ppat.1000713
5. Hager CL, Ghannoum MA (2018) The mycobiome in HIV. Curr Opin HIV AIDS 13 (1):69–72. https://doi.org/10.1097/COH.0000000000000432
6. Perera M, Al-Hebshi NN, Perera I, Ipe D, Ulett GC, Speicher DJ et al (2017) A dysbiotic mycobiome dominated by Candida albicans is identified within oral squamous-cell carcinomas. J Oral Microbiol 9(1):1385369. https://doi.org/10.1080/20002297.2017.1385369
7. Baraniya D, Chen T, Nahar A, Alakwaa F, Hill J, Tellez M et al (2020) Supragingival mycobiome and inter-kingdom interactions in dental caries. J Oral Microbiol 12(1):1729305. https://doi.org/10.1080/20002297.2020.1729305
8. Diaz PI, Hong BY, Dupuy AK, Strausbaugh LD (2017) Mining the oral mycobiome:

methods, components, and meaning. Virulence 8(3):313–323. https://doi.org/10.1080/21505594.2016.1252015

9. Diaz PI, Dongari-Bagtzoglou A (2020) Critically appraising the significance of the oral mycobiome. J Dent Res 2020:22034520956975. https://doi.org/10.1177/0022034520956975
10. Gaitanis G, Magiatis P, Hantschke M, Bassukas ID, Velegraki A (2012) The Malassezia genus in skin and systemic diseases. Clin Microbiol Rev 25(1):106–141. https://doi.org/10.1128/CMR.00021-11
11. Dupuy AK, David MS, Li L, Heider TN, Peterson JD, Montano EA et al (2014) Redefining the human oral mycobiome with improved practices in amplicon-based taxonomy: discovery of Malassezia as a prominent commensal. PLoS One 9(3):e90899. https://doi.org/10.1371/journal.pone.0090899
12. Bandara H, Panduwawala CP, Samaranayake LP (2019) Biodiversity of the human oral mycobiome in health and disease. Oral Dis 25 (2):363–371. https://doi.org/10.1111/odi.12899
13. He X, McLean JS, Edlund A, Yooseph S, Hall AP, Liu SY et al (2015) Cultivation of a human-associated TM7 phylotype reveals a reduced genome and epibiotic parasitic lifestyle. Proc Natl Acad Sci U S A 112(1):244–249. https://doi.org/10.1073/pnas.1419038112
14. Vartoukian SR, Moazzez RV, Paster BJ, Dewhirst FE, Wade WG (2016) First cultivation of health-associated Tannerella sp. HOT-286 (BU063). J Dent Res 95 (11):1308–1313. https://doi.org/10.1177/0022034516651078
15. Kruger W, Vielreicher S, Kapitan M, Jacobsen ID, Niemiec MJ (2019) Fungal-bacterial interactions in health and disease. Pathogens 8 (2):70. https://doi.org/10.3390/pathogens8020070
16. Huseyin CE, Rubio RC, O'Sullivan O, Cotter PD, Scanlan PD (2017) The fungal frontier: a comparative analysis of methods used in the study of the human gut mycobiome. Front Microbiol 8:1432. https://doi.org/10.3389/fmicb.2017.01432
17. Ali N, Mac Aogain M, Morales RF, Tiew PY, Chotirmall SH (2019) Optimisation and benchmarking of targeted amplicon sequencing for mycobiome analysis of respiratory specimens. Int J Mol Sci 20(20):4991. https://doi.org/10.3390/ijms20204991
18. Weiss S, Amir A, Hyde ER, Metcalf JL, Song SJ, Knight R (2014) Tracking down the sources of experimental contamination in microbiome studies. Genome Biol 15 (12):564. https://doi.org/10.1186/s13059-014-0564-2
19. Salter SJ, Cox MJ, Turek EM, Calus ST, Cookson WO, Moffatt MF et al (2014) Reagent and laboratory contamination can critically impact sequence-based microbiome analyses. BMC Biol 12:87. https://doi.org/10.1186/s12915-014-0087-z
20. Schoch CL, Seifert KA, Huhndorf S, Robert V, Spouge JL, Levesque CA et al (2012) Nuclear ribosomal internal transcribed spacer (ITS) region as a universal DNA barcode marker for Fungi. Proc Natl Acad Sci U S A 109 (16):6241–6246. https://doi.org/10.1073/pnas.1117018109
21. O'Brien HE, Parrent JL, Jackson JA, Moncalvo JM, Vilgalys R (2005) Fungal community analysis by large-scale sequencing of environmental samples. Appl Environ Microbiol 71 (9):5544–5550. https://doi.org/10.1128/AEM.71.9.5544-5550.2005
22. Nash AK, Auchtung TA, Wong MC, Smith DP, Gesell JR, Ross MC et al (2017) The gut mycobiome of the Human Microbiome Project healthy cohort. Microbiome 5(1):153. https://doi.org/10.1186/s40168-017-0373-4
23. Hoggard M, Vesty A, Wong G, Montgomery JM, Fourie C, Douglas RG et al (2018) Characterizing the human mycobiota: a comparison of small subunit rRNA, ITS1, ITS2, and large subunit rRNA genomic targets. Front Microbiol 9:2208. https://doi.org/10.3389/fmicb.2018.02208
24. Op De Beeck M, Lievens B, Busschaert P, Declerck S, Vangronsveld J, Colpaert JV (2014) Comparison and validation of some ITS primer pairs useful for fungal metabarcoding studies. PLoS One 9(6):e97629. https://doi.org/10.1371/journal.pone.0097629
25. Blaalid R, Kumar S, Nilsson RH, Abarenkov K, Kirk PM, Kauserud H (2013) ITS1 versus ITS2 as DNA metabarcodes for fungi. Mol Ecol Resour 13(2):218–224. https://doi.org/10.1111/1755-0998.12065
26. Tang J, Iliev ID, Brown J, Underhill DM, Funari VA (2015) Mycobiome: approaches to analysis of intestinal fungi. J Immunol Methods 421:112–121. https://doi.org/10.1016/j.jim.2015.04.004
27. Liu CM, Kachur S, Dwan MG, Abraham AG, Aziz M, Hsueh PR et al (2012) FungiQuant: a broad-coverage fungal quantitative real-time PCR assay. BMC Microbiol 12:255. https://doi.org/10.1186/1471-2180-12-255

28. Kim SK, Hong SJ, Pak KH, Hong SM (2019) Analysis of the microbiome in the adenoids of Korean children with otitis media with effusion. J Int Adv Otol 15(3):379–385. https://doi.org/10.5152/iao.2019.6650
29. Bolger AM, Lohse M, Usadel B (2014) Trimmomatic: a flexible trimmer for Illumina sequence data. Bioinformatics 30 (15):2114–2120. https://doi.org/10.1093/bioinformatics/btu170
30. Camacho C, Coulouris G, Avagyan V, Ma N, Papadopoulos J, Bealer K et al (2009) BLAST+: architecture and applications. BMC Bioinformatics 10:421. https://doi.org/10.1186/1471-2105-10-421
31. Bolyen E, Rideout JR, Dillon MR, Bokulich NA, Abnet CC, Al-Ghalith GA et al (2019) Reproducible, interactive, scalable and extensible microbiome data science using QIIME 2. Nat Biotechnol 37(8):852–857. https://doi.org/10.1038/s41587-019-0209-9
32. Caporaso JG, Kuczynski J, Stombaugh J, Bittinger K, Bushman FD, Costello EK et al (2010) QIIME allows analysis of high-throughput community sequencing data. Nat Methods 7(5):335–336. https://doi.org/10.1038/nmeth.f.303
33. Kuczynski J, Stombaugh J, Walters WA, Gonzalez A, Caporaso JG, Knight R (2011) Using QIIME to analyze 16S rRNA gene sequences from microbial communities. Curr Protoc Bioinformatics Chapter 10:Unit 10 17. https://doi.org/10.1002/0471250953.bi1007s36
34. Edgar RC (2013) UPARSE: highly accurate OTU sequences from microbial amplicon reads. Nat Methods 10(10):996–998. https://doi.org/10.1038/nmeth.2604
35. Chong J, Liu P, Zhou G, Xia J (2020) Using MicrobiomeAnalyst for comprehensive statistical, functional, and meta-analysis of microbiome data. Nat Protoc 15(3):799–821. https://doi.org/10.1038/s41596-019-0264-1
36. Dhariwal A, Chong J, Habib S, King IL, Agellon LB, Xia J (2017) MicrobiomeAnalyst: a web-based tool for comprehensive statistical, visual and meta-analysis of microbiome data. Nucleic Acids Res 45(W1):W180–W188. https://doi.org/10.1093/nar/gkx295
37. McMurdie PJ, Holmes S (2012) Phyloseq: a bioconductor package for handling and analysis of high-throughput phylogenetic sequence data. Pac Symp Biocomput 2012:235–246
38. McMurdie PJ, Holmes S (2013) Phyloseq: an R package for reproducible interactive analysis and graphics of microbiome census data. PLoS One 8(4):e61217. https://doi.org/10.1371/journal.pone.0061217
39. Henson BS, Wong DT (2010) Collection, storage, and processing of saliva samples for downstream molecular applications. Methods Mol Biol 666(1940-6029 (Electronic)):21–30. https://doi.org/10.1007/978-1-60761-820-1_2
40. Callahan BJ, McMurdie PJ, Holmes SP (2017) Exact sequence variants should replace operational taxonomic units in marker-gene data analysis. ISME J 11(12):2639–2643. https://doi.org/10.1038/ismej.2017.119

Chapter 16

Measuring Effects of Dietary Fiber on the Murine Oral Microbiome with Enrichment of 16S rDNA Prior to Amplicon Synthesis

Lea M. Sedghi, Stefan J. Green, and Craig D. Byron

Abstract

The oral cavity houses a diverse consortium of microorganisms corresponding to specific microbial niches within the oral cavity. The complicated nature of sample collection limits the accuracy, reproducibility, and completeness of sample collection of the dentogingival microbiome. Moreover, large variability among human oral samples introduces inexorable confounds. Here, we introduce a method to study the dentogingival microbiome using a murine model that allows for greater control over experimental variability and permits collection of the dentogingival microbiome in an intact state and in its entirety.

As an example of this approach, this chapter provides a workflow to explore the effect of dietary fiber consumption on the murine dentogingival microbiome. Mice are fed diets corresponding to Fiber, Sugar, Fiber + Sugar, and Control groups for 7 weeks. A whole-mandible extraction technique is described to isolate the mandibular dentogingival surfaces. 16S rRNA gene analysis is coupled with removal of unwanted host DNA amplification products to allow an investigation of the dental microbiome in the presence of increased fiber in terms of microbial taxonomic abundance and diversity.

Key words 16S rRNA, Dentogingival microbiome, Oral microbiome sample collection, Murine oral microbiome

1 Introduction

The oral cavity is a highly dynamic microbial environment that houses a diverse array of distinct microenvironments. Such microenvironments within the oral cavity include the non-shedding occlusal, lingual, buccal dental and inter-dental surfaces, the gingival and subgingival surfaces, the epithelial surfaces of the mucosal membrane, saliva, and the dorsal surface of the tongue [1, 2]. Microbial communities that inhabit these niches are exposed to different environmental challenges, such as masticatory

Supplementary Information The online version of this chapter (https://doi.org/10.1007/978-1-0716-1518-8_16) contains supplementary material, which is available to authorized users.

Guy R. Adami (ed.), *The Oral Microbiome: Methods and Protocols*, Methods in Molecular Biology, vol. 2327,
https://doi.org/10.1007/978-1-0716-1518-8_16, © Springer Science+Business Media, LLC, part of Springer Nature 2021

challenges, daily hygiene practice, salivary flow, or, if subgingival, gingival crevicular fluid flow [3–6]. Moreover, exogenous microorganisms are introduced to the oral cavity by open mouth breathing, dietary intake, and host contact [3]. The geographical variability within the oral cavity and the near-constant environmental challenges that its microbial communities encounter creates distinct microenvironments within the oral cavity that are highly site-specific [7–9]. Studies of the oral microbiome often seek to characterize microbial communities in distinct niches within the oral cavity, such as dental or gingival surfaces, that may correspond to specific states of health and disease (i.e., dental caries or periodontal disease) [6, 8, 10–12]. Saliva, plaque, and gingival crevicular fluid samples are most frequently collected from human subjects [10, 13–17]. While such human studies are invaluable to studying the oral microbiome, they are subject to many confounding factors including large differences in the pre-existing microbiota among individuals, patient diets, medications, and hygiene practice [18, 19]. Current sampling techniques are often utilized that do not capture the targeted microbial community of interest [20]. For example, while salivary fluid collection is commonly utilized to obtain a representative sample of the oral microbiota, this technique does not account for the stagnant nature of the dentogingival surfaces that harbor resilient and diverse microbial biofilm communities [7, 15, 21]. In addition, many sampling techniques that utilize brushes or scrapers neglect to account for collection of the entire targeted microbial community [22–24]. Such variability makes it difficult to design patient-based studies of the oral microbiota, and this is even more relevant among studies that require patients to adhere to strict dietary or hygiene regimens that cannot be continuously and objectively observed [19]. In vivo murine models are excellent alternatives to study the oral microbiota in a controlled manner [18, 25, 26]. Murine models are able to recapitulate the dynamic nature of the oral cavity and also capture the relationship of the oral cavity to other disease states and to the host inflammatory response [18, 25, 26]. An added advantage of murine models is the ability to collect specific intraoral sites in their entirety and in isolation from other sites within the oral cavity [27]. However, a difficulty associated with studies of the oral microbiota that utilize murine models lies in the complicated nature of sample collection and obtaining an adequate and representative microbial sample [28]. Here, we describe a successful whole-jaw extraction technique to collect and analyze the dentogingival microbiome in its entirety away from other sites within the murine oral cavity. This extraction technique is demonstrated in a study of dietary fiber and the interaction of sugar and fiber on the murine dentogingival microbiome, in terms of microbial taxonomic abundance and diversity [27]. Microbial community structure is profiled using 16S ribosomal RNA (rRNA) gene amplicon sequencing, and a simple solution to remove unwanted murine host DNA amplification products is also described.

2 Materials

2.1 Animal Usage and Sample Collection

1. Three-week-old CD-1 mice.
2. Shoebox rodent housing.
3. 2 oz cups of DietGel 76A (Clear H_2O).
4. Sucrose.
5. Lignin.
6. Microdissection scissors.
7. Periosteal elevators.
8. −80 °C Freezer.

2.2 Characterization of the Dentogingival Microbiome

1. DNeasy Powerbiofilm Kit (Qiagen).
2. Vortex.
3. PCR machine.
4. 515f/806r primer pair.
5. BluePippin device (Sage Science).
6. 2% Agarose, PippinHT, 100–600 bp gel cassettes.
7. AMPure XP beads.
8. phiX spike-in (Illumina).
9. Illumina MiniSeq.
10. Sequencing primers (Fluidigm).

2.3 Statistical Analysis

1. EdgeR software.

3 Methods

3.1 Animal Usage and Sample Collection

1. Three-week-old female CD-1 mice ($N = 28$) are housed in standard shoebox rodent housing and randomized into four groups ($N = 7$) (*see* **Notes 1** and **2**).
2. Mice are initially raised on control diets (DietGel 76A) for 10 days prior to beginning assigned dietary regimens.
3. At day 10, randomize animals into four dietary groups ($N = 7$ each), including Control, Control + Sugar (Sugar), Control + Fiber (Fiber), and Control + Sugar + Fiber (Sugar + Fiber) groups.

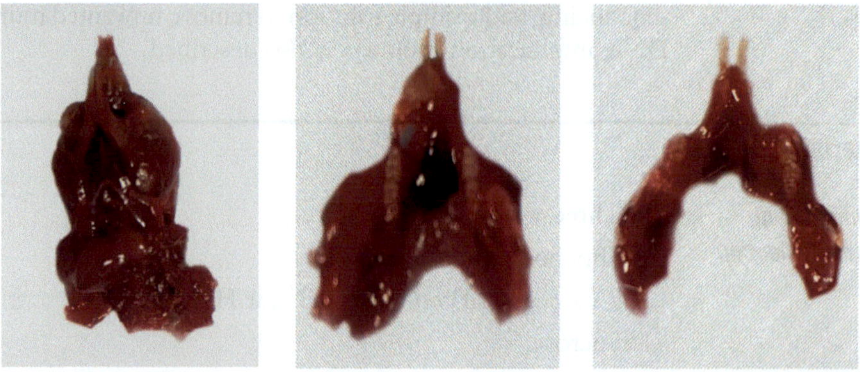

Fig. 1 The leftmost panel depicts a whole mouse cranium without the integument from a ventral perspective. The middle panel depicts an isolated mandible from the dorsal perspective with host tissue attached. The rightmost panel depicts an isolated mandible following removal of attached host tissue and is representative of a sample with intact dentogingival surfaces subjected to DNA extraction

4. To create the various diets, the control diet is supplemented with either sucrose (6.5 g/2 oz) to obtain the Sugar diet, lignin (4.2 g/2 oz) to obtain the Fiber diet, or sucrose (6.5 g/2 oz) and lignin (4.2 g/2 oz) to obtain the Sugar + Fiber diet. For the control group DietGel 76A is continued (*see* **Notes 3** and **4**).
5. Mice are fed once daily for 60 days (*see* **Note 5**).
6. At day 60, mice are euthanized using CO_2 overdose as recommended by veterinarian.
7. Under a laminar flow cell culture hood, each animal is processed with a microdissection protocol as follows:

3.2 Microdissection

1. The cranium is removed with scissors.
2. The cranial integument is next removed by separating the dermal and hypodermal layers.
3. Using microdissection scissors and periosteal elevators, all bilateral masticatory muscles are severed as near to their distal insertion points as possible.
4. External to the mandible, the masseter and temporalis are dissected free. Internally, the medial and lateral pterygoids are dissected free.
5. Next, the tongue and its extrinsic suprahyoid muscles are severed.
6. Finally, the extracapsular and intracapsular ligaments surrounding the temporomandibular joint (TMJ) were dissected until the mandibular condyles could be gently disarticulated from the mandibular fossa of each temporal bone (*see* Fig. 1).

7. Careful attention is given to not disturb the dentogingival surfaces as the lower jaw of each specimen is removed and gently cleaned of most muscle, epithelia, and connective tissue.
8. Micro-dissected mandibles are stored at −80 °C until DNA extraction.

3.3 Characterization of the Dentogingival Microbiome

1. DNA is isolated from extracted mandibles using the DNeasy PowerBiofilm Kit. The entire jaw is added to the extraction tube for processing. Extractions are carried out according to the manufacturer's instructions with the following modification: cell lysis is conducted by performing bead-beating by vortexing the extracted jaws for 5 min total time, with breaks on ice to cool samples. Jaws are removed following this step to prevent tissue from blocking the filter.
2. The V4 variable region of the microbial 16S ribosomal RNA (rRNA) gene is PCR-amplified using the 515f/806r primer pair, using a two-stage "targeted amplicon sequencing" (TAS) protocol, as described previously [29, 30]. During the second stage amplification, Illumina sequencing adapters and a sample-specific unique 10-base index must be incorporated (*see* **Note 6**).
3. Following second-stage amplification, libraries (i.e., second-stage PCR products without purification) are pooled in equal volume and the pool is subject to size selection (350–450 bp) using a BluePippin device and employing 2% Agarose, PippinHT, 100–600 bp gel cassettes. The size-selected library pool is recovered from the PippinPrep system and purified with AMPure XP beads using a 0.6× ratio [31] (*see* **Notes 7** and **8**).
4. The size-selected pooled library, with a 20% phiX spike-in, is loaded onto an Illumina MiniSeq mid-output flow cell. Sequencing is initiated with custom Fluidigm sequencing primers, according to the Fluidigm AccessArray for Illumina user guide (*see* **Note 9**).
5. After sequencing is completed, the number of sequencing clusters per sample is determined by demultiplexing conducted automatically on the Illumina Basespace cloud computing environment. These numbers are used to generate a second pool of second-stage PCR libraries. In the first pool, libraries are pooled in equal volume; in the second pool, the objective is to generate equimolar pooling based on the initial sequencing results. Input volumes for each sample into a single pool are adjusted based on sequencing yield with the aim to generate identical output from the second sequencing run. The second pool of libraries is subsequently processed through the same pipeline of PippinPrep size selection and AMPure cleanup

(**step 3**) and loaded onto an Illumina MiSeq v2 (500 cycle) flow cell using the same loading conditions described in **step 4**. The purpose of this approach is to use a low-output sequencing run to guide a pooling strategy and also allow for the simultaneous size-selection of many libraries while still generating similar numbers of reads per sample. Using this approach, we found that when a set of 28 samples was sequenced, a total of 970,621 clusters were generated, with a median number of clusters per sample of 35,315 and a range of 27,921 to 41,502 clusters per sample [27] (*see* **Note 10**).

6. Raw sequence data can be processed using QIIME workflow with GreenGenes v13 8 [32, 33], as described previously, with minor modifications. Such modifications included discarding quality and primer trimmed merged sequences shorter than 225 bases and rarefaction to a depth of 25,000 sequences per sample. The output of the QIIME pipeline is a series of biological observation matrices (BIOMs) at taxonomic levels from phylum to genus [34].

3.4 Statistical Analyses

3.4.1 Beta Diversity Analyses

1. To determine overall differences in the populations based on taxa, Bray-Curtis indices are calculated with default parameters in R using the vegan library. In our experience, the data are $\log_{10}(x + 1)$ transformed for best results. The resulting dissimilarity indices are modeled and tested for significance with the dietary groups using the ADONIS test. Additional comparisons of each dietary factor, separately, are also performed using ANOSIM. Plots are generated in R using the ggplot2 library [35].

3.4.2 Differential Analysis of Microbial Taxa

1. Differential analyses of taxa, as compared with dietary factors, can be performed using the software package edgeR on raw sequence counts which is available through various interfaces [36]. Prior to analysis, the data should be filtered to remove any sequences that are annotated as chloroplast or mitochondria in origin, as well as to remove taxa that accounted for less than 0.1% of the total sequence counts. Data can be normalized using a trimmed mean of M-values (TMM).
2. Normalized data is then fit using a negative binomial GLM, using Sugar and Fiber covariates, and statistical tests are performed using a likelihood ratio test. Adjusted p values (q values) are then calculated using the Benjamini-Hochberg false-discovery rate (FDR) correction. Significant taxa are determined based on an FDR threshold of 5% or 10% (0.05 or 0.1).

4 Notes

1. Mice are caged without substrates on which to chew (i.e., typical wire-top lids or water bottle sippers) and are kept under controlled environmental conditions (temperature 26 ± 0.5 °C, 12/12 h light/dark cycle).
2. Mice cages are cleaned and refreshed weekly over the course of the study. A small amount of litter (~15 mL or 1 Tbs) from each cage is mixed together during the time of cage cleaning. This intermixed old litter is then distributed into the new shoebox cages (15 mL each). Therefore, any between-group differences in the composition of the oral microbiota could be attributed to dietary differences alone, as opposed to habitat-specific microbial communities.
3. DietGel 76A is added to a microwave-safe bowl and heated briefly for 10 s in a microwave prior to adding sugar and/or fiber. Sugar and fiber are mixed into the warmed food using a plastic fork until the mixture is completely homogenized. The homogenized mixture should be carefully added back into the DietGel76A container prior to feeding to mice.
4. The control treatment is also microwaved to account for the effect of heating.
5. Mice should be given free access to feed.
6. Although the MiniSeq mid-output kits are rated for 300 cycles total, sequencing reads can be increased to 153 bases (2 × 153) to generate additional sequence length for merging purposes. This length can be sustained due to the use of the Fluidigm AccessArray barcoded primers, which contain a single unique 10-base barcode per primer pair. Other sequencing approaches, which use dual barcoding strategies, may not have enough excess sequencing reagents to tolerate 2 × 153 base sequencing. Despite the limited amount of overlap, the forward and reverse V4 amplicon reads generated when using the 515F/806R primer set can still be merged with 2 × 150 sequencing using the software package PEAR [37]. This initial sequencing effort, used to determine the relative contribution of each sample to the pool for the purposes of re-balancing the pool, can also be conducted on the Illumina MiSeq Instrument using a Nano flow cell. The 300-cycle Nano flow cell is the lowest cost sequencing kit available and generates 500,000–1 M clusters, which is generally sufficient for assessing the distribution of 384 samples.
7. Size selection is performed to remove nonspecific amplification products, including host mitochondrial 16S rRNA gene amplicons and host 18S rRNA gene amplicons which are generated

due to high host-microbe DNA ratios in samples. High host DNA is expected in the DNA extraction protocol described in this study.

8. Based on the distribution of reads per barcode generated by the preliminary low depth sequencing run, the amplicons are re-pooled to generate a more balanced distribution of reads and are subjected to another PippinPrep size selection.
9. Using the Fluidigm AccessArray barcoding system, up to 384 samples can be pooled and sequenced simultaneously. The approach described herein allows size selection to be performed on all samples simultaneously and requires only two runs through the PippinPrep device.
10. The re-pooled libraries can be loaded onto a MiSeq V2 flow cell (500 cycles) to generate the final data for analysis, though sequencing can be performed on any Illumina sequencer and kit that generates reads of sufficient length for merging forward and reverse reads, and sufficient reads for proper characterization of each sample.

Acknowledgments

This protocol was developed and tested at Mercer University (Macon, GA, USA) and was supported by Mercer University grant #213019. Bioinformatics analysis in the project described was piloted at the UIC Research Informatics Core, supported in part by NCATS through Grant UL1TR002003.

References

1. Williams NB (1963) Microbial ecology of the oral cavity. J Dent Res 42:509–520
2. Aas JA, Paster BJ, Stokes LN, Olsen I, Dewhirst FE (2005) Defining the normal bacterial flora of the oral cavity. J Clin Microbiol 43:5721–5732
3. Proctor DM, Shelef KM, Gonzalez A, Davis CL, Dethlefsen L, Burns AR et al (2020) Microbial biogeography and ecology of the mouth and implications for periodontal diseases. Periodontol 2000 82:26–41
4. Proctor DM, Fukuyama JA, Loomer PM, Armitage GC, Lee SA, David NM et al (2018) A spatial gradient of bacterial diversity in the human oral cavity shaped by salivary flow. Nat Commun 9(1):681
5. Dutzan N, Abusleme L, Bridgeman H, Greenwell-Wild T, Zangerle-Murray T, Fife ME et al (2017) On-going mechanical damage from mastication drives homeostatic Th17 cell responses at the oral barrier. Immunity 46:133–147
6. Uzel NG, Teles FR, Teles RP, Song XQ, Torresyap G, Socransky SS et al (2011) Microbial shifts during dental biofilm re-development in the absence of oral hygiene in periodontal health and disease. J Clin Periodontol 38:612–620
7. Welch JLM, Rossetti BJ, Rieken CW, Dewhirst FE, Borisy GG (2016) Biogeography of a human oral microbiome at the micron scale. Proc Natl Acad Sci U S A 113:E791–E800
8. Hall MW, Singh N, Ng KF, Lam DK, Goldberg MB, Tenenbaum HC et al (2017) Interpersonal diversity and temporal dynamics of dental, tongue, and salivary microbiota in the healthy oral cavity. npj Biofilms Microbiomes 3:1–7
9. Bowen WH, Burne RA, Wu H, Koo H (2018) Oral biofilms: pathogens, matrix, and

polymicrobial interactions in microenvironments. Trends Microbiol 26:229–242
10. Belstrøm D, Sembler-Møller ML, Grande MA, Kirkby N, Cotton SL, Paster BJ et al (2017) Microbial profile comparisons of saliva, pooled and site-specific subgingival samples in periodontitis patients. PLoS One 12:e0182992
11. Costalonga M, Herzberg MC (2014) The oral microbiome and the immunobiology of periodontal disease and caries. Immunol Lett 162:22–38
12. Duran-Pinedo AE, Chen T, Teles R, Starr JR, Wang X, Krishnan K et al (2014) Community-wide transcriptome of the oral microbiome in subjects with and without periodontitis. ISME J 8:1659–1672
13. Barros SP, Williams R, Offenbacher S, Morelli T (2016) Gingival crevicular fluid as a source of biomarkers for periodontitis. Periodontol 2000 70:53–64
14. Zekeridou A, Mombelli A, Cancela J, Courvoisier D, Giannopoulou C (2019) Systemic inflammatory burden and local inflammation in periodontitis: what is the link between inflammatory biomarkers in serum and gingival crevicular fluid? Clin Exp Dent Res 5:128–135
15. Marsh PD (1994) Microbial ecology of dental plaque and its significance in health and disease. Adv Dent Res 8:263–271
16. Kolenbrander PE, Palmer RJ Jr, Rickard AH, Jakubovics NS, Chalmers NI, Diaz PI (2006) Bacterial interactions and successions during plaque development. Periodontol 2000 42:47–79
17. Haffajee AD, Teles RP, Patel MR, Song X, Veiga N, Socransky SS (2009) Factors affecting human supragingival biofilm composition. I. Plaque mass. J Periodontal Res 44:511–519
18. Hajishengallis G, Lamont RJ, Graves DT (2015) The enduring importance of animal models in understanding periodontal disease. Virulence 6:229–235
19. Kim D, Hofstaedter CE, Zhao C, Mattei L, Tanes C, Clarke E et al (2017) Optimizing methods and dodging pitfalls in microbiome research. Microbiome 5:52
20. Santigli E, Koller M, Klug B (2020) Oral biofilm sampling for microbiome analysis in healthy children. J Vis Exp 130:56320. https://doi.org/10.3791/56320
21. Akcalı A, Lang NP (2018) Dental calculus: the calcified biofilm and its role in disease development. Periodontol 2000 76:109–115
22. Luo T, Srinivasan U, Ramadugu K, Shedden KA, Neiswanger K, Trumble E et al (2016) Effects of specimen collection methodologies and storage conditions on the short-term stability of oral microbiome taxonomy. Appl Environ Microbiol 82:5519–5529
23. Göhler A, Samietz S, Schmidt CO, Kocher T, Steinmetz I, Holtfreter B (2018) Comparison of oral microbe quantities from tongue samples and subgingival pockets. Int J Dent. https://doi.org/10.1155/2018/2048390
24. ESNM (2020) Drugs are an important confound in human microbiome studies. https://wwwgutmicrobiotaforhealthcom/drugs-important-confound-human-microbiome-studies/. Accessed 18 June 2020
25. Gootenberg DB, Turnbaugh PJ (2011) Companion animals symposium: humanized animal models of the microbiome. J Anim Sci 89:1531–1537
26. Oz HS, Puleo DA (2011) Animal models for periodontal disease. J Biomed Biotechnol. https://doi.org/10.1155/2011/754857
27. Sedghi L, Byron C, Jennings R, Chlipala GE, Green SJ, Silo-Suh L (2019) Effect of dietary fiber on the composition of the murine dental microbiome. Dent J (Basel) 7:58
28. Hernández-Arriaga A, Baumann A, Witte OW, Frahm C, Bergheim I, Camarinha-Silva A (2019) Changes in oral microbial ecology of C57BL/6 mice at different ages associated with sampling methodology. Microorganisms 7:283
29. Green SJ, Venkatramanan R, Naqib A (2015) Deconstructing the polymerase chain reaction: understanding and correcting bias associated with primer degeneracies and primer-template mismatches. PLoS One. https://doi.org/10.1371/journal.pone.0128122
30. Naqib A, Poggi S, Wang W, Hyde M, Kunstman K, Green SJ (2018) Making and sequencing heavily multiplexed, high-throughput 16S ribosomal RNA gene amplicon libraries using a flexible, two-stage PCR protocol. In: Raghavachari N, Garcia Reyero N (eds) Gene expression analysis: methods in molecular biology. Springer, New York
31. Nelson MC, Morrison HG, Benjamino J, Grim SL, Graf J (2014) Analysis, optimization and verification of Illumina-generated 16S rRNA gene amplicon surveys. PLoS One 9(4): e94249
32. DeSantis TZ, Hugenholtz P, Larsen N, Rojas M, Brodie EL, Keller K et al (2006) Greengenes, a chimera-checked 16S rRNA gene database and workbench compatible with ARB. Appl Environ Microbiol 72:5069–5072

33. Caporaso JG, Kuczynski J, Stombaugh J, Bittinger K, Bushman FD, Costello EK et al (2010) QIIME allows analysis of high-throughput community sequencing data. Nat Methods 7:335–336
34. McDonald D, Clemente JC, Kuczynski J, Rideout JR, Stombaugh J, Wendel D et al (2012) The biological observation matrix (BIOM) format or: how I learned to stop worrying and love the ome-ome. Gigascience 1(1):7. https://doi.org/10.1186/2047-217X-1-7
35. Wickham H (2016) ggplot2: elegant graphics for data analysis. Springer, New York
36. McCarthy DJ, Chen Y, Smyth GK (2012) Differential expression analysis of multifactor RNA-Seq experiments with respect to biological variation. Nucleic Acids Res 40:4288–4297
37. Zhang J, Kobert K, Flouri T, Stamatakis A (2014) PEAR: a fast and accurate Illumina Paired-End reAd mergeR. Bioinformatics 30:614–620

Chapter 17

Antibiotic Conditioning and Single Gavage Allows Stable Engraftment of Human Microbiota in Mice

Zhigang Zhu, Thomas Kaiser, and Christopher Staley

Abstract

Mice transplanted with human microbiota are essential tools for studying the role of microbiota in health and disease, striving for the development of microbiota-modulating therapeutics. Traditionally, germ-free mice have been the principal option for establishing human microbiota-associated (HMA) mouse models, leading to significant insights into the composition and function of the human microbiota. However, there are limitations in using germ-free mice as recipients of human microbiota, including considerable resource allocation to establish and maintain the model and incomplete development of their immune system and physiological functions. Thus, antibiotic-treated, non-germ-free mice have been developed as an alternative to satisfy the growing demand for an accessible HMA mouse model. Several methods have been described for creating "humanized" mice. These protocols vary in their key components, mainly antibiotic conditioning and frequency of oral gavage. To address this practical challenge and formulate a simple and repeatable protocol, we established a HMA mouse model with antibiotic-treated conventional and specific-pathogen free (SPF) C57BL/6J mice, revealing that a single oral gavage allows stable engraftment of the human microbiota. In this chapter, we present our simple protocol for antibiotic conditioning to prepare mice for stable engraftment of human gut microbiota.

Key words Antibiotics, Dysbiosis, Fecal microbiota transplantation, Gut microbiota, Human microbiota-associated mice, Humanized mice, Mouse model, Oral gavage

1 Introduction

The human intestinal tract is densely populated by trillions of microbes, whose collective genome, the *microbiome*, contains approximately 150 times the number of genes found in the human genome [1, 2]. The microbiota plays a key role in host physiology and nutrition [3], immune function [4], and neurodevelopmental outcomes [5]. However, alterations in composition and function of the human gut microbiota, termed *dysbiosis*, have been implicated in multiple human diseases, including inflammatory bowel disease [6], obesity [7], hypertension [8], and even colorectal cancer [9]. Our knowledge of how microbial factors impact various diseases is critical for the development of

Guy R. Adami (ed.), *The Oral Microbiome: Methods and Protocols*, Methods in Molecular Biology, vol. 2327, https://doi.org/10.1007/978-1-0716-1518-8_17,

microbiota-modulating therapeutics. This information is, to a large extent, obtained through use of translational and complementary animal models, which allow interventional testing and have the potential to overcome the technical and ethical limitations of human studies.

To date, human microbiota-associated (HMA) mice have been used extensively in gut microbiome research to study microbial-related pathologies [10–13]. Establishment of a HMA mouse model typically involves transfer of a defined consortium of bacteria or the entire intestinal microbiota collected from volunteers or patients into germ-free or conventionally raised mice. This allows subsequent monitoring of disease phenotypes [11, 14]. Traditionally, germ-free mice have been the principal option for humanization studies, leading to significant insights into the role of the gut microbiota in several human diseases such as autism [10], obesity [11, 15–17], inflammatory bowel disease [12], and colorectal cancer [13, 14]. However, there are limitations in using germ-free mice to establish a HMA mouse model. Maintenance of germ-free facilities requires considerable resources, and often, strict operational procedures are needed to maintain sterile conditions. In addition, germ-free animals are not available for many genotypic mouse models [18], and without the stimulation of host-specific microbiota, germ-free mice have underdeveloped immune and digestive systems [4]. Therefore, the reproducibility and translational utility of germ-free mice to human pathophysiological conditions is uncertain.

Conventional, antibiotic-treated mice have been developed as an alternative to germ-free mice to establish HMA mouse models, where the removal of indigenous mouse microbiota is achieved by antibiotic treatment, followed by human microbiota transfer [18–21]. This approach allows successful engraftment of the human fecal microbiota, resulting in remarkable changes in cecal microbiota and metabolite profiles in the recipient mice [18]. Significant insights have been gained related to host-microbe interactions in the setting of metabolic disease when performing transfer of gut microbiota from obese human donors to antibiotic-treated, specific-pathogen free (SPF) mice [20]. Thus, this model represents a promising tool to establish a HMA mouse model and study the relationship between human diseases and intestinal microbiota.

Several groups have developed individualized protocols for creating the HMA mouse models with SPF or conventional, antibiotic-treated mice, but protocols vary in key components including antibiotic conditioning procedures and frequency of oral gavage [18, 22–24]. Different types of antibiotics were administrated either alone or in the form of antibiotic cocktails, and their dosage and length of treatment vary among studies. For instance, when ciprofloxacin (30 mg/kg body mass) was used as the sole antibiotic for treatment prior to fecal microbiota transplantation

(FMT), only a minor portion of the human gut bacterial community was established in the recipient mice [23]. In contrast, the engraftment level of the donor microbiota was significantly improved with an antibiotic cocktail consisting of 50 mg/kg vancomycin, 100 mg/kg neomycin, 100 mg/kg metronidazole, and 1 mg/kg amphotericin-B [18]. On the other hand, this method required weekly oral gavage of human donor microbiota for 12 weeks, making the entire process very labor-intensive [18]. To address this practical challenge, a comparative study focusing on the frequency of FMT over 4 weeks revealed that one FMT following bowel cleansings enabled stable transfer of human microbiota to mice [24]. This observation is comparable to our protocol, which allows stable engraftment of human microbiota following a single oral gavage [21]. This chapter is dedicated to introducing our simple protocol for antibiotic conditioning to prepare mice to allow stable engraftment of human gut microbiota following a single oral gavage.

2 Materials

2.1 Mice

1. C57BL/6J mice, females and males, age 36–42 days (Housed conventionally or under SPF conditions) (*see* **Note 1**).

2.2 Antibiotics

1. Systemically absorbed antibiotics: ampicillin, cefoperazone sodium salt, and clindamycin hydrochloride. Store at 4 °C.
2. Nonabsorbable antibiotics: ertapenem sodium, neomycin sulfate, and vancomycin hydrochloride. Store at 4 °C.
3. Antibiotic solutions of either systemically absorbed or nonabsorbable antibiotics (1 mg/ml each antibiotic) are made in drinking water and delivered in 100-ml glass sipper or standard cage water bottles. Store at 4 °C and use within 7 days of making the solution.

2.3 Human Fecal Samples

1. Prior to donor accrual, studies should receive institutional approval. All donors should provide written informed consent.
2. Qualified, consented donors are asked to collect their stool into a plastic toilet hat and then subsample into 30 ml self-standing, polypropylene, skirted, conical-bottom fecal containers with attached screwcap with spoon.
3. The tubes are immediately transferred, unamended, to a −20 °C or −80 °C freezer for storage (indefinite). −80 °C storage is preferable for long-term (>1 week) storage.

2.4 Human Fecal Microbiota Preparation

1. Human fecal sample (thawed on ice).
2. N_2 gas to minimize the incorporation of air while processing the fecal material.
3. Sterile phosphate-buffered saline (PBS).
4. Autoclavable commercial sized blender (250 ml maximum volume). We use a single-speed 50–250 ml Waring blender. To adjust blender speed, we use a variable transformer to adjust voltage.
5. Stainless steel laboratory sieves (autoclaved prior to use): 2.00 mm (USA 10 Mesh), 1.00 mm (USA 18 Mesh), 0.50 mm (USA 35 Mesh), 0.25 mm (USA 60 Mesh), collecting pan.
6. 10% bleach is maintained in plastic tub for immediate decontamination of equipment.
7. Refrigerated centrifuge (50 ml tube capacity, 4500 × *g* speed, 4 °C required).
8. Glycerol (pharmaceutical grade).

2.5 Cell Counting

1. Fluorescent nucleic acid stain (we use SYTO Green Fluorescent stain).
2. Petroff-Hauser counting chamber.
3. Fluorescence microscope.

2.6 Oral Gavage

1. Oral gavage needle [e.g., 20 G × 25 mm (2 mm tip diameter), straight].

3 Methods

3.1 Fecal Microbiota Preparation and Quantification

The fecal microbiota preparation involves resuspension of human fecal samples following homogenization in a blender under N_2 gas, sieving to remove solid particles, and concentration in PBS. The cell density of the slurry is quantified using a fluorescent dye followed by amendment to 10% glycerol for cryopreservation in approximately 1 ml aliquots. All equipment should be autoclaved or sterilized in 10% bleach prior to use. Fecal preparations should be prepared in a biosafety cabinet.

1. Thaw the fecal sample on ice and weigh the thawed fecal material.
2. Transfer to a standard commercial blender purged with N_2 gas (40 psi).
3. Add sterile PBS (5 ml/g feces).
4. In a covered blender, homogenize the fecal slurry by blending at low speed for 30–60 s. Repeat up to three times, as needed to

achieve a homogenous solution. Allow the slurry to settle for 5 min. We adjust the voltage to 25–30 V to achieve a low speed.

5. Pass the slurry sequentially through the 2.0, 1.0, 0.5, and 0.25 mm sieves and collect in a sterile pan. After passing through, transfer sieves to 10% bleach for at least 10 min.
6. Aliquot the slurry evenly into an even number of conical tubes (e.g., 50 ml tubes). Note the volume of slurry in each tube.
7. Centrifuge the slurry at 4500 × *g*, 4 °C for 15 min. Discard the supernatant into 10% bleach or a biohazard receptacle.
8. Resuspend the pellet in a volume of PBS matching the starting volume, as noted in **step 6**.
9. Centrifuge the slurry at 4500 × *g*, 4 °C for 15 min. Discard the supernatant into 10% bleach or a biohazard receptacle.
10. Combine pellets and dilute 1:1 (vol:vol) in PBS. Maintain on ice.
11. Transfer 100μl of the fecal solution to a microcentrifuge tube and serially dilute to 10^{-3}.
12. Stain the 10^{-3} dilution using the fluorescent nucleic acid stain according to the manufacturer's instructions. Adjust the slurry dilution until appropriate counts can be obtained. Determine the average ($\overline{x}$) of five medium boxes (0.2 mm × 0.2 mm). Determine the concentration: cells mg/ml = Average ($\overline{x}$) × chamber volume (ml) × dilution factor. If using the 10^{-3} dilution, cells mg/ml = $\overline{x} \times (1.25 \times 10^{6}) \times 10^{3}$.
13. Amend the slurry to a final concentration of 10^{10} cells mg/ml diluted in PBS, amended with 10% glycerol. Store as 1 ml aliquots in cryovials at −80 °C.

3.2 Antibiotic Conditioning

Mice are allowed a brief acclimation period, then receive alternating cocktails of systemically absorbed and nonabsorbable antibiotics to ablate the indigenous microbiota (*see* **Notes 2** and **3**). Then, they receive a single gavage of prepared donor fecal material. During antibiotic administration, mouse weights should be monitored every 2–3 days to ensure health and antibiotic solutions should be replenished if needed. Throughout the protocol, mice are maintained under ambient housing conditions (e.g., a 12:12 dark:light cycle with *ad libitum* standard chow) and may be cohoused with other mice receiving the same donor material (*see* **Note 4**).

1. Prepare antibiotic solutions as described in Subheading 2.2. One-liter solutions (1 g each antibiotic) can be stored at 4 °C and used to replenish antibiotic-conditioned water during each round of administration. Cocktails should be freshly made for each administration.

2. Allow mice to acclimate for at least 2 days prior to administration of antibiotics.
3. Deliver the systemically absorbable cocktail to the mice, *ad libitum* as their only source of drinking water, for 7 days (*see* **Note 5**).
4. Replace with normal drinking water for 2 days.
5. Deliver the nonabsorbable cocktail to the mice, *ad libitum*, for 7 days.
6. Replace with normal drinking water for 2 days.
7. Deliver the systemically absorbable cocktail to the mice, *ad libitum*, for 7 days.
8. Replace with normal drinking water for 2 days.
9. Thaw donor material on ice prior to gavage. Gavage each mouse with 100μl prepared donor fecal material.

3.3 Experimental Considerations

3.3.1 Experimental Design

Due to cage-mate interactions including coprophagy, the microbiota of mice in a single cage will homogenize [25]. As a result, individual mice reflect technical replicates while different donor samples reflect biological replicates. Experimental arms should be tested against control mice that receive gavage with drinking water to mimic experimental stress. Our group uses two control groups: one that receives the antibiotic cocktail without water gavage and a second that receives only the water gavage without antibiotics (*see* **Notes 6–8**).

3.3.2 Confirmation of Engraftment

While not an essential element of the protocol, we confirm microbiota engraftment using next-generation sequencing. We extract DNA from the donor sample(s) and fecal pellets collected from mice prior to beginning antibiotics to represent human and mouse microbiota configurations. DNA from fecal pellets subsequent to gavage can then be used to track the engraftment. We use next-generation amplicon sequencing of the 16S ribosomal (r)RNA gene (we use the V4 region on Illumina systems) to characterize the microbiome and investigate engraftment using the SourceTracker software [26] (*see* **Notes 9** and **10**).

4 Notes

1. Our earlier work validated this using SPF, female mice [21]. We have subsequently validated this model using male and female C57BL/6J mice housed under SPF and conventional conditions and observed similar levels of engraftment between housing conditions using the same donor (Fig. 1, ANOVA $P = 0.289$ and 0.191 for males and females, respectively).

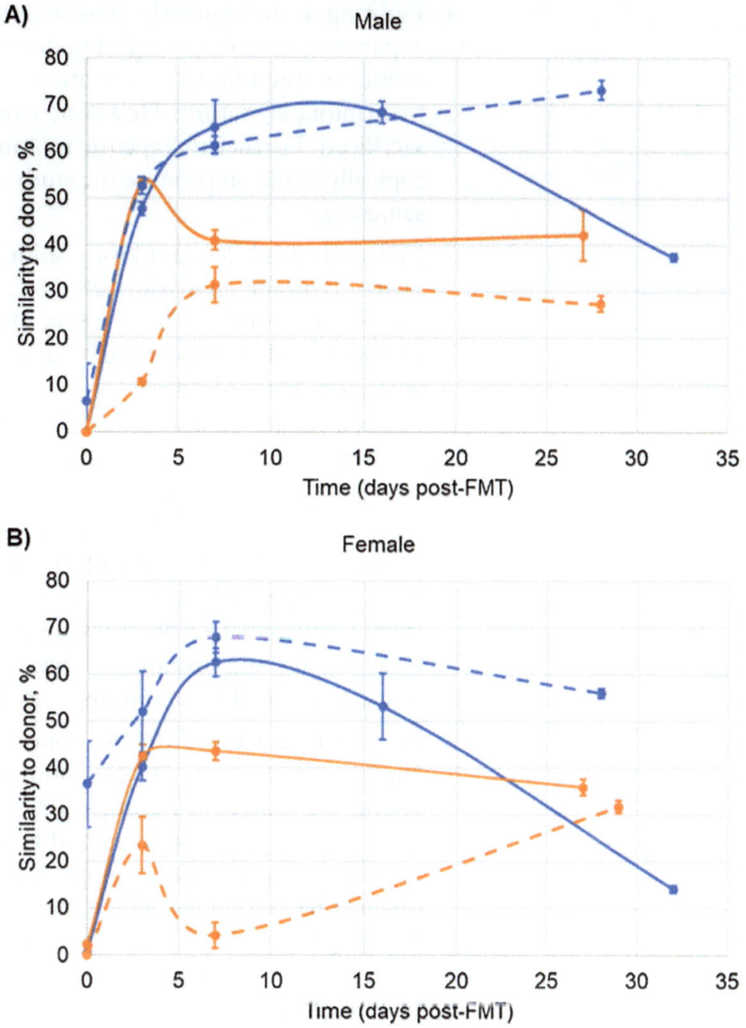

Fig. 1 Mean donor engraftment, as determined using SourceTracker in (**a**) males ($n = 4$) and (**b**) females ($n = 5$). Colored lines reflect individual donors. Solid lines represent conventionally housed mice and dashed lines represent SPF housing. Error bars indicate standard error of the mean

2. We previously tried single-course administrations using either antibiotic cocktail [21]. While this permitted early engraftment of donor human microbiota, it was not durable and the human signature decayed rapidly.

3. Anecdotally, the order the cocktails were provided does not appear to affect the success or stability of engraftment. When nonabsorbable antibiotics were accidentally administered first, followed by the systemically absorbable cocktail and a second administration of the nonabsorbable cocktail, similar results were achieved in SPF mice.

4. Fighting is infrequently observed among our mouse trials. Separating mice due to fighting does not appear to significantly influence microbiota engraftment or durability of the human microbiota signature. However, wounded animals should be sacrificed. Further therapeutic treatment with antibiotics (even topically) will compromise the study design and any mice in the same cage.
5. Mice will drink less antibiotic solution than normal drinking water (typically approximately 20–25 ml every 2–3 days in a cage of 4–5 mice). The systemically absorbed cocktail is less well tolerated. If dehydration is a concern, the cocktails can be amended with saccharin (1 mg/ml) to improve drinking. Balb/C mice do not tolerate antibiotic cocktails without saccharin.
6. We have tested the phenotypic effects of transferring microbiota from lean and obese (body mass index >35 mg/kg^2) humans into the mice under SPF and conventional conditions. We noted that, when housed conventionally, C57BL/6J mice (males and females) do show a phenotypic response in both body weight and insulin sensitivity that is not observed when housed under SPF conditions (Fig. 2).
7. Fecal matter from different donors appears to have variable engraftment success. The second donor tested showed poorer engraftment across all groups (*see* Fig. 1, ANOVA $P < 0.0001$ in both sexes). We hypothesize this is due to a very low proportion of the community comprised of members of the bacterial phylum Bacteroidetes, which may provide a necessary scaffold for engraftment. Compositional features associated with greater or poorer engraftment are being actively investigated by our group and others.
8. We also found that mouse genotype influenced phenotypic shifts following successful human microbiota transfer. The obese phenotype was not observed in Balb/C mice when using the same donor that was associated with weight gain in C57BL/6J mice (Fig. 2). Furthermore, this result was reproducible in second trial using C57BL/6J mice.
9. We extract DNA from approximately 25 mg of fecal sample or individual mouse pellets (approximately 10 mg) using the DNeasy PowerSoil DNA Isolation kit (Qiagen) on the automated QIAcube system using the inhibitor removal technology (IRT) protocol. DNA is amplified using the 515F/806R primer set [27] targeting the V4 hypervariable region of the 16S rRNA gene. Paired-end sequencing is done at a read length of 250–300 nt using the Illumina MiSeq or HiSeq2500 platforms. We have not noticed variability in data

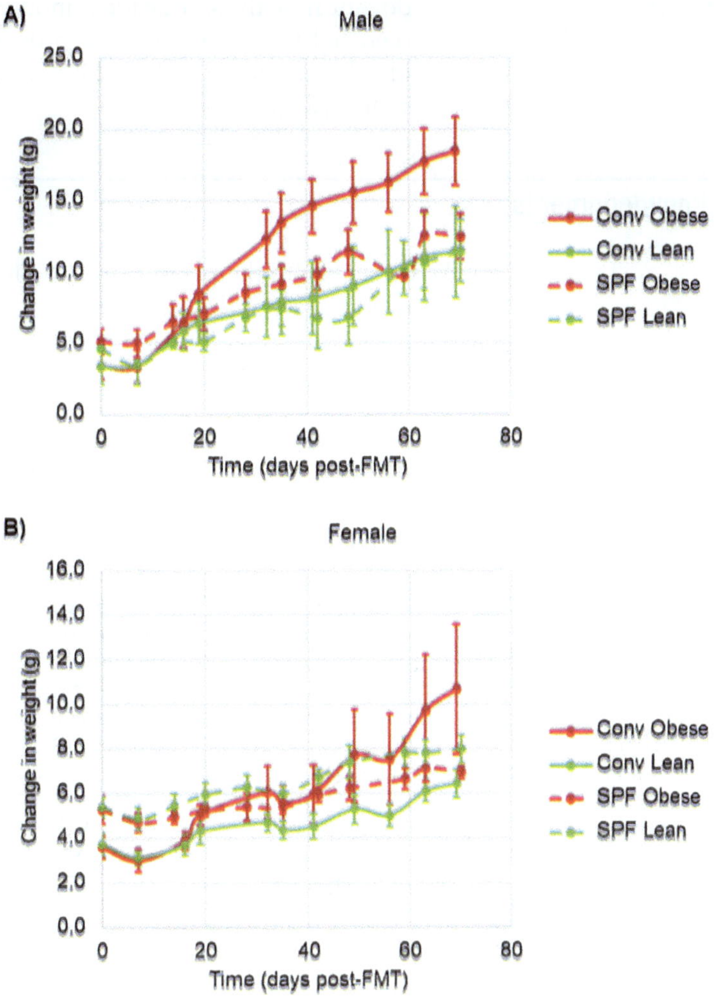

Fig. 2 Mean change in weight from baseline in (**a**) males ($n = 4$) and (**b**) females ($n = 5$). The same obese and lean donors were used for both conventionally housed (solid lines) and SPF-housed (dashed lines) mice. Error bars indicate standard error of the mean

quality or sample composition resulting from batch effects or sequencing platform.

10. We process our sequence data using mothur software [28]. Reads are paired-end joined, trimmed for quality, and aligned against the SILVA database [29]. Operational taxonomic units are classified at 99% similarity using the furthest-neighbor algorithm and classified using the Ribosomal Database Project [30]. We and our collaborators obtain highly correlated results using other freely available software and taxonomic databases. We use SourceTracker [26] to determine source allocations using at least triplicate samples to represent each source (technical replicates from a single fecal

donation or three individual mouse pellets). We have recently reported that SourceTracker provides a conservative estimate of the numbers of OTUs contributed from a single source [31].

Acknowledgments

We would like to thank Dr. Harika Nalluri for her input and editing of the final draft of this chapter.

References

1. Ley RE, Peterson DA, Gordon JI (2006) Ecological and evolutionary forces shaping microbial diversity in the human intestine. Cell 124:837–848
2. Qin J, Li R, Raes J, Arumugam M, Burgdorf KS, Manichanh C et al (2010) A human gut microbial gene catalogue established by metagenomic sequencing. Nature 464:59–65
3. Bäckhed F, Ley RE, Sonnenburg JL, Peterson DA, Gordon JI (2005) Host-bacterial mutualism in the human intestine. Science 307:1915–1920
4. Chung H, Pamp SJ, Hill JA, Surana NK, Edelman SM, Troy EB et al (2012) Gut immune maturation depends on colonization with a host-specific microbiota. Cell 149:1578–1593
5. Lu J, Lu L, Yu Y, Cluette-Brown J, Martin CR, Claud EC (2018) Effects of intestinal microbiota on brain development in humanized gnotobiotic mice. Sci Rep 8:1–16
6. Dupont AW, Dupont HL (2011) The intestinal microbiota and chronic disorders of the gut. Nat Rev Gastroenterol Hepatol 8:523–531
7. Aron-Wisnewsky J, Prifti E, Belda E, Ichou F, Kayser BD, Dao MC et al (2019) Major microbiota dysbiosis in severe obesity: fate after bariatric surgery. Gut 68:70–82
8. Li J, Zhao F, Wang Y, Chen J, Tao J, Tian G et al (2017) Gut microbiota dysbiosis contributes to the development of hypertension. Microbiome 5:14
9. Zou S, Fang L, Lee M-H (2018) Dysbiosis of gut microbiota in promoting the development of colorectal cancer. Gastroenterol Rep 6:1–12
10. Sharon G, Cruz NJ, Kang DW, Gandal MJ, Wang B, Kim YM et al (2019) Human gut microbiota from autism spectrum disorder promote behavioral symptoms in mice. Cell 177:1600–1618.e17
11. Ridaura VK, Faith JJ, Rey FE, Cheng J, Duncan AE, Kau AL et al (2013) Gut microbiota from twins discordant for obesity modulate metabolism in mice. Science 341:1241214
12. Nagao-Kitamoto H, Shreiner AB, Gillilland MG, Kitamoto S, Ishii C, Hirayama A et al (2016) Functional characterization of inflammatory bowel disease-associated gut dysbiosis in gnotobiotic mice. Cell Mol Gastron Hepatol 2:468–481
13. Sobhani I, Bergsten E, Couffin S, Amiot A, Nebbad B, Barau C et al (2019) Colorectal cancer-associated microbiota contributes to oncogenic epigenetic signatures. Proc Natl Acad Sci U S A 116:24285–24295
14. Wong SH, Zhao L, Zhang X, Nakatsu G, Han J, Xu W et al (2017) Gavage of fecal samples from patients with colorectal cancer promotes intestinal carcinogenesis in germ-free and conventional mice. Gastroenterology 153:1621–1633.e6
15. Bäckhed F, Manchester JK, Semenkovich CF, Gordon JI (2007) Mechanisms underlying the resistance to diet-induced obesity in germ-free mice. Proc Natl Acad Sci U S A 104:979–984
16. Turnbaugh PJ, Ley RE, Mahowald MA, Magrini V, Mardis ER, Gordon JI (2006) An obesity-associated gut microbiome with increased capacity for energy harvest. Nature 444:1027–1031
17. Bäckhed F, Ding H, Wang T, Hooper LV, Koh GY, Nagy A et al (2004) The gut microbiota as an environmental factor that regulates fat storage. Proc Natl Acad Sci U S A 101:15718–15723
18. Hintze KJ, Cox JE, Rrompato G, Benninghoff AD, Ward RE, Broadbent J et al (2014) Broad scope method for creating humanized animal models for animal health and disease research through antibiotic treatment and human fecal transfer. Gut Microbes 5:37–41

19. Lundberg R, Toft MF, Metzdorff SB, Hansen CHF, Licht TR, Bahl MI et al (2020) Human microbiota-transplanted C57BL/6 mice and offspring display reduced establishment of key bacteria and reduced immune stimulation compared to mouse microbiota-transplantation. Sci Rep 10:1–16
20. Rodriguez J, Hiel S, Neyrinck AM, Le Roy T, Pötgens SA, Leyrolle Q et al (2020) Discovery of the gut microbial signature driving the efficacy of prebiotic intervention in obese patients. Gut:1–13
21. Staley C, Kaiser T, Beura LK, Hamilton MJ, Weingarden AR, Bobr A et al (2017) Stable engraftment of human microbiota into mice with a single oral gavage following antibiotic conditioning. Microbiome 5:87
22. Manichanh C, Reeder J, Gibert P, Varela E, Llopis M, Antolin M et al (2010) Reshaping the gut microbiome with bacterial transplantation and antibiotic intake. Genome Res 20:1411–1419
23. Wos-Oxley M, Bleich A, Oxley AP, Kahl S, Janus LM, Smoczek A et al (2012) Comparative evaluation of establishing a human gut microbial community within rodent models. Gut Microbes 3:234–249
24. Wrzosek L, Ciocan D, Borentain P, Spatz M, Puchois V, Hugot C et al (2018) Transplantation of human microbiota into conventional mice durably reshapes the gut microbiota. Sci Rep 8:6854
25. Nguyen TLA, Vieira-Silva S, Liston A, Raes J (2015) How informative is the mouse for human gut microbiota research? Dis Model Mech 8:1–16
26. Knights D, Kuczynski J, Charlson ES, Zaneveld J, Mozer MC, Collman RG et al (2011) Bayesian community-wide culture-independent microbial source tracking. Nat Methods 8:761–U107
27. Caporaso JG, Lauber CL, Walters WA, Berg-Lyons D, Huntley J, Fierer N et al (2012) Ultra-high-throughput microbial community analysis on the Illumina HiSeq and MiSeq platforms. ISME J 6:1621–1624
28. Schloss PD, Westcott SL, Ryabin T, Hall JR, Hartmann M, Hollister EB et al (2009) Introducing mothur: open-source, platform-independent, community-supported software for describing and comparing microbial communities. Appl Environ Microbiol 75:7537–7541
29. Pruesse E, Quast C, Knittel K, Fuchs BM, Ludwig W, Peplies J et al (2007) SILVA: a comprehensive online resource for quality checked and aligned ribosomal RNA sequence data compatible with ARB. Nucleic Acids Res 35:7188–7196
30. Cole JR, Wang Q, Cardenas E, Fish J, Chai B, Farris RJ et al (2009) The ribosomal database project: improved alignments and new tools for rRNA analysis. Nucleic Acids Res 37: D141–D145
31. Staley C, Kaiser T, Vaughn BP, Graiziger C, Hamilton MJ, Kabage AJ et al (2019) Durable long-term bacterial engraftment following encapsulated fecal microbiota transplantation to treat *Clostridium difficile* infection. MBio 10:e01586–e01519

INDEX

Guy Adami (ed.), *The Oral Microbiome: Methods and Protocols*, Methods in Molecular Biology, vol. 2327, https://doi.org/10.1007/978-1-0716-1518-8,

T

V

W

MIX
Papier aus verantwortungsvollen Quellen
Paper from responsible sources
FSC® C105338

If you have any concerns about our products,
you can contact us on
ProductSafety@springernature.com

In case Publisher is established outside the EU,
the EU authorized representative is:
Springer Nature Customer Service Center GmbH
Europaplatz 3, 69115 Heidelberg, Germany

Printed by Libri Plureos GmbH
in Hamburg, Germany